Edinburgh
Information Technology Series

MACHINE TRANSLATION TODAY :
THE STATE OF THE ART

S. Michaelson and Y. Wilks
Series Editors

Edited by
MARGARET KING

MACHINE TRANSLATION TODAY : THE STATE OF THE ART

PROCEEDINGS OF THE
THIRD LUGANO TUTORIAL
LUGANO, SWITZERLAND
2–7 APRIL 1984

EDINBURGH UNIVERSITY PRESS

© Margaret King 1987
Edinburgh University Press
22 George Square, Edinburgh

Set in Linotronic Times by
Speedspools, Edinburgh and
printed in Great Britain by
Redwood Burn Ltd, Trowbridge

British Library Cataloguing
  in Publication Data
King, Margaret
  Machine translation today.
  —(Edinburgh information
  technology ; 2)
  1. Machine translating
  I. Title.   II. Series
  418′.02   P308

ISBN 0 85224 519 x

CONTENTS

CONTRIBUTORS

Sophie Ananiadou
Centre for Computational Linguistics, UMIST, Sackville Street, Manchester, England

Christian Boitet
GETA, University of Grenoble, 38402 Saint-Martin-d'Heres, France

Beat Buchmann
Dalle Molle Institute for Semantic and Cognitive Studies, 54 route des Acacias, 1227 Geneva, Switzerland

Jean-Luc Cochard
Department of Applied Mathematics, Federal Institute of Technology, Ecublens, 1015 Lausanne, Switzerland

Anne de Roeck
Department of Computer Science, University of Essex, Wivenhoe Park, Colchester, England

Jean-Philippe Guilbaud
GETA, University of Grenoble, 38402 Saint-Martin-d'Heres, France

Pierre Isabelle
Department of Linguistics, University of Montreal, Montreal, Canada

Roderick Johnson
Centre for Computational Linguistics, UMIST, Sackville Street, Manchester, England

Margaret King
Dalle Molle Institute for Semantic and Cognitive Studies, 54 route des Acacias, 1227 Geneva, Switzerland

Jan Landsbergen
Philips Research Laboratories, 5600 JA, Eindhoven, The Netherlands

Heinz-Dieter Maas
Institute for Applied Information Processing, Eurotra-D, Martin Luther Strasse 14, 6600 Saarbruecken, Germany

Alan Melby
Department of Linguistics, Brigham Young University, Provo 84604, Utah, USA

Sergei Perschke

Commission of the European Communities, DG XIII, Jean Monnet Building, Plateau du Kirchberg, Luxembourg

Dominique Petitpierre

Dalle Molle Institute for Semantic and Cognitive Studies, 54 route des Acacias, 1227 Geneva, Switzerland

Michael Rosner

Dalle Molle Institute for Semantic and Cognitive Studies, 54 route des Acacias, 1227 Geneva, Switzerland

Geoffrey Sampson

Department of Linguistics and Phonetics, University of Leeds, Leeds, England

Patrick Shann

Dalle Molle Institute for Semantic and Cognitive Studies, 54 route des Acacias, 1227 Geneva, Switzerland

Jonathan Slocum

Microelectronics and Computer Technology Corp., 9430 Research Blvd., Austin, Texas, USA

Susan Warwick

Dalle Molle Institute for Semantic and Cognitive Studies, 54 route des Acacias, 1227 Geneva, Switzerland

Eric Wehrli

Department of Linguistics, UCLA, Hilgard Ave. 405, Los Angeles, Cal. 90024, USA

Peter Wheeler

Logos Computer Systems Deutschland GMBH, Lyoner Strasse 26, 6000 Frankfurt am Main 71, Germany

PREFACE

This book, based on a week-long tutorial held in Lugano, Switzerland in spring of 1984, reflects the renewal of interest in machine translation systems. In the late 1950s and early 1960s, much energy, enthusiasm (and research funding) went into work on machine translation. However, the effort met with only limited success. At least in part, this was because the expectations both of the research workers themselves and of their funding agencies were unrealistically high. They had not fully realised the complexity of language and the consequent difficulties of treating natural language with computational means, and therefore fell into the trap of believing that general purpose machine translation could be achieved simply by working hard. Although the research workers themselves quite quickly began to realise that it was not so simple, they were too late to avoid being condemned for not being able to live up to the rash promises initially made, with the result that towards the mid-1960s, machine translation came to be seen as a utopian dream, requiring years of preliminary fundamental research.

Unfortunately, even though research on machine translation could be made, at least in large measure, to go away, simply by making it difficult to find funding, the need for translation itself could not so conveniently be made to disappear. Indeed, the need for translation has been growing steadily over the last twenty years, partly because of a political will towards multi-lingualism in polities where different parts of the population speak different languages, such as the European Economic Community, Canada and Switzerland, partly because of the economic forces springing from the needs of trade (importers have become increasingly unwilling to accept documentation in a language other than their own), and partly because of the growing need to communicate across language barriers in a world which is shrinking rapidly.

Thus, it was inevitable that even after a comparatively short time, attention should again turn towards machine translation as a way of resolving pressing translation needs. By the late 1970s, several systems had proved their worth, and were being used on a regular basis.

More recent years have seen ever-growing interest. A number of new projects have been started, some of them (for example, in France and in Japan) within the framework of national research programmes aimed at rapid technological development.

The pattern of this book follows the pattern of historical development quite closely.

Part I aims at providing the reader with a knowledge of the background of work in machine translation, explaining some of the problems involved and introducing some fundamental terminology. The first two chapters give the historical background in some detail, providing a critical analysis rather than simple narration of facts. The second two chapters are concerned with linguistics in machine translation: chapter 3 concentrates on relating linguistic theory to the development of machine translation over time, whilst chapter 4 argues that the results of recent work in theoretical linguistics can be applied more or less directly to machine translation. The final two chapters of this section set out other approaches. Chapter 5 describes and discusses work in artificial intelligence relevant to machine translation. Chapter 6 takes the rather iconoclastic view that much can be achieved with relatively simple theoretical tools.

Just as every machine translation system embodies in some way or another a linguistic theory, even more clearly does every machine translation system require software. Different kinds of software are involved: first there is the software underlying the core of the machine translation system itself. Then there is support software used to create an environment around the core system. This in its turn can be sub-divided into the environment needed for developing a system, and that required when a system is in operation use. Clearly, these two are likely to be very different, since in the first case the users of the system are the linguists and computer scientists involved in the construction of the system, whilst in the second the users are the translators and revisors involved in using the system to produce actual translations. Part II introduces the reader to some of the issues involved in designing and building these different kinds of software. The first two chapters are introductory. Chapter 7 introduces and explains some of the software related terms frequently encountered in the description of machine translation systems, whilst chapter 8 describes a particular data structure which has, in recent years, very frequently been used in machine translation applications, in particular in several of the systems described in the closing section. Chapter 9 directly addresses the question of a suitable environment for an operational system, looking at the question from the point of view of the translator using the system. Chapter 10 describes the evolution of research in one particular group, covering both the underlying philosophy as well as direct discussion of software issues and of the relation between linguistic work and software. Chapter 11 discusses the question of a software environment within which to carry out research on machine translation, where software tools need to be tailored to match the different linguistic solutions being tested.

The final Part aims at giving the reader an overall picture of the state of the art. Chapter 12 serves as an introduction, giving an overview of some of the systems not discussed in later chapters and picking out some of the chief characteristics of current systems. Of the later chapters, each describes a particular system, and can be left to introduce itself.

A footnote should be added on the bibliography. Since it is rather substantial, it has been structured to reflect different aspects of the general problem area, in order to help the reader to find his way around it. There are separate sections on machine translation literature up to 1973 and after 1973, on software, on linguistics (including computational linguistics) and on artificial intelligence. References in the text are given with a number which corresponds to the number of the reference in the bibliography, as well as with the year of publication and the author's name.

M.K., Geneva, 11 July 1985

Acknowledgements

The initial work on this book was done in preparation for a Tutorial on Machine Translation, organised by the Dalle Molle Institute for Semantic and Cognitive Studies of the University of Geneva, the third in a series of tutorials held in Lugano under the general title of the Lugano tutorials. The authors and organisers would like to thank the Commune of Lugano for their continuing support and for their kindness in providing a beautiful setting for a week's intensive work.

As always, very many thanks should also go to Signor Dalle Molle, the founder of the Institute and the origin of all that has followed, and to Franco Boschetti who has been our constant support.

The editor would like to record her personal gratitude to Anneke De Roeck, who did much of the initial organisation of the tutorial, to Martine Vermeire for consistent and untiring help, and to Beat Buchmann, both for his work during the tutorial and for his invaluable help in producing this book: without him neither tutorial nor book would have happened.

Finally, special thanks should go to Jonathan Slocum, who generously provided us with a lot of material for the bibliography.

PART ONE: INTRODUCTION

EARLY HISTORY OF
MACHINE TRANSLATION

Historical surveys of a given field of academic endeavour usually remain on a purely descriptive level – and this one is no exception – although it is often difficult retrospectively to separate fact from opinion. This is particularly true for the field of machine translation (MT) which, during the period from its beginnings up to 1966, has been marked by both a wealth of written material (some 10 000 pages of notes, mimeographs, research papers and reports) and a wide variety of approaches, false starts, and surrenders in the face of the complexity of the task. The claims made for individual systems/approaches and the hopes raised as to their future development, too, show the same kind of variation and divergence, especially when pronounced by researchers directly involved in a given project and by researchers working on other 'competing' projects. Nonetheless, an attempt has been made to present a balanced and impartial view of the pre-1966 world of MT.

Pre-1954

Surely, if one were to consult classical scholars, passages could be found in ancient writings that could conceivably be construed and interpreted as containing the seeds of mechanical translation from one language into another. However, serious and quite detailed technical proposals for MT were first put forward in 1933, when George Artsruni, a French engineer, and P. P. Smirnov-Trojanskij, a Russian scholar, independently took out patents for the translation machines they had invented. In retrospect, it is well worth having a closer look at Trojanskij's invention as it incorporated many of the fundamental design features and principles of MT which are as valid today as they were then.

Trojanskij recognised three stages in the translation process,

3

which, in modern terminology, could be equated with analysis, transfer and generation. In analysis, the source language would be transformed into a canonical form (base forms or roots) plus a set of labels indicating the forms' syntactic functions; in transfer, the source language canonical form would be replaced by the corresponding target language canonical form; lastly, in generation, the target language proper would be reconstructed/generated from the canonical form and the associated labels. Transfer would be mechanical; analysis and generation were conceived of as being done manually, a scheme which in effect meant that Trojanskij's translation machine was nothing more than an automatic dictionary look-up mechanism. A model implementation of the machine in the form of a moveable tape with vertically arranged word equivalents from different languages underlined its conception as a scroll-type word selection mechanism.

Apart from the tripartite distinction between analysis, transfer and generation, Trojanskij also fathered the notion of 'intermediary language' (IL). He maintained that independently of their individual lexical and grammatical forms, all languages had a common logical content which allowed translation from any language into any other language via a universal logical intermediary language. The canonical form labelled with syntactic information was to be considered a first approximation to such an intermediary language.

It was thirteen years later and at a time when the newly invented digital computers became more widely available that A. D. Booth, a British scholar, took up the idea of creating an automatised computerised dictionary look-up mechanism as a first, and purely heuristic, approach to machine translation. The idea was based on the fact that a computer could store a sufficiently large amount of data, in this case lexical items, to allow a rather fast kind of 'word-for-word dictionary look-up translation' like the one a human equipped with a bilingual dictionary but with no or only little knowledge of the target language would produce. Consequently, in 1947, A. D. Booth and D. H. V. Britten, working at Princeton University, New Jersey, developed a programme for detailed bilingual dictionary coding and implemented it on a computer. A year later, in an attempt to push automation of the translation process a step further, R. M. Richens, also a British scholar, introduced the concept of split versus full dictionary entries. Together with Booth, he developed a dictionary coding scheme in which word roots and affixes appeared as separate entries with their respective target language equivalents and, where pertinent, relevant grammatical information. An example (taken from [9] Delavenay, 1963, 42) may serve to illustrate this scheme:

The Spanish form *comprarlo* and its French equivalent would be coded in the following manner:

compr –	*achet*
– *ar* –	– *er* – (inf.)
– *lo*	*le* (art.)/*le* (pron.) *lui*

Upon encountering the form *comprarlo* as its input, the dictionary look-up mechanism would try to find a direct match for the whole form with a dictionary entry. Failing such a match, the look-up mechanism would search for a match of the longest possible word segment, always working from left to right, until all segments had found their match – and their translation.

The reasons for having such split-form dictionary entries – whether they be roots and affixes or syllables – were linguistic and computational. On the linguistic side, the idea of identifying and coding such 'minimal semantic units' was, at least in 1948, a reflection of the then prevailing opinion that syntax was of only minor importance for the comprehension of language and that translation could be achieved by replacing the *invariant semantic units* of one language with the corresponding semantic units of another language. On the computational side, it was a reflection of the constant preoccupation of all MT researchers at that time with questions of machine performance and capabilities such as storage capacity and memory access time; it is obvious that a morphological analysis of dictionary entries as carried out by Booth and Richens drastically reduced the number of dictionary entries and hence also the demands made on the machine with regard to storage capacity.

The translations produced by the Booth and Richens programme were word-for-word or rather root-for-root and affix-for-affix translations maintaining the word order of the source language and, in the case of multiple meanings, supplying all of the alternative target language equivalents. To make up for the lack of any syntactic analysis, the translations contained indications of the units' grammatical meaning(s) such as (art.), (pron.) etc. (see example above). Later, in the years to follow, the two researchers, by then working at Birkbeck College in London, experimented with several schemes for microglossary construction to cover the specialised word meanings of particular scientific fields and introduced these microglossaries into their programme in order to reduce the number of possible meanings of individual technical terms. Yet, despite all these various back-ups to compensate for the absence of an underlying grammatical analysis, the translations were such that, at the first MT conference held at MIT, Mass., in 1952, 'some of the participating linguists indicated in private conversations that the samples of automatic dictionary output were unintelligible to them' ([39] Reifler 1954, 25).

However, intelligibility and transmission of information for scanning purposes was what W. Weaver, Director of the Natural Sciences Division of the Rockefeller Foundation, expected to be the minimal results of word-by-word translation when, in a memorandum called *Translation* (1949), he took stock of the work done by Booth and Richens. In his memorandum, which was circulated to some two hundred interested scholars, Weaver related the problem of machine translation to the problems of code breaking, for which digital computers had been successfully employed during the Second World War. He noted: 'It is very tempting to say that a book written in Chinese is simply a book written in English which was coded into the "Chinese code". If we have useful methods for solving almost any cryptographic problem, may it not be that with proper interpretation we already have useful methods for translation?' ([51] Weaver 1949, 22). He thought that the concept of statistical invariance found in cryptography and asserted by Shannon's information theory and the concept of logical invariance (as expressed by the German logician Reichenbach) held for all natural languages. It should, in his view, therefore be possible to identify linguistic universals and some logical structure common to all languages; clearly, such a 'universal' language would facilitate the task of MT enormously. In the light of these theories, Weaver suggested undertaking statistical semantic studies as to the minimal context necessary for the disambiguation of multiple word meanings, one of the major and most pressing problems that needed resolving in MT.

The circulation of Weaver's memorandum sparked off a flurry of activity in MT research. A number of one- or two-man part-time research teams sprung up in various American universities and institutes. Among the scholars who had become interested in MT was E. Reifler, a sinologue working at Washington University, Seattle, who, in 1950, proposed the first serious scheme of how the mechanical processing of texts should be organised in an MT system. He introduced the notions of pre-editing and post-editing: In pre-editing, the known grammatical ambiguities and the multiple-word-meaning problems would be solved by means of inserting diacritics (i.e. special signs) into the source text; this would be done manually by a human pre-editor. In post-editing, remaining problems of word-sense ambiguities and problems of target language word order would have to be taken care of by a human post-editor whose task it would be to turn the raw machine output into a readable translation. However, in the light of new developments, this scheme was given up again two years later after the first MT conference.

Another of the early MT enthusiasts was the Israeli logician Y. Bar-Hillel, who went down in MT history as having been the first scholar being engaged full-time in MT research. It was he who, with the sponsorship of the Rockefeller Foundation, organised the first conference on MT in 1952, a conference that, for the first time, brought together eighteen linguists and computer engineers interested in the problems of mechanical translation.

From the first MT conference to the
Georgetown University MT experiment: 1952–54

At this conference, at MIT, the participants gave a number of papers on previous or ongoing research in MT or related fields. Many of the early MT concepts and ideas introduced in these papers deserve closer attention as in some form or other they have been around in MT ever since.

One of the ideas put forward was that the 'advances' on the computational side now allowed for some kind of 'syntactic analysis' to be mechanised. Syntactic analysis in those days basically meant automatic recognition of syntactic word classes and syntactic functions of word blocks (e.g. word blocks functioning as nouns, verbs, adjectives, or adverbs). Such an analysis was to lead to a block-by-block translation which was seen as representing a considerable improvement over the mere word-for-word translation. In this connection, Bar-Hillel presented what he called 'operational syntax', a scheme using symbolic logic to recognise syntactic connections. Once these syntactic relations – read 'word blocks' – had been identified, the logical symbols (symbolic fractions) used in the program, according to Bar-Hillel, allowed the rearrangement of the source language syntax into the target language syntax on the basis of simple arithmetic operations. It is reported that the linguists and the computer engineers alike were very impressed by and enthusiastic about the 'exceedingly simple arithmetic operations' ([41] Reynolds 1954, 51).

Dictionary construction and in particular the concept of microglossaries for the non-technical vocabulary of particular scientific subfields also received much attention at the conference. V. A. Oswald and S. L. Fletcher, of UCLA, stressed that such microglossaries drastically reduced the number of word sense ambiguities in the general language portions of scientific texts. They maintained that 'translations so obtained conveyed the meaning of the original article with correlations of meaning better than 90 per cent, on the assumption that the problems of syntax and contextual modification had previously been solved' ([41] Reynolds 1954, 52).

It would seem though that contextual modification is the very touchstone against which the usefulness of microglossaries has to be measured; for, in the majority of cases, the lexical items concerned are not *types* in any given text, but *tokens* whose meaning has to be inferred/modified from the context. However, given the pressing problem of word-sense disambiguation and the still limited memory space of the computers available, most participants of the conference welcomed the concept of microglossaries as being *the* avenue to explore in the future.

In the light of the evidence that ambiguities of syntactic word class were amenable to fully mechanised solutions (see 'syntactic analysis' above) and that problems of multiple word meanings could to a large extent be solved by the use of microglossaries, Reifler, in a discussion about pre-editing and post-editing, modified his previously (1950) proposed scheme slightly as to the division of labour between man and machine: '. . . the optimum may be reached in an arrangement in which a pre-editor signalizes certain types of grammatical information, the machine abstracts some other types of grammatical information and on the basis of this information from two sources determines certain types of incident non-grammatical meaning and reshuffles the word order. A post-editor then solves the residual semantic problems on the basis of (the) output context . . .' ([41] Reynolds 1954). Later, Reifler gave up the pre-editing/post-editing scheme altogether.

Related to the pre-editing/post-editing scheme were two further suggestions, namely 'writing for MT', i.e. writing the texts to be translated in an MT-digestable form (cf. chapter 12, Titus), and constructing 'model target languages', i.e. normalising the forms and structure of the target languages. Such a set-up would have the indubitable advantage that the machine would no longer have to translate from irregular language into irregular language but would simply have to ensure a mechanical correlation between regularised languages.

In a similar vein, L.E. Dostert of Georgetown University, Washington, introduced the concept of a 'pivot language' for use in general MT (MT from one into many languages). The idea behind it was to become independent of individual language pair systems by translating the given source language into a pivot language – say

English – and then from that pivot language into any one of the output languages desired. Clearly, such a set-up would considerably increase the flexibility within and among different MT systems.

The various proposals made at this conference have been discussed here at some length, so as to introduce the reader to some of the leading concepts and principles of MT not only valid in 1952 but far beyond. The conference closed with the participants agreeing to press ahead on two fronts: investigations of word frequency and word meanings to be carried out with a view to constructing a number of detailed microglossaries, and studies of 'operational syntax' to be undertaken with a view to computer implementation of the programmes thus specified. As several accounts of the conference report, it closed 'on a note of optimism regarding the potentialities now known to be physically present in the concept of mechanical translation' ([41] Reynolds 1954, 55), an ending that comes as no great surprise given Bar-Hillel's powerful, yet so simple arithmetic algorithm for 'syntactic analysis', Oswald's ninety-percent meaning correlation achieved by the use of microglossaries, and Booth's report of having produced 'usable' word-for-word machine translations in twenty languages including such exotic ones as Albanian, Arabic, Indonesian, and Japanese. Also small wonder that J. W. Perry of MIT concluded that 'mechanical translation was not only feasible but far closer to realizations than possibly the audience recognized' ([41] Reynolds 1954, 48).

After the conference, research in MT proceeded in a number of American universities along the general lines of the programme of work agreed upon at the conference. At Georgetown University (GU), L. E. Dostert, a newcomer to MT in 1952, yet no novice as to general problems of translation – he had invented simultaneous interpreting – started up an MT project with a small group of other linguists, with the declared aim of creating a pilot system to prove to the world, and in particular to potential funding agencies, not only the feasibility but also the practicality of MT. With a target public of various US government agencies (US Office of Naval Research, CIA, US Air Force) interested in collecting intelligence material, the choice of language pair and the direction of translation for this pilot project was almost predetermined: from Russian into English. On 7 January 1954, the time had come for the by now famous Georgetown University experiment.

With the assistance of IBM, Dostert and P. Garvin had programmed (key-punched) a computer to translate forty-nine *selected* Russian sentences with a vocabulary of two hundred and fifty words into English, thereby using only six 'syntactic' rules. The test sentences to be translated were of a very restricted syntactic nature: only simple declarative sentences were used, no negations were included, verb inflections were restricted to the third person singular or plural, only a few prepositions were included. For those readers who know Russian, a few samples of the test sentences used are given below (transliterated) ([512] Zarechnak, 1979, 22–3):

1. *Pryigotovlyayut tol*
2. *Tol pryigotovlyayut yiz* uglya*
3. *Tol pryigotovlyayetsya yiz uglya*
4. *Boyets pryigotovlyayetsya k boyu*
5. *Kachyestvo uglya opryedyelyayetsya kaloryiynostjyu*
6. *Tol pryigotovlyayetsya yiz kamyennogo uglya*
7. *Byenzyin dobivayut yiz nyeftyi*
8. *Byenzyin dobivayetsya yiz nyeftyi*
9. *Ammonyit pryigotovlyayut yiz syelyitri*
10. *Ammonyit pryigotovlyayetsya yiz syelyitri*

The six 'syntactic' rules employed in the experiment were also highly constrained. They basically dealt with two major problems of the translation process, namely, problems of lexical selection and problems of lexical arrangement in the target language. However, the scope of the rules determining lexical selection and arrangement was purely *local* in that their search limits were defined as only one step (one unit) to the left or one to the right, or both. This, by definition, restricted the rearrangement of units to simple inversion of only *continuous* units; discontinuous rearrangements were excluded. Which rules were activated and what decisions they took with regard to lexical selection and arrangement was dictated by codes or so-called diacritics attached to the lexical entries in the dictionary; the system was thus entirely driven by the dictionary, and this is how it was done:

Each dictionary entry had attached to it a two-digit or three-digit numerical code: the first digit indicated whether the particular entry was a so-called decision-point entry or a decision-cue entry. A decision-point entry meant that the computer had to make a decision as to lexical choice or word arrangement at this point, thereby using the cues coded with the adjacent decision-cue entries.

The second digit, the so-called program-initiating diacritics, indicated what kind of decision had to be taken at this point by activating one of the six 'syntactic' rules. There were six program-initiating diacritics (ranging from 0 to 5), i.e. one for each rule. The diacritics and the corresponding rules, expressed in words, were ([28] Macdonald 1963, 4):

0. The order of the original text is to be followed.
1. There is to be a difference of order in the translation from the order in the original, and an inversion is necessary.
2. There is a problem of choice; the choice depends on an indication which follows the word under consideration.
3. There is a problem of choice; the choice depends on an indication which precedes the word under consideration.
4. A word appearing in the original text is to be dropped, and no equivalent will appear in the translation.
5. At a point where there is no equivalent word in the original text, a word is to be introduced into the translation.

The above list of rules again shows that the search space, both in lexical search (rules 2 and 3) and in 'syntactic' scanning (rules 0, 1, 4 and 5) was limited to a single step to the right or to the left. Also, the rules nicely show what kind of 'syntax' the system could deal with; straight-forward word-for-word translation (rule 0), inversion of two adjacent units (rule 1), deletion of an element (rule 4) and insertion of an element (rule 5); the last rule was used to insert articles into the English translation.

The third digit, finally, was used with some decision-cue entries to mark which of two possible choices had to be made.

Thus, schematically, a dictionary entry had the following format:

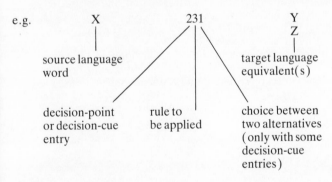

The translation process as such can be represented diagrammatically (figure 1.1).

The experiment was heralded as an unqualified success and was seen as absolute proof of the feasibility of MT. The germination of the MT seeds was over, they had now started to blossom:

'The notion of . . . fully automatic high quality mechanical translation, planted by overzealous propagandists for automatic translation on both sides of the Iron Curtain and nurtured by the wishful

Entry: Russian text (sentence).

dictionary look-up activated

Word look-up on a sentence-by-sentence basis.
Coupling of code attached to dictionary entries with matching words in the sentence.

dictionary look-up produces

Linear sequence of words with attached codes (and TL equivalent(s))
e.g.   A 111   B 23   C 25   D 14

translation algorithm activated

The program-initiating diacritics (second digit) activate the respective rules performing lexical search and 'syntactic' scanning; at decision-point entries, decisions as to lexical choice and linear word arrangement are made on the basis of neighbouring decision-cue codes.
The translation algorithm, i.e. the activation of the rule, presumably works linearly from left to right looking first for decision-points (this could not be verified in the literature).

translation algorithm produces

English translation.

Figure 1.1. Georgetown University experimental translation process.

thinking of potential users, blossomed like a vigorous weed' (Oettinger 1963, cited in [637] Dreyfus 1972, 3). And the dollars started to flow: over twenty million dollars were invested in MT over the next ten years by the various US government agencies.

But how is this experiment that raised so much hope and enthusiasm for the future prospects of MT to be seen in retrospect?

In a review of the Georgetown University experiment more than a decade later, Garvin, the linguist who had developed the 'linguistic' strategy underlying the system, characterised it in the following way ([14] Garvin 1967, 48):

1. The scope of the program was clearly specified.

Honest as this statement may be, one could express the same fact also in different words: The selected sentences, the dictionary entries and 'syntactic' rules specially tailored for these sentences created an ideal closed linguistic system thus guaranteeing success of the experiment.

2. The translation algorithm was based on the collocation of decision-points and decision-cues, rather than directly on the linguistic factors involved.

If this admission of a feeble linguistic basis and Dostert's conclusions about the experiment, namely that 'the problem of machine translation is primarily one of linguistic analysis . . .' ([28] Macdonald 1963, 5) are to be made consistent, then the only possible interpretation of 'linguistic analysis' is that of 'studying *local lexical* co-occurrences', for this was in fact the only linguistic input for the coding of decision-point and decision-cue diacritics. What is more, the fact that an algorithm based on lexical co-occurrences worked for a carefully selected set of sentences, where the individual words occurred as *tokens,* would not seem to warrant the extrapolation of such information to the point of coding it with word *types,* which would be the way in which the entries would have to be regarded if the algorithm were to have any claim to generality. Trying to put the burden of both paradigmatic and syntagmatic decisions on the shoulders of the lexicon was, at least at that time, clearly not so much the result of insights into and an understanding of sentence syntactic relations as an expression of the 'brute force' approach belief in the power of the word-for-word lexical, empirical strategy.

What, then, did the Georgetown University experiment show? It certainly did *not* prove the feasibility of MT; if it proved anything, it showed that certain non-numerical decisions could be formalised and computerised. It did, however, do two things; it provided an essentially correct formulation of the problem of MT: 'The machine translation problem is basically a decision problem' ([14] Garvin 1967, 48), and it gave quite a stimulus to the search for solutions of this decision problem; the gauntlet had been thrown down, the race was open.

*From the Georgetown University experiment
to the ALPAC report: 1954–66*

For a new and up-coming field of research to become recognised as an independent 'scientific' discipline by the outside scientific community, it has almost become a prerequisite that there should be a special journal devoted to it. In 1955, W. Locke and V. Yngve, two researchers working at MIT, started publishing the journal of *Mechanical Translation* thus providing a discussion forum for the now rapidly growing field of MT. This was followed by the publication of the first book on MT called *Machine Translation of Languages* ([27] Locke and Booth 1956), a collection of state-of-the-art papers of that time. In the same year, the first *international* conference on MT was held at MIT bringing together researchers from Britain, the USA, and Canada; the USSR, which started rapidly becoming one of the major centres of MT research, was represented by a written communication from D. Panov of the USSR Academy of Sciences.

From 1956 onwards, the dollars (and roubles) really started to flow. Between 1956 and 1959, no less than twelve research groups became established at various US universities and private corporations and research centres. In the USSR, too, the 1954 Georgetown University experiment had a great impact so that, after a first experiment in English–Russian MT in late 1955, a number of research groups sprang up, almost all of them within the Academy of Sciences in Moscow. In Britain, finally, MT research was undertaken in two universities, Birkbeck College (University of London) and Cambridge University.

The kind of optimism and enthusiasm with which researchers tackled the task of MT may be illustrated best by some prophecies of Reifler, whose views may be taken as representative of those of most MT workers at that time: '. . . it will not be very long before the remaining linguistic problems in machine translation will be solved for a number of important languages' ([40] Reifler 1958, 518), and, 'in about two years (from August 1957), we shall have a device which will at one glance read a whole page and feed what it has read into a tape recorder and thus remove all human co-operation on the input side of the translation machines' ([40] Reifler 1958, 516). At the time of writing in 1984, there are only a few operational MT systems, none of which would claim to have solved all linguistic problems, and there is still no automatic print reader for the purposes of MT.

The research efforts undertaken in both East and West during the decade 1955–65 give a lively picture of the various approaches adopted by individual groups and of the evolution of the concepts and ideas in MT as a whole. Four not always hard-and-fast distinctions can be identified as to the approaches and evolutionary tendencies in MT: the empirical versus the theoretical, word-for-word versus syntax, comprehensive MT-systems versus the fragmental approach, and first generation versus early second generation. As there is a good deal of intersection between these somewhat artificial binary distinction pairs, they will not be discussed separately, but globally.

The lack of adequate formalised theories, in both linguistics and computation, forced the early MT researchers to adopt a *brute-force empirical approach*. Initially, this meant crude word-for-word translation with no or only minimal *local* word reshuffling, the underlying idea of this approach being that, as more computer power in storage capacity and access time would become available, the remaining problems could be solved by sheer force. It is perhaps of interest to note that research on word-for-word translation continued from the first MT experiments right up to 1959 in at least two American universities, Harvard and Washington. Whilst yielding little in terms of linguistic insights, the word-for-word approach nonetheless had two beneficial effects: With its enormous demands on computer storage it led to the development of large capacity, fast-access memory devices such as the photoscopic disc. Also, it provided the field of MT with large computerised and operational dictionaries of several hundred thousand entries. In the USSR particularly, a good deal of the early research efforts went into the compilation of dictionaries and glossaries for particular scientific fields, often based on statistical material such as word frequency counts etc. The word-for-word approach and the empiricism behind it is perhaps best expressed by Yngve: 'Since a strict sentence-for-sentence translation is entirely impractical, and since word-for-word translations are surprisingly good, it seems reasonable to accept word-for-word translation as a first approximation and then see what can be done to improve it' ([38] Pendergraft 1967, 309).

A number of researchers (including Yngve) fairly quickly recognised that improvement of the word-for-word output was only possible if some kind of syntactic analysis was performed and incorporated into the MT system. However, with 'syntactic' analysis to serve mainly as a means of syntactic word-class disambiguation, 'syntactic' analysis was only done at a local word-centred level through 'syntactic' scanning of the words in the immediate neighbourhood of the word to be disambiguated; syntax, in this *direct first-generation approach,* was never used to build up a syntactic representation of the entire sentence. At its worst, syntax was no more than a matter of statistical co-occurrence information; at its best, it was a kind of local immediate constituent analysis (IC-analysis) based on the grammatical concepts of structural linguistics. The kind of empirical local IC-analysis envisaged by the RAND research group may serve as a representative example of what 'syntax' meant in late 1950 MT linguistics: '. . . it is planned to investigate hundreds of thousands of Russian consecutive word pairs in order to arrive at a revealing classification of such pairs for the purpose of reducing syntactical ambiguity, to be followed by an investigation of word triplets, etc.' ([3] Bar-Hillel 1960, 110).

What the work on dictionary compilation and early 'syntactic' analysis had in common was the fact that it was, almost without exception, narrowly corpus-based and thus empirical to a degree that allowed little or no generalisations to be made over the whole of the particular language (or language pairs) thus treated. A. Brown, the head of one of the four Georgetown University research groups, may be mentioned here as the epitome of the empirical approach: he developed his MT programme on a sentence-by-sentence basis solving the translation problems one after the other as they came along, reportedly with great ingenuity. With MT being conceived of as a pure engineering problem by most researchers, ingenuity and clever programming rather than linguistic understanding were the qualities sought after in an MT researcher.

All this taken together – a local, empirical approach, a certain linguistic naivety, and 'clever' programming – resulted in the development of quite huge, ad hoc and non-modular translation programmes with nonetheless very limited linguistic coverage, in which unsystematic linguistic descriptions of two languages and algorithmic specifications were inextricably intertwined. Systems exhibiting these characteristics (undirectionality of translation, inseparability of linguistic data and processing algorithm, and 'one-problem-one-solution' univariantness) have become known as *first-generation* systems.

However, with the advent of more formalisable linguistic theories, in particular with N. Chomsky's formal models of linguistic description (1957), a few MT researchers recognised that MT was in desperate need of a broad and sound theoretical linguistic foundation. Among them were Z. Harris, the former teacher of Chomsky and intellectual father of the concept of transformations in linguistic theory, as well as Yngve who, only a few years after his positive pronouncement on the merits of the word-for-word approach, now held the opinion that a complete syntactic and semantic analysis of both source and target language was a necessary prerequisite for MT of a worthwhile quality. Advocating rather long-term theoretical research, these few researchers were the first exponents of a tendency that became more and more pronounced: away from the empirical heuristic approach towards more theoretical linguistic foundations and away from aiming at a comprehensive operational MT system towards adopting a fragmental approach concentrating research efforts on problems of syntactic parsing.

In the USSR where, during the first years of active MT research, researchers had taken a similar empirical approach and had produced a number of rudimentary bilingual translation programmes – mostly on paper for lack of computing facilities – a corresponding shift towards more theoretical approaches, though in this case logico-mathematical rather than linguistic, became quite discernable in the late 1950s.

Such tendencies, in both East and West, becoming more and more obvious at the same time as the first empirically-based MT systems became semi-operational and were presented to a larger public (demonstrations of the Georgetown University system in 1959 and 1960, demonstration of the IBM system in 1960), were a clear indication that the initial euphoria was slowly giving way to disillusion as it started to dawn on at least some of the researchers that MT, and in particular high quality MT, was not just round the corner but miles away. Bar-Hillel's 1960 critical review and evaluation of past MT research and his pessimistic forecasts for future MT developments can therefore not just be branded, as in the past, as being the work of a confirmed MT destroyer; it was rather the natural consequence and reflection of doubts and disillusion that had grown from within the MT community itself.

Bar-Hillel, an early MT enthusiast who had turned into an MT realist (rather than pessimist), took stock of the organisational side of MT research and noted that MT had become a 'multimillion dollar affair' ([3] Bar-Hillel 1960, 92) with several hundred researchers working on problems directly or indirectly concerned with MT. In this connection he later criticised the fact that the adoption of an empirical approach to MT and the competitive and distrustful rather than co-operative atmosphere among the different MT groups had led to duplication of work and had thus resulted in the waste not only of effort and time but also of money.

In the sections on the more practical aims of MT and its theoretical foundations, Bar-Hillel tried to prove that fully automatic high-quality translation (FAHQT) was unattainable, not only in the near future but *in principle*. Arguing on the basis of the by now famous example ([3] Bar-Hillel 1960, 158) of 'The box was in the pen', he maintained that for a system to be able to disambiguate the word 'pen' in such a sentence and thus to produce FAHQT-output, it needed to have access to the wealth of world knowledge that the human being had at his disposal; this, to him, was a theoretical and practical impossibility. This, however, did not mean that MT was condemned to death; if taken as a practical, commercial tool, MT and MT research had its future in two possible ways: either rather crude MT systems could be employed in 'the very many situations where less than high quality machine output is satisfactory', such as intelligence gathering, or MT-systems could be integrated into a machine-post-editor partnership, in which case the problem of MT became that of determining 'the region of optimality in the continuum of possible divisions of labor' ([3] Bar-Hillel 1960, 95). Surely, such views and prophecies are not those of an incurable pessimist but those of a 1960 MT realist; indeed, some of these views are as valid today as they were two decades ago.

In the early 1960s, it was realised that the brute-force approach to MT did not bear the expected fruit of high-quality machine translation – not even with the deployment of massive computer power – and that systems based on this approach would soon reach a complexity and quality ceiling beyond which improvements were only possible by means of fundamentally restructuring the very bases of the systems. This caused many MT researchers to look for more solid and sound foundations of MT, thereby gradually shifting from the empirical to a more theoretical approach. This shift went hand in hand with the increased activity in theoretical linguistics sparked off by Chomsky and theoretical advances in the field of computation. Thus, a number of the US MT research teams started studying and experimenting with modern linguistic theories and models such as transformational grammar (MIT), finite state grammar and predictive analysis (National Bureau of Standards), stratificational grammar (Berkley), and formational theory (Texas). Other MT researchers drifted off into pure, theoretical linguistic studies thought to be the necessary prerequisite for further advance in MT thus effectively anticipating the conclusions of the 1966 ALPAC report.

In the USSR, too, the shift was towards more fundamental theoretical research though with a clear preference for logico-mathematical and set-theoretical models, and this mainly in the pursuit of a universal logical structure that could serve as an intermediary language for MT.

In Europe, a handful of scientists, possibly influenced and animated by the 1961 First International Conference on Machine Translation of Language and Applied Language Analysis held in England, became interested in MT and started up their own projects, notably at the University of Grenoble and at the University of Milan. And all this at a time when disenchantment and disillusion started to spread in the US MT community. However, the European teams, like the Montreal team set up in 1962, had learnt from the mistakes of their American colleagues: from the start, they based their projects on clear theoretical foundations (Tesnière's dependency theory in Grenoble, correlational grammar in Milan, transformational theory in Montreal) thus bringing in the period of *early second-generation approaches*. Their design characteristics were:

1. Maximum independence of analysis and generation, and thus at least potential multilinguality.
2. Maximum separability of algorithm and linguistic data,
3. Polyvariantness with regard to problem solutions.
4. Indirect translation via a level of transfer and possibly using an intermediary language.

Almost identical principles of system design had already been formulated in a programmatic way by D. Panov ([36] Panov 1956) as the line of attack to be followed by Russian MT research. However, due to the lack of adequate theories of linguistics and computation, they could not be adhered to then; in the early sixties they found eager followers.

In sum, this period of MT research had seen two major tendencies: first, a clear shift away from the empirical to a more theoretical approach as the researchers started to look for sounder linguistic foundations for MT as a whole. Secondly, as a direct consequence of the above tendency, a shift away from the construction of comprehensive MT systems to a fragmental approach as the researchers became aware of the complexity of the underlying linguistic problems to be solved. Whilst both these tendencies were undoubtedly beneficial for both linguistics and MT, at least in the long run, they were also one of the reasons for there not being a really operational MT system producing output of acceptable quality when it came to the crunch: the investigation of the Automatic Language Processing Advisory Committee (ALPAC) in 1964–6.

After having funded MT research for ten years with an overall budget of some twenty million dollars, the funding agencies wanted to see concrete results. In 1964, the US National Academy of Sciences set up ALPAC whose task it was to investigate the results obtained so far and, on the basis of this evaluation, to give advice as to further funding.

The six-man committee approached its task from a purely commercial point of view: in order to establish a framework within which MT could be assessed, it studied such questions as quality and cost-effectiveness of human translation, time and money required for scientists to learn Russian, amounts spent for translation within the US Government and the needs and demands for translations and translators. It is quite obvious that the first-generation systems available at the time of the investigation, with their feeble linguistic basis and the resulting deficiencies in their output, could stand up neither to the cost-effectiveness criterion measured in hard dollars nor to the still entertained expectations that MT would deliver perfect, high (if not highest) quality translation. Both these criteria are nicely expressed in the committee's evaluation of the Georgetown University MT system: 'The contention that there has been no machine translation of general scientific text is supported by the fact that when, after eight years of work, the Georgetown University MT project tried to produce useful output in 1962, they had to resort to postediting. The postedited translation took slightly longer to do and was more expensive than conventional human translation' ([1] ALPAC 1966, 19). Given such criteria and, in all fairness, given the state of MT linguistics in the early 1960s, there was only one possible conclusion for the ALPAC to draw: '. . . we do not have useful machine translation. Further, there is no immediate or predictable prospect of useful machine translation' ([1] ALPAC 1966, 32). The committee therefore recommended that research funds should be channelled into more long-term fundamental research in theoretical and computational linguistics – a shift in approach that had already started within the MT community itself.

The immediate effects of the by now notorious ALPAC report on the field of MT were disastrous. MT research was brought to a practical standstill for almost a decade, and this for two reasons: MT was devoid of all its funding and it had been branded with a stigma which it took the academic community years to overcome.

However, the more long-term effects of the ALPAC report have to be judged in a more positive light: The newly available funds in the fields of theoretical and computational linguistics allowed linguistic theory to catch up with the pretensions of proposed applications – to the unquestionable benefit of post-1975 and present-day MT.

Conclusions

Looking back at fifteen years of early MT research, at least two things can be said: first, that perhaps the early MT researchers had tried to run before they could walk and had thus preprogrammed their fall; but also that, without their pioneering work, linguistics, computational linguistics and natural language processing in general would perhaps not be what they are today. Second, that early MT, despite its mistakes and shortcomings, has produced a wealth of insights which present-day MT researchers would be well advised not to sneeze at but to take full advantage of.

AN OVERVIEW OF
POST-ALPAC DEVELOPMENTS

A quiet decade followed the ALPAC report, but this was not the only effect the report had on the field of MT. Its influence can still be seen in the issues it addressed and the place it assigned MT within the field of computational linguistics. With the conclusion that '. . . fully automatic high-quality machine translation was not going to be realized for a long time' ([1] ALPAC 1966, 25), the goals of work in MT had to be re-defined. Some obvious criteria used to differentiate systems today stem from a more refined look at the terms contained in the now well-known abbreviation FAHQMT. How 'fully automatic' a system is will depend on how much manual preparation or human interaction is foreseen in the process. To what degree the output is 'high-quality' can be reformulated as what is acceptable output with respect to a given application. As to MT, a distinct computational model of 'translation' has been developed in line with current natural language processing techniques.

Partly due to the ALPAC report, MT is not considered an integral part of mainstream computational linguistics. Its recommendation that 'linguistics should be supported as science, and should not be judged by any immediate or foreseeable contribution to practical translation' ([1] ALPAC 1966, 34) effectively eliminated MT as a possible candidate for testing new theories in computational linguistics. Although the report acknowledged that the field of MT 'had produced much valuable linguistic knowledge and insight' ([1] ALPAC 1966, 24), it proceeded to criticise all MT work on the basis of out-dated, short-term development projects. The separation of work on MT – viewed as a practical exercise – from other topics of research in the academic sector is still somewhat prevalent.

Assigning MT systems the status of mere development projects has been underscored by the commercialisation of MT systems, some of which are still based on the mid-1960s' state of the art. Moreover, most current projects in the field are both research and development projects with a view to large-scale applications specific to translation needs in the commercial sector. The growing demand for MT, claimed to be 'nonexistent' in the ALPAC report, has become an important factor in the renewal of interest in the field. This overview is meant to provide a more realistic view of the possibilities and purposes of MT today.

After ALPAC – A brief historical review

Over the past ten years a number of informative surveys and histories on MT have been published. A handbook of MT by Bruderer gives an inventory of existing systems including sample output and some relevant technical information ([114] Bruderer 1977). A good description of the linguistic models underlying MT systems can be found in a paper by Hutchins ([211] Hutchins 1979). Hutchins has also published an excellent report on the evolution of MT ([212] Hutchins 1982). An insightful review of basic developments and major problems in current MT is contained in a paper by R. Johnson ([217] Johnson 1980). A very general survey of the field emphasising the translator and the need for translation was presented at the international computational lin-guistic conference (COLING) in 1984 ([416] Slocum 1984). And finally, an in-depth study of the major MT systems can be found in a report published at UMIST ([496] Whitelock and Kilby 1983). This list of documents is by no means exhaustive.

Although it is generally acknowledged that the ALPAC report's recommendation to fund basic research in computational linguistics rather than MT was appropriate, the report itself was felt to be biased and short-sighted in its condemnation of the entire field of MT. A more accurate review of the state of the art of MT in the 1960s must be looked for in other documents.

Disagreement with the contents of the report was so strong in the academic community that a lengthy document was prepared in reply to ALPAC in an attempt to invalidate it and thereby reduce its impact. The *Commentary on the ALPAC report* reviewed the work in the field of MT from a more scientific point of view, including an informed look at the work of human translators and the (potential) need for mechanical aids ([34] Pankowicz 1967). The commentary carefully documents the mistakes and misrepre-sentations in ALPAC through a series of letters from experts in the corresponding fields and a critical analysis of the various sections of the report.

In general, two different attitudes towards MT can be distinguished during the period immediately after ALPAC, i.e. research of a more fundamental nature and an attempt to incorporate insights from the field of linguistics versus development of operational systems based on what was well known. As may be expected, private companies were producing operational systems whereas university centres were concentrating on new methods of building MT systems based on more theoretical foundations.

The operational systems were based on the 'direct' approach, i.e. the systems were designed on the basis of a direct mapping from one language to another. SYSTRAN, for example, was developed during this time and became operational at the end of the 1960s (cf. chapter 13). Another system was LOGOS, designed to translate Vietnamese into English for the US Air Force; the project was abandoned at the end of the Vietnam war. Supported by private sources, LOGOS Corporation continued work on an MT system based on developments in the field of MT and the availability of more sophisticated word-processing systems; it currently sells a German–English system implemented on a Wang office computer ([427] Staples 1983) (cf. chapter 12).

At universities, work on an 'indirect' approach to translation continued, based on theoretical developments in the field of linguistics. Influenced by the notion of 'deep structure' representations and the notion of 'universals' in language, this period can be characterised by the attempt to develop an 'interlingua' model for machine translation. Translation seen as a two-stage process – mapping the source language into a 'deep structure' – an interlingua – from which the target language could then be generated was 'indirect' in that an intermediate representation was established between the languages.

In Grenoble, CETA, the centre for automatic translation established in 1961, continued its work ([477] Vauquois 1975). The Grenoble approach was based on a dependency model to represent relationships between lexical items. The translation strategy was first to build a phrase structure for surface syntactic configurations and then to assign a 'deep structure' in the form of predicates and arguments (see chapter 3 for further discussion). In Texas the basic linguistic model was influenced by Chomsky's theory of transformational grammar to define the 'deep structure' ([26] Lehmann et al. 1972–5). Both groups abandoned this approach in favour of a more realistic 'transfer' approach which allowed for an explicit mapping of translational equivalents between language pairs in the transfer phase.

A few other university centres were established during this period, e.g. in Germany a group was set up in Saarbruecken (cf. chapter 14) and in Canada at the University of Montreal (cf. chapter 15).

In general, the decade following ALPAC was a time of relative non-activity in the field of MT. However, in the fields of computational linguistics and artificial intelligence, theoretical issues also relevant for MT were being researched. Workers in MT closely followed the developments, incorporating them into their own work on research prototypes where applicable.

By 1975, a few tentative results in the field of MT, more well-developed techniques for natural language processing and a growing interest from the commercial sector – based in part on the relative success of operational systems and in part on growing pressure for cheaper translation – brought about a new enthusiasm for machine translation.

A number of conferences were organised in the mid-seventies reflecting the renewed interest and enthusiasm in MT, including a 1976 conference organised by the US Government ([461] FBIS Seminar 1976). A quotation from R. Troike in his talk on 'The Future of MT' may serve to illustrate the mood: 'There is a widespread myth among linguists that machine translation – or, properly, machine-aided translation – which was the object of intense effort and research a decade and a half ago, was found to be a failure and has since been abandoned. Nothing, in fact, could be further from the truth . . . The time has now come for a new effort in MT to be undertaken. Properly conducted, such an effort would not only improve the quality and efficiency of translation, but would add to our knowledge of substantive universals and semantics, as well as deepen our understanding of particular languages. MT can make an important contribution to the building of the information base on which the growth of linguistic theory must depend, at the same time that it produces a result of great practical value.'

A more realistic view of machine translation and its place in computational linguistics is evident in this quotation. Significant improvements in natural language processing techniques helped to make MT a viable field of research once again. Practical results, such as the TAUM–METEO system, encouraged investment in developmental projects.

The work of the decade following ALPAC helped establish a

firm basis for a second generation of MT systems. A translation model specific to the concerns of computational treatment of natural language had emerged. The separation of linguistic data and algorithms was recognised as a principle for system design. The representation of linguistic forms (morphology and syntax) was better understood and specific areas where more research was necessary had been recognised (e.g. semantics and extralinguistic knowledge).

A number of new centres for MT have been established since 1975 in both the public and private sector. In the United States, most work in MT is limited to the private sector, e.g. ALPS, LOGOS, SYSTRAN, WEIDNER. In Europe, the activities are basically centred on universities, e.g. GETA – University of Grenoble, SUSY – University of Saarbruecken, and EUROTRA – spread across universities all over Europe. A renewed interest from the European private sector can be seen in the two MT projects in Holland, i.e. Philips (cf. chapter 18) and BSO ([503] Witkam 1983).

In Japan work in MT can be found in all sectors – public, private and university – where machine translation is seen as a main component of basic application systems in the fifth-generation project. Companies have defined systems for internal use and also participate in joint projects with university groups ([334] Nagao 1983).

The following sections provide a review of the state of the art in MT today.

On translation

On the basis of what MT systems had produced (up to 1964) and an evaluation of translation needs in the United States, the ALPAC report recommended funding in two distinct areas: 'basic research in computational linguistics' and 'the improvement of translation' ([1] ALPAC, 34). These two areas may serve as themes to discuss the basic considerations underlying the system design. Developments in MT are based on results coming from the fields of linguistics, computer science and artificial intelligence. A discussion of 'improvement of translation' must take into account the different uses and types of applications of MT systems based on the interaction with human translators and the need for and possibility of machine translation.

The translation problem specific to MT is two-fold: (1) how to analyse the source language text, assigning to it a representation which captures the 'meaning' it conveys and (2) how to generate the target language text, guaranteeing that this same information is present. Exactly how a human translator performs this task is little understood; nevertheless, the types of information used in this process are clear. The first type is information about the individual words in the text (lexicography), the second is about how these words are combined (syntax) and the third is about the meaning they convey (based on semantics, pragmatics, stylistics and real-world knowledge). All three play a role in translation and are further related to the text type and subject area (as a discipline referred to as Language for Special Purposes).

For a human translator at least part of the process is conscious as seen in the courses offered at translators' schools. In an MT system, this information must not only be explicitly stated, but also the relationship between the different kinds of information and a correspondence between source and target language must be established. The generally agreed upon standard model is referred to as the 'transfer' model, with its three stages: analysis, transfer and generation, as introduced in chapter 1.

Although the basic unit of translation is often a lexical item, the translation choice is based on the role it has been assigned in the source language representation and takes into account the possible structures and relations defined for the target language representation. Source language constituent structures and relations are also mapped on to a target language representation.

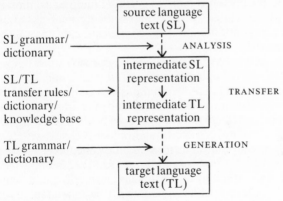

Figure 2.1. Transfer model.

The transfer model can be depicted schematically as in figure 2.1. The three stages may be further subdivided into steps treating e.g. morphology, surface and 'deeper' syntactic relations, semantic relations, and may include different devices to account for considerations of a pragmatic and stylistic nature. The basic text for translation is usually no more than one sentence. Safety-net procedures, i.e. when a complete structural analysis is not achieved, may include translation of the largest constituents identified or, in the worst case, simply the words.

The target language(s) may dictate the type of information aimed at in the intermediate representation. In a bilingual system, only information necessary for disambiguation with respect to the target language may be included (e.g. TAUM, METAL). In a multi-lingual system, a semi-abstract (or multilingual) representation of data, interpretable for a pre-defined set of languages may be defined (e.g. EUROTRA). The attempt to define representations and attributes that are abstract enough to be valid for various languages is a more realistic interlingual approach to MT. It also overlaps with current directions in AI concerned with meaning representation of a well-defined world.

The interlingual approach in MT was an attempt to represent the 'underlying meaning' of a text at a level that abstracted away from the language-specific surface realisation. This direction in MT work was influenced by earlier developments in linguistics and by psychological considerations on meaning and translation (see chapter 3). One major difficulty with the interlingual approach – aside from the complexity of defining such an abstract model – was that language-specific attributes necessary for defining translation equivalents on the lexical and structural level were neutralised in the interlingual representation, thereby complicating the task of generation considerably.

In AI projects, a type of interlingual approach to MT has also been pursued (cf. chapter 5). However, the emphasis was on generating a text from a conceptual description instead of finding lexical equivalents and resulted in a kind of paraphrase system. Another use of interlingual techniques can be seen in TITUS. It is an example of a highly restricted system which is closed with respect to the form of the input and the semantic concepts it treats (see chapter 12).

The field of terminology also provides an interlingual technique for MT. As a technique which applies to a well-defined set of lexical items, it can be incorporated into the lexicon of any MT system. Standardised technical terms can be defined in terms of concepts, hence ensuring correct translation. Other interlingual techniques may be used in transfer for areas where an adequate model exists and when explicit access to language-specific attributes is not necessary.

In principle, the transfer model allows for any number of languages to be added to a system. In practice, the choice of what information is represented will limit the possible target languages, especially once this information has been encoded in e.g. the analysis-module and the dictionaries. The extensibility of the system will depend, in part, on the relative independence of the transfer phase. That is, does this module have a well-defined input and output and to what extent does it influence/interact with analysis and generation. Many modern systems, e.g. LOGOS, METAL, were developed on the basis of only one language pair; to what degree they are extensible to other languages remains an open question. EUROTRA is an example of a multi-lingual system design which has specified from the outset a data model and a linguistic representation for several languages.

The transfer model as a basic principle for MT design reflects a modular approach to the problem of translation. By the very definition of 'transfer' from one language to another, the importance of contrastive linguistics is apparent. Lexical as well as structural equivalents must be described and implemented. Work in this discipline may provide insights into the process of human translation although this is not a primary goal for the field of MT.

The ALPAC report criticised output from an MT system on the basis that it had to be 'post-edited'. Yet every human translator knows that translated documents may be subjected to numerous revisions depending on the intended public and the practical use of the document. The notion of 'fully automatic high-quality translation' must clearly be evaluated according to a more refined set of criteria. MT will not replace qualified translators in any near future but it can serve to aid in parts of the translation process and to meet different kinds of translation needs.

According to the three aspects addressed in the term FAHQMT, i.e. 'automatic', 'quality', and 'machine translation', a classification of the different goals and applications can be discussed. With regard to automating the process of translation, a classification can be developed according to which part of the process is done by humans or machines, e.g. a text may be manually prepared for automatic processing or it may be manually revised after an initial translation. This notion interacts with the question of which parts of the translation process itself may be mechanised, ranging from dictionary consultation and translation of words in text to complete translation without any possibility for human intervention. The different approaches are classed together under the heading machine-assisted translation (MAT) (figure 2.2).

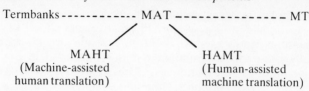

Figure 2.2.

Computerised dictionaries are the most common mechanical aid for human translators. A number of commercial systems, e.g. ALPS and WEIDNER, provide these facilities in a sophisticated word-processing environment specifically designed for translation needs. (For further discussion, see chapters 9 and 12.)

The large termbanks developed recently – containing multi-lingual translations of standardised terminology in numerous technical areas – are an attractive type of dictionary to be incorporated into MT systems. Given the complexity of encoding lexical information and simply the amount of time required for dictionary development, a goal in MT, as in other NLP applications, is to make use of what already exists, avoiding reduplication where possible. However, the integration of standard mono-lingual and bi-lingual dictionaries into MT systems poses important problems due to differences in format and in the type of data humans use as opposed to information necessary and accessible for computational processes. From one MT system to another, the information varies greatly according to the strategies employed.

The success of SYSTRAN, one of the oldest MT systems, is largely due to the extensive and continued development of the dictionary component. Words are encoded with their specific contexts and are constantly augmented via the users. The sheer size of the dictionary provides the basis for understandable, if not elegant, translation. The user environment for dictionary maintenance is an important consideration for any operational MT system.

At the other end of the classification scale is the fully automatic MT system. The one example of such a system is TAUM–METEO (see chapter 15). Used for the translation of weather bulletins in Canada, the system automatically recognises texts it cannot translate and sends them to a human translator. No revision is necessary for automatically translated texts.

Machine-assisted translation (MAT) is what is usually meant when talking about computer translation; all current systems require human intervention at some stage of the process. As indicated in the scheme above, MAT can be subdivided into (1) human-assisted MT, i.e. the human prepares a text for automatic treatment (pre-editing) or interacts with the system during processing and (2) machine-assisted human translation, a more realistic view of most current MT systems whose output is usually revised (post-edited).

A text that is manually prepared for automatic translation may be done by a human who has only source language knowledge. Pre-editing can mean re-writing a text according to a pre-defined syntax (as in TITUS) or pre-analysing by marking constituents with specific information. In the CULT system, for example, syntactic and semantic indicators are inserted into the text before translation; after translation some manual revision is also necessary. This system has been successfully used to translate mathematics and physics papers from Chinese to English for a number of years ([283] Loh et al. 1978).

Interactive systems require human intervention throughout the translation process. The human is called on to resolve ambiguities as they arise. The first experimental system, based on a transfer approach, was the MIND system developed in the early seventies ([601] Kay 1973). For example, during analysis of the sentence 'I saw the girl with a telescope', the machine would ask whether 'with telescope' referred to the verb 'see' or to the object 'girl'. Another example of an interactive system is MAHA, developed in Japan by Mitsubishi as an aid for writing business letters ([180] Fukushima 1983). MAHA supplies the user with a number of model Japanese sentences to choose from according to subject matter and stylistic preference. A translation is then generated, based on the user's choice and on an analysis of the user's input. This type of system is only possible due to the restricted type of text and subject field.

The question of how much post-editing any text requires will vary according to its intended use. Given that current MT systems provide understandable translations, almost all of which need revision, the question of quality must be judged relative to the purpose of the output.

Information acquisition is an argument often put forward for the development of MT – a 'rough' translation is better than no translation at all. Some of the Japanese systems currently under development reflect this concern; one system translates only titles of scientific documents, another only abstracts.

The major goal for developmental projects is to provide a translation of a text which is accurate and can be understood by the human translator who will revise it. Commercial systems stress the reduction in cost and manpower that an MT system provides. Partial automation of the translation process is felt to be an important step towards meeting the growing translation needs and reducing the cost of translation.

The increase in translation demands can be traced to three major sources. The need to keep up with international developments, especially in technical fields, has encouraged the development of MT as a means of information acquisition. Another source of expansion in the field of MT is the need for adequate translation to assure the dissemination of information. For example, a Japanese exporter of technical products must ensure that the documentation is accurate and understood by the recipients. A third explanation, by no means independent of the other two, is the political emphasis on 'multi-linguality'. One obvious example here is the right to documentation in the national language for all members of the European Economic Community (cf. chapter 19).

In order to meet these needs, two approaches to system development have been pursued: limited applications to help reduce the problems an MT system must solve; and general purpose systems, although unable to provide perfect output, as an aid to translation. Most of the systems presented in this book are general purpose, in some cases restricted with respect to subject area and text type. Limited systems such as TITUS can be described according to the artificial limitations they impose on a text with respect to syntactic structures and semantic domains. Sublanguage research into possible MT applications has concentrated on 'natural' restrictions found in certain text types and subject areas ([604] Kittredge and Lehrberger 1982).

Linguistic and computational considerations

Like other systems which treat natural language, an MT system is based on the ability to represent, access and process large amounts of linguistic and extralinguistic knowledge. The transfer model (as described above) provides current system designers with a framework in which to combine results from the disciplines of generative linguistics, artificial intelligence and computer science. The role of these three disciplines in MT system design have been addressed in other chapters of this book (cf. the chapters on linguistics, software, and AI). This section will therefore only briefly discuss some of the problems which an MT system must address, taking into account partial solutions that current implementations provide.

The advances in MT systems which serve to distinguish second from first generation systems fall into the three major categories of linguistics, knowledge representation and processing techniques. As to linguistics, the need to base the system on a formal model of language had been recognised even before the ALPAC report was published. The need to incorporate semantic information and world knowledge had also been recognised with the now well-known statement from Yngve in 1964 that 'Work in mechanical translation [had] come up against a semantic barrier . . .'. Different means of representing meaning were pursued, for example in the interlingual approach and by studying results from the field of AI where the focus is on the relationship between text and knowledge structures.

The first major advance in processing techniques was to separate the data from the algorithms, i.e. the linguist could write linguistic rules without reference to the processes that applied them. The type of programming languages has also been investigated in different directions. Whereas Q-systems (TAUM-METEO) were a general tool developed for the linguist, at GETA the emphasis is on developing constrained tools specific to a linguistic process. Although standard programming languages have been used in many systems, e.g. Fortran in SUSY, Lisp in METAL, a current trend is to develop more problem-oriented languages for natural language processing and specifically machine translation (cf. chapter 11).

The problems MT systems address are to a large extent well-known in the field of computational linguistics ([588] Grishman 1984 and [625] Winograd 1983). The problems of natural language representation, the treatment of ambiguity and how to reduce complexity and ensure efficiency are not specific to MT. Prototype MT systems can serve as a testbed for natural language experiments in each of the fields independently; operational MT systems test 'real' language treatment on a large scale.

According to the three-phase process of analysis, transfer and generation, specific types of linguistic problems can be identified. Reflecting a modular approach, problems specific to a single language are, in principle, limited to analysis and generation; the transfer phase addresses the bi-lingual problems specific to translation. This separation is, however, not always possible, nor necessarily efficient. For example, mono-lingual criteria applied in analysis to determine word-senses may not be sufficient for the transfer phase. The verb 'to know' may seem unambiguous in analysis but the lexical equivalent in e.g. French can be either 'connaître' or 'savoir' and a choice must be made in transfer. The assignment of tasks according to analysis and transfer is a useful organising principle, realised differently in each system according to the languages, the computational model, etc.

The problem of ambiguity is central to all MT systems. For 'direct' systems, which basically translated words, without the larger structural context, the problem was overwhelming. Current MT systems have recognised the need for a strong semantic component to aid in the disambiguation process at the various levels.

In analysis, ambiguities may arise due to homography, ellipsis, anaphora, etc. Where no general solution exists (e.g. conjunction), practical MT systems must rely on an ad hoc treatment of the phenomena. When ambiguity is not resolved or not recognised in analysis, the transfer component may not be able to compute the correct translational equivalent. A few examples will serve to illustrate the problem of ambiguity.

For anaphora:

He dropped the bottle on the floor and it broke.

In order to ensure correct gender agreement in a language like French or German, it must be clear that the pronoun 'it' refers to the bottle and not the floor.

For homonymy:

He went to the board.

The translation will obviously depend on the meaning of board, i.e. the blackboard or the board of directors.

The example of the 'board' is a clear case of monolingual lexical ambiguity; the distinction between this type of ambiguity and multiple lexical equivalents is, however, not always so clear. The following examples serve to illustrate some specific lexical problems for translation.

For multiple lexical equivalents (source language English): *corner* may be translated in Spanish as *rincon* (inside) or *esquina* (outside). The transfer dictionary may state the following context for correct transfer: 'in the corner' = *rincon* and 'on the corner' = *esquina*. These types of entries for transfer imply a very large and explicit bilingual dictionary. A further example is: *put*→ German *stellen* (put something that is upright, e.g. a bottle) and *put*→ German *legen* (put something that is flat, e.g. a piece of paper).

In order to resolve these problems systematically, a strong semantic component is necessary. For lack of any formal semantic model applicable to translation, current MT systems concentrate on defining specific semantic criteria in the form of features used as selectional restrictions to resolve ambiguity.

Representing linguistic and extra-linguistic information in a way that is adequate with respect to content and to the form in which it is stated is another important issue in natural language processing (NLP). MT, as a process going from source language strings to representations and then to target language strings is faced with the problem of the complexity of manually describing all of the structural and semantic characteristics of the texts and controlling their interaction in the automatic process. First generation systems, relying on rather unstructured lexical information quickly reached a threshold for improvement due to unforeseeable interactions between words in differing contexts. The more modular approach currently pursued (e.g. separation of data and algorithms, better understood linguistic levels, the transfer phase and high-level programming languages) has improved the situation. However, the growing awareness of the complexity of natural language and the need to draw on information from numerous domains including common sense and encyclopaedic knowledge and the vaguely understood domains of pragmatics and discourse has added to the difficulties.

The linguistic knowledge ranges from information about the surface syntactic constructions of objects in the text to the underlying relations of a conceptual nature. The linguist must be able to state syntactic and semantic expectations for 'shallow' and 'deeper' levels in order to build representations and state the relationship between them. For areas where no formal model exists, explicit reference to individual lexical items and surface word-order may be necessary. To compute translational equivalents, access to all the 'levels' is often called for, especially when the aim is to keep the transfer component as simple as possible, i.e. a lexical substitution process.

The constituent structure is represented as a phrase structure or dependency tree in most MT systems; this is augmented by case frames, co-indexing devices, etc. In place of a formal representation of meaning, semantic attributes such as 'concrete', 'animate', and 'machine-like' are used for ambiguity resolution. For example, in the sentence

The printer was running all day

the verb will be interpreted as either 'locomotion' or 'operation' on the basis of the attribute assigned to printer, i.e. either 'animate' or 'machine-like'. In this example, the ambiguity can only be resolved according to a larger context (a reference in the preceding text to computers) or knowledge about the subject field (a text on physical activity versus a description of an automated office). The use of AI techniques to state preferences about expectations can serve to resolve this type of ambiguity.

Another example of AI representation methods can be found in work on the MT system SALAT in Heidelberg ([198] Hauenschild et al. 1978). The SALAT machine translation system experimented with deduction processes for pronoun resolution, disambiguation of alternative analyses and for choosing correct lexical equivalents by means of logical formulae (see chapter 5 for discussion).

Two major problems for computational treatment of natural languages are the complexity of how the human being communicates with the machine and the efficiency of the actual processing. (For a description of computational techniques and a more technical discussion of the concerns specific to the computer scientist, see the chapters on software.)

One method of overcoming the ambiguity problem and the inability to adequately characterise certain aspects of natural language in MT was the interactive approach. The major problem with this method, aside from the difficulty of identifying the appropriate places for human interaction and the formalisation of the questions put to the human being, was the inefficiency in real-time it introduces. The translation process is slowed down to such an extent that it is not usually considered a practical alternative for general purpose MT systems.

In fully automatic systems as well, machine efficiency is still a problem. The large grammars that define the representations require a powerful and efficient computational component. The large body of dictionary knowledge that must be stored and accessed poses problems similar to other database applications.

Communication between the grammar writer and the machine, during development of the system, and between the user and the MT system, once it has become operational, is another important concern in MT. A full-scale system is written by a group of people, posing problems of consistency and integrity with respect to the grammar rules, the dictionary entries and the interaction between the components. Once an MT system is operational, a well-developed user environment for, for example, dictionary update is necessary. Commercial systems, in particular, have invested a great deal of energy in developing an appropriate environment for the user.

Two major concerns for practical MT systems are adequacy with respect to a given application and extensibility of a system. For operational systems, expansion may be defined in terms of vocabulary expansion, new text types and coverage of different languages. For research projects which work on experimental prototypes, extensibility means developing a full-scale system. The development of a prototype which tests new and more powerful techniques does not guarantee that it can serve as a basis for an operational system. Similarly, new techniques developed in other fields of NLP and AI are not necessarily directly applicable to or consistent with current MT systems. With the exception of Japan, where MT is considered an integral part of fifth generation work, research in NLP does not directly address problems in MT.

Most commercial systems are based on 'old technology' due to the time and expense involved in constructing a full-scale MT system. Only a small portion of the work goes towards system design; most of the time is devoted to writing grammars and encoding dictionary information. Since most MT projects have been funded as both research and development projects, with a rather short- term goal of building an operational system, flexibility and progress in the field has been slowed down in comparison to other computational linguistic projects.

Conclusions

The status of work in the field of MT has only slowly overcome the devastating effects of the ALPAC report. The political and economic pressures for practical MT have helped make research in the field a useful and interesting enterprise in the commercial and academic sector.

One future direction in MT is to test large-scale implementations of well-established principles in computational linguistics and MT. METAL is an example of such a second-generation system, only recently released to the public. The second direction is to pursue more fundamental long-term research into problems specific to machine translation and into the more general problems of natural language processing, using MT as a testbed.

The developments in the field of past and current systems are reflected in this book. These systems attest to the importance and potential for advance in the field. Although FAHQMT remains a long-term goal, the possibility of a third generation of MT systems, incorporating more powerful semantic methods, is foreseeable in the near future.

LINGUISTIC THEORY AND
EARLY MACHINE TRANSLATION

Whatever other factors machine translation (MT) and linguistics may or may not share as disciplines, they have in common an interest in human language although they approach it from different points of view. But does that observation capture all there is to their relationship? After all, MT is essentially a practical enterprise concerned with converting texts from one language into another by means of computers. Most of us would agree that tasks of such a practical nature benefit from reference to a framework in order to guarantee consistency in the approach. Linguistics concerns itself with providing such frameworks for the characterisation of human languages. Can MT benefit from guidance by linguistic theory?

It is an interesting question to what extent linguistic theories have influenced MT in the past, and in particular, whether the spectacular rise of generative linguistics has had an effect on the approaches adopted by MT systems. The area is of course a large one but can be restricted. First of all, unless an MT system incorporates at some stage a static representation which describes the text being translated, it is difficult to assess what linguistic information has been used in the translation process. This effectively reduces the area to be investigated to indirect MT systems – as opposed to direct ones which encode that information in a procedure from which it can be extracted only with great difficulty. Secondly, the study will be limited to the period between 1955 and 1966, eleven years which witnessed the appearance of the generative approach in linguistics and during which the accent in MT shifted from direct to indirect systems. Thirdly, the interest in the rise of generative linguistics warrants the exclusion of all MT systems not explicitly concerned with syntax.

MT in search of some foundations

The influence of linguistic considerations on the very origin of MT is highly questionable. It seems that people first started to think about automatic translation in the early 1950s, when many maths departments suddenly gained access to expensive computers which, because of their cost, had to be put to good – i.e. full time – use. As a consequence, the people who knew how to run the machines started to look for new applications, overconfident in their estimate of the kind of tasks computers could perform. Expectations of what computers could do for humanity boomed. MT emerged as one of those tasks, and the discipline has paid in later years for the over-confidence which marked its early days.

But it would be wrong to suggest that MT led an existence only to the point where it kept a computer going. Some people had a genuine interest in the field and its theoretical aspects and they tried to give it a framework of its own, or at least tie it in with established disciplines like mathematics, information theory or even psychology. Their efforts resulted in a number of assumptions concerning the why and how of MT, some of which have survived into the present. Oddly enough, none of these pioneers were linguists and their contributions reflect their particular specialities.

The acknowledged father of MT is Warren Weaver, a mathematician who had been involved in cryptography during the Second World War and who had established a reputation in information theory. In 1949 he circulated a memorandum among his colleagues to solicit their interest in MT (see chapter 1). In it, he reflects that when looking at a text in Chinese he is, in a way, looking at a text in English written down in some strange code. Since there is a proof available in information theory that every code is decipherable, and since languages can be viewed as ciphers for each other, it follows that every text is translatable.

As Mounin ([30] Mounin 1964) points out, Weaver's metaphor is easily attacked. Human languages cannot appropriately be represented as codes for one another. First, the relation between a text and its coded version is invariant and mathematically calculable from the form (the surface) of the original text.

The same cannot be said of the relation between human languages, or texts which are translations of one another. Secondly (a more practical point), it is not necessary to know a language in order to be able to decipher a code for it. On the other hand, it is impossible to translate a text into a language one does not speak.

But in spite of the clumsy metaphor, Weaver has indirectly established one of the principles of MT: for any text expressed in one language there is a text in any other language which is its translation and such that the relation between the two texts can be calculated by a machine.

As to how to translate, Weaver proposes to descend to the common basis of all human linguistic communication, which he thinks can be formalised in a true universal language. The translational model which he introduces can be represented schematically as in figure 3.1.

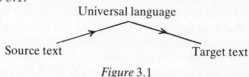

Figure 3.1

Because of his profession, Weaver's reasoning must be placed against the background of information theory. Communication is defined as the transmission of information from a source to a recipient by means of a vehicle in which the message is encoded and the properties of which must be known to both parties (say, a language). Information is seen as quantifiable and the amount of information conveyed stands in a direct relation to the properties of the language it is expressed in, the capacity of the channel through which the message is sent and the length of the transmission. Given this background it is understandable that Weaver should equate meaning and information and that he should treat meaning as an equally quantifiable concept.

Although Weaver does not describe in detail what this universal language would look like, the resulting model for MT can, even today, count a fair number of followers. It assumes that all human languages are used to express messages the content of which can be represented in a universal formalism which is not prejudiced with respect to the characteristics of those languages. In other words, the 'true universal language' which Weaver proposes as the basis for MT is the same as what has since been called the interlingua. Although it seems uncertain whether anyone will ever be able to define such a formalism, the model for translation which it entails is still a familiar one.

Information theory provided the basis for another approach to MT. Weaver and Shannon had developed a scheme for human communication which adheres to the principles outlined in the previous paragraphs and which can be roughly sketched as in figure 3.2.

Source $\xrightarrow{\text{Message}}$ Destination

Figure 3.2

Their research concentrated on the 'Message': the relation between the quantity of information conveyed and the nature and
distribution of the elements that can carry it in a particular coding
system. This model was taken up by Richards and Yngve who
adapted it to the requirements of translation and MT respectively.
Richards postulated that a theory for translation presupposes a
theory of meaning based on the study of 'comprehension'. He
believed that coding (expressing a message in a particular language) assumes reference to the situations which have given rise to
the message and to what the current message has in common with
similar messages covering similar situations. The link between situation and utterance, though, is unique for each language. Yngve
accepts this general point of view but expands on it by remarking
that it is difficult to pin down what 'meaning' is in more detail, let
alone formalise it. The only thing he believes one can be sure of is
that, if both source and recipient speak the same language, then a
message will be understood by the recipient as conveying what it
was intended to convey at the source. Since 'meaning' is an inconve-
nient notion for formalisation and since the link between situation
and utterance depends on the language of expression, translation
must involve a mapping between coding systems (languages). The
resulting model is illustrated in figure 3.3.

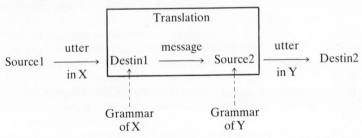

Figure 3.3

The important change here is the recognition that 'comprehension' has to do with the link between situation (meaning?) and
utterance as provided for by the coding schemes prescribed by a
particular language. The existence of describable universal mean-

ing is not challenged but it is judged to be too vague to be of immediate use.

This model can be looked at as the immediate predecessor of indirect-transfer MT systems, although in 1933 Trojanskij had already proposed this approach, as reported in chapter 1. Unfortunately, Trojanskij formulated his proposal twenty years too early and by the time a more general interest in MT had emerged his work had apparently been forgotten.

All approaches taken to MT investigated so far were based on information theory, but the link between the two disciplines was not unanimously accepted among the pioneers. Quite a violent attack on these views came from the psychology corner and has been worded by Ceccato ([6] Ceccato 1961).

As a psychologist Ceccato held the conviction that language cannot be seen independently from thought. Thought has a structure, which languages reflect and which can be discovered from clues found in signs and sounds and their order. The distribution of these indicators is different for each language. This view has a number of consequences. First of all one cannot establish a direct correspondence between languages: word-for-word translation is clearly absurd. Secondly, translation via an intermediary language is possibly even more absurd as the representation can only duplicate whatever information is already present in the source and target texts. Ceccato believes that similarities between languages cannot be systematic but must be accidental and trying to capture them is a futile exercise. Thirdly, because sentences are by no means units of thought, they are useless as units of translation.

According to Ceccato, translation requires that the translator reconstruct the thought processes reflected in the source text and derive the target text on the basis of those processes and the signs and rules for their distribution as dictated by the target language. Since translation involves active thought on behalf of the translator, any hopes for MT can be abandoned straight away as long as the effort is directed towards: 'linguistic studies whose object is . . . formal similarities found in words of various languages, or, to be more precise, found in the particularly fortunate cases of languages that happen to show such similarities' ([6] Ceccato 1961, 15). As an alternative he proposes that more time be spent investigating human thought processes.

From the above quotation it becomes clear that Ceccato has no confidence in work done in linguistics. One can even describe his attitude towards linguistic theory contributing to MT as hostile. This rejection is based on the opinion that languages can and do differ from each other in random ways (a view held even by some linguists at the time; cf. the Sapir–Whorf hypothesis) and results in a denial of what many linguists would nowadays see as the very foundation of linguistic theory: searching and accounting for language universals.

Ceccato's views conclude the investigation into the origin of the various methodologies advocated for indirect MT, and one can safely conclude that the influence of linguistics on the different approaches taken has been sparse or absent. On the other hand, linguistics as a discipline is not particularly oriented towards accounting for translation so that the above conclusion may not be particularly surprising. Indirect MT, though, presupposes that source and target text are assigned a representation (possibly the same one) which makes explicit the information which is necessary for the translation. Most of that information will be linguistic. Can one not assume that the mappings between texts and representations would be drawn from work done in linguistics, a discipline which not only seeks to describe human languages but which even uses similar mechanisms to represent the structural properties of texts?

Mainstream linguistics I: structuralism

Most of the very early MT systems worked word for word and used linguistic information rather sparsely. In fact, most of the operations were of the word-on-the-left, word-on-the-right type. In the introduction this lack of linguistic awareness has been ascribed to the fact that the people creating and working with these systems usually did so because they knew how to operate computers. But Georges Mounin ([30] Mounin 1964) gives another reason. He believes the situation may be explained by the fact that mainstream linguistics at the time was not very helpful for the purposes of MT.

The most widely accepted linguistic theory in the early- and mid-1950s was – as it had been ever since 1930 – structuralism. It was rooted in empiricism, a view of science, scientific methodology and theory construction which postulated that all knowledge is either explicitly acquired or is derived from experience. This implies that generalisations must be linked to observation. Structuralist linguistics culminated in the work of Bloomfield who defined the relation between a corpus and its theoretical description as based on the principle that 'the only useful generalisations about language are inductive generalisations' ([609] Newmeyer 1980, 4). The goal of linguistic theory thus became to discover a grammar from a corpus of data by performing a number of well-defined operations on it. The starting point for all investigation had to be the physical record of the flow of speech, since this was the only type of data considered suitable for objective description. The description of language would proceed from there to the discovery of other linguistic phenomena in a strict order: phonemics first, then morphemics, syntax and finally discourse. This ordering of levels was important: morphemes can only be discovered thanks to an adequate description of the phonemic level, etc. It is therefore out of the question that morphemic, leave alone syntactic or discourse information should interfere with the description of the corpus at the phonemic level. The principle behind the approach rests in a concept of hierarchy in the building blocks of language. One starts by describing units, then groups of those units, then groups of groups of those units and so on.

Structural linguistics had considerable success, especially in the fields of phonemics and morphemics. As a matter of fact, the linguistic community was quite pleased with itself, not only about what they had achieved already, but equally about what they thought might yet be accomplished. They were confident that the procedures which had been developed for corpus analysis were complete apart from filling out the details. Once that was done, these principles could be fed into a computer, run on a corpus, and a grammar for the language would fall out.

Early MT systems were influenced by this approach from two unexpected angles. First, empiricism as a methodology. The MT system at RAND corporation, for instance ([18 and 19] Hays 1961) worked from a set of rules for translation which were extracted from a corpus. The system was expected, in a first instance, to deal with that corpus only. It was hoped that by extending the corpus and gradually adjusting the set of rules one would arrive at a set of instructions sufficient to translate any text from source to target language. Secondly, maybe less important, confidence. People were convinced that MT no longer held problems and that in a matter of months a working system could be marketed and exploited commercially. The example here is the word-for-word MT effort at Georgetown University.

But these similarities do not go deep. In many ways, the assumptions underlying indirect MT as described in the previous section cannot be reconciled with structuralism. First of all, Weaver suggests that the basis of human communication is not only universal, it is also describable. Structuralism in general shies away from assigning 'meaning' a significant role in their procedures. Bloomfield even warns linguists that to give a scientifically accurate definition of 'meaning' would involve a scientifically accurate description of every aspect of the speaker's world: an unattainable goal ([609] Newmeyer 1980, 9). Secondly, both Weaver's model and Richards and Yngve's model presuppose a mapping from text to representation. The latter even explicitly assumes a set of grammar rules for both source and target languages. Nevertheless, they do not insist that grammar be discovered via a particular procedure applied to a corpus. In designing their model they do not insist on the same strictly empiricist methodology. Maybe Mounin was right in suspecting that the aims of structural linguistics and MT were not easily married.

Mainstream linguistics II: generativism

Despite the optimistic atmosphere in structuralist circles in the early fifties the golden days were drawing to a close as the generativist approach to linguistics emerged. Newmeyer ([609] 1980, 36) reports that by 1950 logicians simply assumed that human language, just like formal languages, was defined by a finite set of recursive rules. In 1953 Bar-Hillel published an article in *Language* in which he sets out to 'present an outline of a method of syntactic description that is new insofar as it combines methods developed by . . . Adjuciewicz on the one hand and American structural linguists on the other' ([2] Bar-Hillel 1953, 47). He states that 'this could be of value in those situations in which a completely mechanical procedure is required for discovering the syntactic structure of a given thing. Such a situation arises, for instance, in connection with the problem of mechanical translation' (ibid.). The notation he proposes is an extension of Adjuciewicz' categorial grammar, a formalism designed to define the well-formedness conditions for a certain class of formal languages. It consisted of a finite vocabulary of primitive categories (say N, S, . . .) which could be used in the construction of complex categories (e.g. S/N, S/N/N, etc.), and a set of cancellation rules which determine how, in a well-formed formula, items of specific categories can be combined. Bar-Hillel's refinements to the syntax of the system amounted to the addition of restrictions on which part of the context could be used by a particular cancellation.

Important here is not the notation, but what Bar-Hillel intended the formalism to achieve. He saw that MT had a need for syntactic structural analysis and emphasised the necessity for an adequate definition of the resulting syntactic representation, as opposed to 'developing a method which a linguist might use to arrive at the analysis of a linguistic corpus' (ibid.).

According to Newmeyer, this contribution is to be regarded as the first published attempt in the US to formulate rules for the generation of sentences in natural language. It is worth noting that the motivation must be sought in an investigation of what is needed for a mechanical treatment of language (such as MT).

But the real blow to structuralism came somewhat later, with the publication of *Syntactic Structures* ([570] Chomsky 1957). Chomsky noticed that natural sciences had cast off the empiricist methodology they adhered to in the thirties and that they had adopted a deductive method of theory formation. He applied the same principles to linguistics, and viewed a grammar as a theory of language, not as a mechanical abstraction over a corpus. A grammar becomes a formal axiomatic system which generates (describes, predicts, characterises) by means of a finite set of rules the infinite number of all but only those sentences which belong to the language. The rules are postulated and are held to be accurate until they are falsified by the data. A theory is thus arrived at deductively: the grammar is hypothesised and is accompanied by an evaluation procedure which defines how well the theory describes the data. 'Discovery' of the grammar is no longer an issue and, by extension, neither is the procedure defined for that purpose. The strict hierarchy of separated levels of description is no longer important.

Many variations on Chomsky's original transformational grammar have arisen over the years, but the deductive generative approach to linguistic description has become widely accepted in the western world. Is it the case that the aims of this new view of linguistics are as incompatible with the needs of MT as the structuralist view had been? The very fact that Bar-Hillel's motivation for attempting to promote a generative methodology explicitly involves his concern with the needs of mechanical language processing seems to suggest that the two must be compatible. But MT remains a practical enterprise: it requires physically observable results which can be evaluated against human translation and which are sufficiently accurate and precise to allow for commercial development. Obviously an MT system would benefit from reference to a complete generative grammar for a particular language, but the fact remains that no such grammar has yet been developed. The relevance of generative linguistics for MT is less practical.

Translation is a one-off activity: translating a text between a particular language pair once is enough. A system designed to translate one corpus (a physically limited number of strings) only is of no great consequence. The interest of generative grammars for

MT lies in the ease with which they can account for infinite sets of sentences by describing them in terms of classes of strings that share a derivational history and that have the same structural properties. Even though no grammar has as yet been defined that accounts for the whole of a particular language, it is possible to describe a subset, still infinite, of those sentences of a language which are instances of particular linguistic phenomena. In order to design an MT system that would exploit these advantages one needs to know what type of text is being translated. In other words, one must redefine the notion of corpus, not as a limited number of sentences, but in terms of the linguistic phenomena which occur in it and which can be described by means of a generative grammar. Furthermore, one must ask what information is necessary to translate any text which includes these phenomena, how it is to be represented, and how it can be obtained. This way of proceeding goes against empiricism in the sense that the corpus is no longer the source from which the grammar is to be extracted by observation but becomes the object which the grammar defines by fiat.

The empirical approach is severely attacked by Bar-Hillel ([3] Bar-Hillel 1960), who judges that the importance of statistical methods for MT has been overestimated and should be considered as a hangover from the statistical theory of communication. He rejects the empirical approach to MT because it behaves as if no other work has been done, citing MT at RAND corporation as an example. On the other hand, he has doubts about the relevance of transformational grammar for MT. In 1960, phrase structure grammars were widely considered inadequate for generating human languages, but Bar-Hillel is sceptical about transformations which he believes to be very complex and ill understood. He points to Yngve's project at MIT as a serious and positive enterprise which does not underestimate the importance of fully automatic high-quality translation (FAHQMT), concentrates on the syntactic and semantic analysis of source and target language and which, although influenced by structuralist methods, takes Chomsky's work into account. Funnily enough, the MIT MT project played a somewhat controversial role with respect to Chomsky's ideas, as will be seen later on. But first let us have a closer look at the MIT system.

MT at MIT

When Bar-Hillel left MIT in 1953 he was replaced by Yngve who immediately embarked upon MT. Yngve ([54, 56] Yngve 1961, 1967) stressed the need for long-term linguistic research to assist

MT, as opposed to the attitude of his colleagues, who believed that the problem was easy. His motivation was that word-for-word translation would not yield the expected results and that many ambiguity problems encountered by those systems could be resolved combinatorially by syntactic analysis of the text under translation. Although he pointed out that satisfactory completion of the research would take a long time, he did specify how the results should be used. The model for MT which he proposes can be summarised as in figure 3.4.

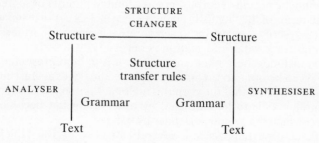

Figure 3.4

The aim was completeness in the sense that the system should cover as much as possible of the syntax of the languages involved. In practice the model had been split up into six parts resulting from a division into three modules (analysis, transfer and synthesis) each with two components (static descriptive knowledge and procedural knowledge or programme). The grammar of source and target languages and the rules for mapping between representations make up the declarative knowledge; the procedural aspect of the model (in bold in figure 3.4) specifies how to use that knowledge in a programme. Yngve's group developed a special programming language (COMIT) in which these procedures could be more easily expressed than in the languages available at the time.

The translation process itself happens in three steps: recognition of a structure for the input text, selection of the structure for the target text (the one which will yield the best translation equivalence), and production of the output text. The set up requires two language dependent representations of input and output texts which Yngve calls 'specifiers'. An example is given in figure 3.5 ([56] Yngve 1967, 482).

The specifiers include information on the constituency structure of the text, categorisation and some syntactic functions. Movement phenomena are represented by allowing for crossing branches in

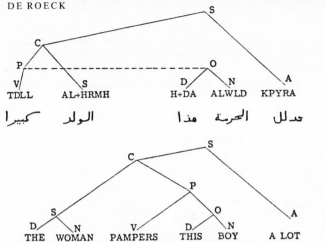

Figure 3.5. Syntactic translation of Arabic to English, showing how the translation is structure-for-structure in spite of word-order differences.

the tree. Matthews ([29] Matthews 1961), one of Yngve's collaborators, explains the status of these specifiers within a linguistic framework. A specifier combined with a grammar defines a sentence. A grammar then is a recursive function consisting of an ordered finite set of rules. The rules can be optional or obligatory and optional rules may be recursive. If one specifies which optional rules apply to a given structure as well as how many times to apply them, then the resulting mapping between representation and string is unique. Matthews points out that analysis amounts to recognition of a sentence as belonging to a class of strings defined by a specifier. Transfer then becomes a mapping between two specifiers which have translation equivalents among the classes of strings which they characterise. Clearly the framework described by Matthews is a generative one. But has it been influenced by views held in linguistics at the time? Can one talk about generative mainstream linguistic theory in the 1960s without reference to transformational grammar?

As a matter of fact, the Yngve group did not rate transformational grammar (TG) very highly. Although some of the terminology they use (optional versus obligatory, etc.) may lead one to suspect that they adopted some of the concepts of TG, they did not use transformations at all but were violently opposed to them. The transformation versus phrase structure battle at MIT even got slightly out of hand.

In 1963 Harman ([589]), at the time a collaborator of Yngve's, published an article in *Language* which he called a defence of phrase structure. In his footnotes he recorded gratitude to Yngve for suggestions and comments and continued to outline the phrase structure grammar used in the MIT MT project. The grammar has two kinds of rules:

A = Z where A is a constituent and Z a (possibly empty) string of constituents.

A = X + ... + Y where A, X and Y are constituents. This rule means: rewrite A as X and, in the resulting derivation string, insert Y after whatever followed A. This produces a structure:

A grammar of this kind can cope with discontinuity and strongly generates structures of the type illustrated in figure 3.6.

Grammar

S = A + B
A = A + D
A = E + ... + F
B = G + H
G = H + ... + I

Structure

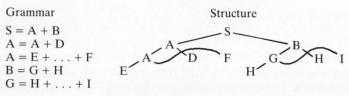

Figure 3.6

Harman admits that grammars of this kind can grow very large and introduces a system of subscripts to reduce their size. He argues that a fully specified set of phrase structure rules would be about as long as a transformational grammar dealing with the same phenomena, that it expresses the same syntactic information, that it does not produce spurious structure and that it is at least as well motivated.

Chomsky ([571] Chomsky 1965) dedicated a long footnote to Harman's publication claiming that the argument is based on a total misunderstanding of the notion of phrase structure grammar. He reduces the issue to a 'confusion' claiming that Harman has not even challenged his assertion of the inadequacy of phrase structure grammar.

From the above it has become clear that the MT group at MIT made a genuine effort to incorporate linguistic research into the field. They developed a framework which is generative in spirit, a property it shared with Chomsky's TG. But that framework was clearly not influenced by mainstream linguistics which at that time was strongly biased towards a transformational approach. The MIT group even entered into open conflict with that aspect of Chomsky's work. The theoretical background to their system can be seen as evolving in parallel with transformational linguistics as a continuation of Bar-Hillel's pioneer work on generative grammar. Their effort to define an alternative to TG by defending phrase structure rules and developing mechanisms to facilitate their use was to be repeated more successfully some twenty years later by Gazdar ([585] Gazdar 1982). At that time, they were fighting a losing battle.

Alternative linguistic theories

Not least because of the publicity it received, TG had such a widespread impact in the western world in the 1960s that it is easily forgotten that other linguistic theories continued to exist after the Chomskian revolution. Europe especially continued to support alternative linguistic traditions. Furthermore, various centres in Europe had an established interest in MT, the most important one being the CETA project at Paris and Grenoble, funded by the CNRS, the French scientific research council. This project is of singular importance for the MT field, not only because it has led a continuous existence up to today, but more specifically with reference to this chapter; it was the first of its kind to develop its own theoretical framework oriented towards the task of translating by machine and guided by decisions concerning what linguistic information is adequate to achieve that task.

Some of the alternative linguistic theories, surviving on the fringes of the major scene, are worth describing in more detail, either because they emerged from a particular MT effort, or because they hold views on criteria for text description which are important as background to the model for translation which emerged from the CETA project.

In 1960 Lamb worked in the MT research team at UCLA at Berkeley. It is not clear whether they ever produced a working system, but in the 1960 National Symposium on Machine Translation, Lamb took a strong line. First of all, he shared Bar-Hillel's rejection of the empirical approach to MT. Secondly, because a system is 'as weak as the weakest rule in it', he believed it was in the researcher's interest to ensure that the rules are firmly based and adequate. As a consequence, he predicted that ninety per cent of the work necessary to build an MT system lies in gathering the necessary linguistic information beforehand ([24] Lamb 1961).

As to what information needs to be gathered, Lamb states that a text has a simultaneous existence on all levels of description. The grammarian has the task of separating out all information belonging to each level during analysis. An element or a combination of elements classified at a particular level may be referred to as the representation at that level. A simple example is given in figure 3.7.

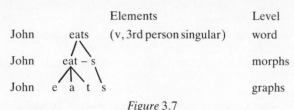

Figure 3.7

Elements at one level derive their existence from combinations of elements on another level. For instance, the elements at the morphemic level are morphs, but those are made up of graphs, i.e. elements at the graphemic level. Between sound and meaning there are several such strata.

Lamb's work later grew to become an independent linguistic theory (stratificational grammar). As defined in Lamb ([24] 1961), the approach is a compromise away from structuralism. He rejects the empiricist methodology of grammar discovery but retains the notion that language can be described at different levels which can be ordered in a hierarchy.

Petr Sgall, one of the Prague structuralists, developed a model of linguistic description based on the view that a grammar is a hierarchy of levels of description which are linked to one another by means of transducers (i.e. abstract machines which take structures or strings as input to produce other structures as output). The highest level, corresponding to the proposition, is generated by a Push Down Store automaton (PDS) (the procedural equivalent of a context-free grammar). The description of the proposition, expressed by means of dependency symbols (as opposed to bracketing), contains information concerning all grammatical categories and the dependencies between them. The representation is fed into a series of PDS transducers which then yield a final surface form.

As a rule, American indirect MT systems take recourse to immediate constituency (IC) trees for the representation of linguistic structure. The exception is the project at RAND where Hays ([18] Hays 1961) spent considerable effort on working out a new formalism, based on existing work in France and in Eastern Europe. His dependency syntax assumes that the presence of words (occurrences) in a sentence depends on the presence of other words in that same string. Each sentence has exactly one occurrence which is independent and which is represented as the root of the dependency tree. The type of structure which emerges from the analysis is illustrated in figure 3.8 on an example sentence 'John eats an apple'.

```
        eats
        /\
 John     apple
            |
            an
```

Figure 3.8

Hays' dependency theory relates to work on predicative valency by Tesnière. His work includes a proof that IC and dependency representations are one-way equivalent: for each dependency tree one can calculate an IC equivalent, but not vice versa.

The fact that the MT system at RAND uses dependency structures is not the only sense in which the theory is important. In 1967 Kulagina and Mel'cuk ([23] 1967) made the interesting observation that probably both IC and dependency representation would be needed for MT. They do not question the usefulness of one over the other, but observe that IC is very well suited to represent scope phenomena. Their statement implies that different linguistic phenomena may well be best represented by different formalisms.

The above linguistic theories concentrate on two points of particular relevance to MT as it evolved in the CETA project. The first point concerns the fact that a text contains phenomena which originate at a variety of different levels of linguistic description and which should be recognised as such. Furthermore, the levels and the information they account for are interdependent. The second point concerns the realisation that different phenomena may require different formalisms for their characterisation.

The CETA MT system

'Nous sommes naturellement conduits a envisager la traduction automatique comme le passage entre deux langages artificiels, le premier étant un modèle de la langue source, le deuxième représentant la langue cible. A ce problème de transfert . . . s'ajoute celui de la fabrication des modèles' ([49] Vauquois 1968). This quotation summarises very well the basic assumptions underlying the CETA model for MT. Vauquois ([48] 1966) remarks that structural linguists have given up the study of semantics and have concentrated on the study of the properties of languages for their own sake. Nevertheless, he points out that any formalism devised for formal syntactic representation needs a semantic interpretation. This notion is generalised in a framework which views linguistic analysis as a hierarchy of levels of description, each defined by a model, and each finding its interpretation at a higher level in the hierarchy. Translation under those terms becomes a mapping, not between strings, but between models of source and target language.

A model is a formal axiomatic system which characterises an artificial language: the set of all objects at a particular level of description. Besides this formal part, each model except for the top one (which gives the ultimate interpretation), contains an interpretative component which provides a link with other models higher up in the hierarchy ([50] Vauquois et al. 1966). Vaquois ([48] 1966) proposes a sequence over three levels, simplified in figure 3.9, where: λ, λ' are human languages; φ, φ' are sentences in λ, λ'; M, M' are words in λ, λ'; L, L' are artificial languages, models of λ, λ'; Z, Z' are terms in L, L'; F, F' are formulae in L, L'; Σ is a set of concepts and provides meaningful interpretation for L and L'; and σ are valid (meaningful?) members of Σ.

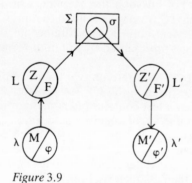

Figure 3.9

The framework defines equivalent texts as those which share an interpretation in Σ. A text is ambiguous if it has more than one interpretation in Σ. The above model for translation also allows for a classification of MT systems. Word-for-word or sentence-by-sentence systems provide a mapping between λ and λ' and transfer systems between L and L'. Those translating via σ make use of an interlingua.

The model for translation described in the previous paragraph is clearly an abstraction of an ideal situation which may not be attainable in practice. The CETA system, though, retains the basic assumption that translations are equivalent texts in the sense that they share the same underlying semantic interpretation. But whereas the ideal semantic model would remain invariant for all languages, the CETA system compromises on this universality. Vauquois ([49] 1968) recognises that a true interlingua remains inaccessible for the time being and proposes translation via a 'langage pivot', an intermediary formalism chosen with care according to the languages the system expects to deal with. A further compromise is made by regarding sentences, and not texts, as the units of translation.

Both analysis and synthesis are carried out in three steps, each corresponding to a level of linguistic description, and for each of which a model has been defined. The morphological model, M1, decides, by means of a finite-state automaton, whether particular morphological segmentations of the input string are admissible in the language. It adds further information to the elements in the sequence (category, number, case, . . .) and delivers syntagms as a result. The syntactic model, M2, relies on a context-free grammar in order to rule out those combinations in the syntagm which do not amount to a legal sentence, yielding a binary IC tree. It also contains transformation rules (i.e. mappings between tree structures) which convert that IC structure into a dependency tree. The resulting representation is marked for syntactic function and has discontinuity phenomena resolved. The sememic model, M3, interprets the syntactic structure in terms of the langage pivot as defined by a sequence of transformation rules. It yields a dependency representation marked for semantic roles.

The CETA MT system and the model for translation it derived from incorporates elements of a variety of linguistic theories, some of which have been described earlier. The framework it provides for analysis is generative in spirit. But the impact of mainstream transformational linguistics is minimal. The group at CETA has undertaken to provide its own theory for translation by subordinating the elements borrowed from various linguistic traditions to the needs of the task at hand. They refrain from continuous reference to any particular linguistic approach but rather follow the principle that translation needs an account of different linguistic phenomena which may be best handled by different formalisms.

Conclusions

This chapter has attempted to provide answers to two questions. First of all, can linguistic theory contribute to the success of MT? As for Mounin's ([30] 1964) suggestion that the structuralist tradition was not particularly helpful in that respect, one cannot but agree. Structuralist linguistics sought to discover a description of human language empirically by means of mechanical operations over a corpus. The goal of MT is to develop working systems which deliver translations of a quality allowing commercial use. Such systems might benefit from reference to description of human language but cannot draw advantage from the empiricism so deeply encrusted in the structuralist methodology. Generative linguistic theories, on the other hand, adopt an approach far more compatible with the aims of MT, in that they provide mechanisms for the characterisation of infinite numbers of instances of linguistic phenomena. These mechanisms can be referred to by MT systems in order to deliver descriptions of source and target texts which include occurrences of those phenomena. The aims of MT as a discipline and of generative linguistic theories are easily married.

The second question is how far linguistic theories have contributed to approaches taken in MT during the period under investigation. The answer here is complex. As to the first efforts to provide MT with a framework, the work of the pioneers seems little influenced by linguistic considerations. Their contributions, resulting in methodological principles which have survived into the present, were rooted in other disciplines, mainly mathematics, information theory and psychology. Their attitude to linguistic theory is mostly neutral, and in some cases rather hostile.

Those indirect MT projects which did make an effort to include linguistic research adopted a generative approach quite early on, but they did not do so because of advances in mainstream linguistics. The Chomskyan revolution, characterised by the rise of transformational grammar, had no direct impact on MT. The generative views emerging in MT were developed in parallel to those in linguistics; although both were based on the work of forerunners like Bar-Hillel, they were motivated differently. The opinions held by people in MT and in linguistics were at times violently opposed.

On the whole, linguistic description in MT seems to occur quite independent of current work in linguistics during the period under scrutiny. If MT groups are interested in linguistic description at all, as a general strategy they develop their own framework. They borrow from various linguistic traditions those elements which seem most suited and try to reconcile them with their model which, as a result, can become quite complex and obscure. It can be concluded that in the period investigated the possible advantages to be gained for MT by reference to theories in linguistics has not really been exploited.

Acknowledgements
I am indebted to Anneke Neijt and Sam Steel for comments and suggestions.

RECENT DEVELOPMENTS IN THEORETICAL LINGUISTICS AND IMPLICATIONS FOR MACHINE TRANSLATION

There seems to be a consensus among both translators and linguists to consider the task of machine translation primarily as a problem of applied linguistics. This view, incidentally, is clearly reflected in the basic architecture adopted by the overwhelming majority of MT systems, which distinguishes three fundamental steps in the translation process: a linguistic analysis of the text in the source language, transfer at some abstract level of linguistic representation and generation of the text in the target language. (For an overview of MT systems, see ([416] Slocum 1984) as well as references cited there.) The linguistic nature of these three basic components of MT systems is obvious enough and does not need to be further stressed.

Given the central role of linguistics in the MT enterprise, one would expect intensive and fruitful collaboration between MT researchers and theoretical linguists. Ideally, such a collaboration should benefit both camps: theoretical linguistics provides MT with models of grammar, rules and representations which are part of the basic tools and concepts that MT researchers need. Also, working within the same framework as theoretical linguists would enable MT researchers to draw directly from theoretical work. In turn, MT research provides an unmatched testing ground for theoretical linguistics, not to mention the tremendous incentive for basic research that such applications would create.

Reality, though, has been somewhat different. Rather than working in close collaboration, MT researchers and theoretical linguists seem to have largely ignored one another. Thus, contacts between the two fields have been rather scarce over the last decade or so, and generally speaking, one might say that the relation between the two disciplines is best characterised by mistrust and scepticism, and, above all, lack of communication (cf. also chapter 3). As a result, the rapid development that linguistic theory underwent since the so-called standard theory of generative transformational grammar of the mid-1960s has had very little impact on MT research. In fact, one cannot fail to be struck by the discrepancy between the kind of grammatical concepts and rules used in the field of MT and those advocated by current grammatical theories.

The lack of exchange and communication between theoretical linguistics and MT has multiple and complex causes, and it is not the purpose of this paper to discuss them in detail. However, it should be briefly pointed out that in view of the history of the relationship, part of the responsibility for the present state of affairs can certainly be attributed to the mutual frustration which resulted from earlier MT research (cf. chapter 1).

The goal of this chapter is to discuss some recent developments in theoretical linguistics, emphasising some of the features which may be of interest for MT, and more generally for natural language processing. One of the objectives of this chapter is to try to revive some interest in theoretical linguistics, and in generative grammar, among people working in the field of MT.

At the outset, we should mention two important limitations of the chapter. First of all, we will only be concerned here with one component of the grammar, that is the syntactic component. Second, only one particular paradigm of theoretical linguistics will be discussed, namely generative transformational grammar. (The main point of this paper is that MT research would greatly benefit from adopting a grammatical framework more in line with current work in transformational grammar (i.e. GB theory). To a large extent, a similar point could be made in favour of some of the current competing syntactic theories, such as 'lexical functional grammar' ([567] Bresnán 1982) or 'generalised phrase structure grammar' ([585] Gazdar 1982, and references given there), although we consider the GB framework more adequate.) The starting point is the standard theory of the mid- and late-1960s, and a certain familiarity with the central concepts of this framework is assumed.

The chapter is organised as follows: the first section is a short review of some basic theoretical assumptions in generative grammar. These notions will serve as background for the remainder of the paper. The second section has an overambitious goal: to summarise in a few pages the results of nearly twenty years of linguistic research. Needless to say, our purpose is not to provide a detailed account of the evolution of generative grammar from the standard theory to 'Government and Binding', but more modestly to give a feeling for the fundamental change in focus which occurred during this period of time. Finally, the third section is concerned with the relevance of the evolution of generative grammar for natural language processing and MT. In particular, it will be argued that even though the major changes in generative grammar were not motivated by computational considerations, it turns out that these changes make the theory computationally much more interesting and also more tractable than previous models of generative grammar.

Theoretical outline

From the very beginning, it has been a basic assumption in generative linguistics that the goal of linguistic theory is to provide an account of the language faculty (as reflected in the competence of the speaker and the hearer) and of the acquisition of language. Thus, the theory must develop a set of concepts rich enough to express the wide range of linguistic data and processes, but at the same time it must be sufficiently constrained and restricted in the number of its options, to account for the fact that language is acquired on the basis of defective data and without direct negative evidence.

The first of these two goals has to do with descriptive adequacy, the second with explanatory adequacy. As Chomsky and Lasnik ([577] 1977, 427) observe. 'There is a certain tension between these two pursuits. To attain explanatory adequacy it is in general necessary to restrict the class of possible grammars whereas the pursuit of descriptive adequacy often seems to require elaborating the mechanisms available and thus extending the class of possible grammars.'

The problem of the acquisition of language in the face of the defective nature of the linguistic data available to the learner has led to the assumption that the language faculty is best characterised as a biological faculty, often referred to as universal grammar (UG).

Biologically equipped with UG, the language-learner has the task of fixing the values of parameters. Borer ([565] 1984), for instance, suggests that UG be viewed as being composed of two major components, one containing those principles and rules which hold universally, such as 'move α', X-bar theory, the binding conditions, and so on. The other component of UG determines the principled ways in which languages may differ from each other with respect to the application of the principles of UG.

Thus, within the GB framework, it is assumed that the core grammar of a language is determined by setting the values of the parameters of UG. Chomsky ([576] 1982) observes that if the parameters are embedded in a theory of UG that is sufficiently rich in structure, then the languages that are determined by fixing their values one way or another will appear to be quite diverse, since the consequences of one set of choices may be very different from the consequences of another set.

From a system of rules to a system of principles

Keeping in mind these remarks, let us turn now to what is sometimes perceived as one of the most interesting aspects in the development of modern theoretical linguistics: the shift from a conception of grammar as a system of rules to a system of principles.

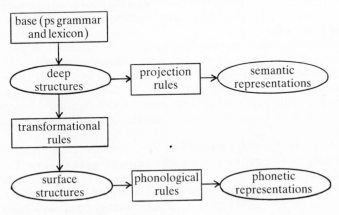

Figure 4.1. The 'aspects' model.

In the standard theory, the grammar consists of (i) a base component which contains a lexicon and a set of phrase-structure rules and (ii) a transformational component. As illustrated in figure 4.1 (where the circles identify the levels of representation and the boxes the rule types), the base component generates a set of abstract phrase-markers, called 'deep structures', by the application of some context-free phrase-structure rules followed by lexical insertion. Notice also that the semantic representations are derived from the level of deep structure by means of some projection rules. It follows that the level of deep structure must contain all the information relevant for a correct semantic interpretation. Another consequence is that transformations are meaning-preserving operations.

The point is that, in this model, the grammar consists essentially of two types of rules, the phrase-structure rules, which determine the level of deep structure, and the transformations, which are (possibly) complex combinations of elementary operations such as insertion, deletion, movement and copying, plus a variety of conditions. Typically, transformations in the standard theory are language-specific. (In their structural description, transformations can refer to specific words or morphemes of the language; similarly, in their structural change they can insert specific words or morphemes. For instance, many formulations of the passive transformation specify the past participle morpheme *en* in the structural description and insert the preposition *by,* etc.) It is also usually the case that specific constructions are derived by particular transformations. Thus, constructions such as passives, raising-to-subject, raising-to-object, existential (e.g. *there is a man at the door*) and so on, are all derived by distinct transformations, which usually bear the name of the construction.

Much of the work done in the 1960s was concerned with the precise formulation of the rules (in particular transformational rules) required to account for an increasing number of constructions. As Stowell ([618] 1981) notes, 'the very complexity and variety of the transformational grammars of individual languages frustrated attempts to develop explanatory theories of language acquisition.'

Most of the complexities in specific grammar rules appeared to be largely idiosyncratic. This was perhaps most obvious for the transformational rules, each of which appeared to require an arbitrary collection of elementary operations and various mysterious conditions preventing individual rules from applying in certain environments. It was obvious, from the perspective of a theory of acquisition, that these complexities could not be directly learned on the basis of experience, since the learning task would have to depend on explicit negative evidence of a very obscure kind. Stowell concludes: 'It was not until work within the Extended Standard Theory had led to a drastic reduction in the expressive power of individual rules that the goal of explanatory adequacy was close to being realized.'

The shift from rules to principles has its origin in the desire to reduce the expressive power of the transformational component. First, Ross ([616] 1967), in his pioneering work, observed that many transformations had similar conditions built in. If these conditions could be factored out, not only would an important generalisation be captured, but the transformations would also be much simplified. He proposed a number of general constraints on extraction, such as the complex noun-phrase constraint or the coordinate structure constraint. Chomsky ([572] 1972) went one step further, showing that some of these generalisations follow from more abstract principles, such as subjacency and the opacity conditions (the specified subject condition and the tensed-S condition).

At approximately the same time, Jackendoff ([595] 1972) and Chomsky ([572] 1972) convincingly argued against the Katz-Postal hypothesis, according to which all semantic interpretation takes place at the level of deep structure. They showed that this hypothesis was much too strong and had to be abandoned. Recall that one of the consequences of the Katz-Postal hypothesis was that transformational rules should not be able to derive an ill-formed surface structure from a well-formed deep structure. As a consequence of this exceedingly strong requirement, transformations had to be formulated in such a way that they would only apply to the relevant structures. It follows that many of the complexities of transformational rules were then justified as a means of preventing the derivation of ungrammatical sentences from semantically already interpreted structures.

The shift to surface interpretation, which was fully achieved with the adoption of trace theory in the mid-1970s makes it possible to invoke general interpretive constraints to rule out overgeneration. The first consequence of this shift was the gradual abandonment of the complex transformational rules, of the complex ordering of rules, of obligatory transformations, replaced by simpler and more general rules (e.g. NP-preposing). The introduction of 'minimal factorisation' and finally the even simpler 'move α' in Chomsky ([575] 1981) places the burden of explanation on general principles rather than on specific rules encoding specific constructions.

The GB model consists of interacting components that can be looked at in various ways. For instance, like earlier versions of the extended standard theory, the GB model can be viewed as a rule system consisting of the three basic parts in (1):

(1)a. Lexicon
    b. Syntax: (i) base component
              (ii) move α
    c. Interpretive components: (i) PF component
                     (ii) LF component

The rules in (1) determine the levels of representation illustrated in figure 4.2.

Alternatively, one can think of the grammar as a system of principles and subtheories, such as the one in (2) below. As

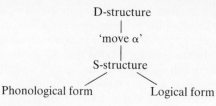

Figure 4.2. Levels of representation.

Chomsky ([575] 1981) points out, this view has become increasingly important in recent years.

(2)a. X-bar theory
  b. θ-theory
  c. case theory
  d. binding theory
  e. government theory
  f. bounding theory
  g. control theory
  h. projection principle

Roughly, the X-bar theory (cf. [572] Chomsky 1972, [596] Jackendoff 1977, [618] Stowell 1981) imposes various constraints on the form of constituent structures. In particular, it states that each phrase of type X (X = Adj, N, V or P) has a node X as head. More generally, the X-bar theory states the endocentric nature of phrases, i.e. a VP contains a V, NP a N, and so on. This theory also brings a solution both to the problem of the definition of the notion of head of a phrase and to the old problem noticed by John Lyons ([606] 1968), namely that the theoretical apparatus of the grammar had no way to prevent 'crazy' rules such as VP→Det Adj P.

The θ-theory is concerned with the assignment of thematic roles such as agent, theme, location, etc. This, of course, is reminiscent of case grammar (not to be confused with case theory!) or valency grammar. The basic idea is that certain lexical items, such as verbs, assign thematic roles (θ-roles) to other categories. Thus, for instance, the verb *persuade* has the property of assigning a certain θ-role to each category for which it is subcategorised, i.e. its direct object and (possibly) its clausal complement. It also assigns a θ-role to its subject, although in this case, the θ-assignment is indirect, being mediated by the VP. Notice that the specific content of θ-roles may play a role with respect to semantic interpretation, but does not in sentence grammar.

An important part of θ-theory is the θ-criterion, which states that the relation between θ-roles and arguments is biunique, that is, each argument is assigned a θ-role and each θ-role as lexically determined must be uniquely assigned to an argument. To give an example, the θ-criterion rules out the presence at D-structure of a referential NP in the subject position of a structure containing *seem* as main verb, as in (3):

(3) [John seems . . .]

(The so-called raising verbs, i.e. *seem, appear, happen,* etc. have the property of not assigning any θ-role to their subject.)

Case theory is concerned with the assignment of abstract case. The central notion, here, is that lexical NPs must have a case. This requirement is enforced by the case filter:

(4) N, where N is lexical and has no case.

Case is assigned by a verb to its direct object, by a preposition to its complement and by the inflection node INFL to the subject. (More specifically, INFL is itself composed of a Tense and an agreement (AGR) component. It is the +AGR INFL node which assigns subject case to the subject.) Thus, the subject of a tensed clause receives a case and, therefore, can be lexical, whereas the subject of an infinitival clause does not receive a case (INFL is −AGR in infinitives) and, therefore, cannot be lexically realised (in the unmarked case).

The binding theory specifies the kind of relations that NPs can entertain with each other. In particular, it is concerned with the relation of pronominals and anaphors to their antecedents. In this sense, it subsumes the opacity conditions of earlier work. A formulation of the binding conditions is given in (5) (from [565] Borer 1984, 12):

(5)A. an anaphor is bound in its governing category
       (anaphors: NP traces, lexical anaphor, PRO)
    B. a pronominal is free in its governing category
       (pronominals: pronoun, PRO)
    C. an R(=referential)-expression is free
       (R-expressions: names, variables)

The notion of government corresponds roughly to the traditional notion that a verb governs its complements, etc. This notion is relevant for subcategorisation, θ-theory, case assignment and binding.

Bounding theory states locality conditions such as subjacency, which constrain the scope of the transformation move α.

The control theory is concerned with the problem of the interpretation of the abstract pronominal PRO subject of infinitival clauses.

Informally, the projection principle states that the θ-assignment properties of lexical items must be represented categorically at each

level of the derivation. Chomsky also introduces a stronger formulation, called the extended projection principle, which adds the
requirement that clauses have subjects.

As observed by Borer ([565] 1984), the distribution of NPs at
S-structure is partially predicted by the interaction of the projection
principle and the θ-theory. Since all θ-positions must be filled at
D-structure, and since movement of an argument to another θ-position results in a violation of the projection principle or of the
θ-criterion, it follows that movement is only possible from a θ-position to a non θ-position.

To return to the question of the status of rules such as phrase-
structure rules in a GB framework, it has been argued by Chomsky
([575, 576] 1981, 1982) and Stowell ([618] 1981) that, contrary to
what was the case in the earlier framework, rules can be derived to
a large extent (maybe entirely) from principles such as X-bar, the
θ-theory, the projection principle and lexical information.
Chomsky ([576] 1982) notes: 'We might still say that there is a base
phrase structure grammar, say, a context-free grammar, which
generates D-structures (or, for that matter, S-structures or LF),
but it is derivative: it need not be specified or learned, but derives
from deeper principles and the specification of parameters that
constitute the actual grammar.'

To illustrate this point, consider the following examples, adapted
from Chomsky ([576] 1982):

(6)a. John was persuaded to leave

b. $[_s [_{NP_i} John]$ was persuaded $e_i [_s PRO_i$ to leave$]]$

Consider now (7), the phrase structure rule expanding VP assumed in earlier work to derive the D-structure underlying (6b):

(7) VP→V NP S

The question is how much of it must be specified in the grammar.
The fact that the head of VP is V need not be specified. It follows
directly from the X-bar theory. Notice, furthermore, that the presence of NP and S is also unnecessary, being redundant with the
lexical information associated with the verb *persuade*. Recall, incidentally, that the projection principle requires that this information
be present at every level of the derivation. Therefore, the S-structure must be something like (6b), with an empty category in direct
object position. The corresponding D-structure is essentially the
same, with *John* in the direct object position and the matrix subject
position unfilled. Chomsky goes on to show that even the order of
NP and S can be derived – presumably from the adjacency requirement for case assignment – and thus need not be specified.

To summarise, within the GB framework, a grammar can be

thought of as a highly modular system of abstract principles and well-formedness conditions which interact – or conspire – in various ways, to provide the kind of complexity required to describe the intricacies of natural languages. There is obviously a certain amount of redundancy among some of these notions (for instance between θ-theory, subcategorisation and case theory), and much of the current work concerns various proposals to eliminate part of the redundancy.

Some implications for MT and natural language processing

In this section, I would like to discuss briefly the relevance of the evolution of generative grammar discussed in the previous section for machine translation and more generally for natural language processing.

As mentioned in the introduction, it has usually been assumed among computational linguists as well as among MT researchers that generative transformational grammar is too complicated to be used as a processing model. This very unfortunate conclusion originated partly in the rather unsuccessful attempts of the late 1960s and early 1970s to implement parsers based on some versions of the standard theory. A detailed discussion of the kind of problems faced by the so-called transformational parsing is given in King ([603] 1983) and references cited there (cf. also ([625] Winograd 1983, 383–7).

The major problems mentioned in King ([603] 1983) can be summarised as follows: (i) determining the surface structures on which the set of inverse transformations must apply; (ii) formulation of inverse transformations; (iii) ordering paradoxes, cycling, etc.

Interestingly, these three basic problems are directly connected with the transformational component of the grammar. More specifically, they have to do with the complexity of the transformational apparatus. As we pointed out earlier, this criticism of the standard transformational theory is not specific to computational linguistics. It is precisely to restrict the excessive descriptive power of the transformational component that triggered the evolution of modern theoretical linguistics which led to the GB model.

Clearly, arguments against generative grammar as a computational model based on problems such as the ones mentioned above become irrelevant with modern versions of generative grammar. One of the most spectacular consequences of the evolution of theoretical linguistics over the last twenty years has been to 'trivialise' the mapping between D-structure and S-structure, so that the transformational component is now reduced to the sole rule 'move α'. In fact, it is very likely that the transformational component could be dispensed with altogether, although the advantage that would result from the abandonment of 'move α' is still unclear. In any case, the fact that D-structure and S-structure are essentially isomorphic appears to be a welcome step both from the point of view of linguistic theory and from the point of view of computation.

Also, part of the problem opposing computational linguistics and generative grammar seems to have been a lack of understanding of what theoretical linguistics is all about. In particular, it was taken for granted that a parser based on generative grammar would have a one-to-one correspondence between rules of the grammar and computational operations, oblivious of the fact that generative grammar is a characterisation model, not a production model, although it is certainly fair to say that the rather systematic presentation of earlier models of grammar as systems of rules helped maintaining the confusion. In fact, as Berwick ([631] 1983) points out, 'there need not be any clean 1–1 rule-computational operation correspondence in order for the "realization" relation to hold. A computational model can abide by the operating constraints of a transformational grammar and yet not mimic its rules.'

An example of such an approach is the parser produced by Marcus ([607] 1980), which operates 'as if it obeyed the operating principles of a transformational grammar'. Another and more recent example is the parser for French described in Wehrli ([620, 621] 1983, 1984), which is an attempt to develop a parsing strategy based on the modular nature of GB-grammars.

Like earlier models of generative transformational grammar, current models offer MT a framework and provide the tools that grammar-writers need to describe particular grammars. However, because of its declarative nature, a theoretical model such as GB does not have the strong bias towards generation which characterised earlier versions of transformational grammar. (It should be emphasised that this bias towards generation has not been intended. Chomsky and his followers have always presented transformational grammar as a characterisation model. The bias should be viewed as a consequence of the choice of a particular formalism, i.e. derivational rules.) The interaction of the modules described in the previous section specifies the properties of well-formed linguistic structures. To put it slightly differently, given the set of all possible phrase structures, the diverse principles and conditions can be used to filter out the well-formed structures. The point is that, as such, the grammar does not specify *how* a particular structure is built. The task of defining parsing or generation strategies is entirely left to the computational linguist. In this respect the GB model is very different from other competing theories such as 'lexical functional grammar' (cf. [567] Bresnan 1982) or 'generalised phrase structure grammar' (cf. [585] Gazdar 1982 and references given there), which are both strictly rule-based.

As an example of how parsing strategies can be designed, which will reflect the principles of a GB grammar, consider the projection principle. By requiring that the θ-properties of a particular lexical item, say X, be categorically represented at every level of representation, this principle greatly limits the choice of structures in which X can occur. In other words, it follows from this principle that structures can be viewed to a large extent as projections of lexical properties. In addition, phrase structures are constrained by the X-bar theory. Crucial, here, is the notion of *head* of a phrase. In accordance with the X-bar theory, the head of a phrase determines the categorical nature of the phrase. As observed in Wehrli ([621] 1984), put together, these two principles suggest a particular parsing strategy, which can be dubbed 'head-driven parsing'. In essence, the head-driven parsing strategy states that a phrase is built on the basis of the lexical specification of its head. As we just saw, the head determines the category of the phrase and also creates some expectations with respect to the kind of complements the parser should find and attach to the phrase. Assuming that the choice of specifiers is largely determined by the nature of the phrasal category, the head-driven parsing module, triggered by the presence of a potential head, will essentially be able to build the entire phrase structure.

Needless to say, the head-driven parsing module that has just been sketched, cannot by itself stand as a parser. It only constitutes one possible module of a parser based on a GB grammar. It is a possible implementation of parts of the projection principle and θ-theory. A GB parser will also have other modules corresponding to the other principles and subtheories of the grammar.

Conclusion

Contrary to what seems to have been a fairly common belief in the field of MT, modern theoretical linguistics and in particular transformational generative grammar has undergone significant changes during the past twenty years. In fact, these changes have been so drastic that the GB model – the latest version of transformational generative grammar – bears very little resemblance to the 'aspects' model of 1965.

Although primarily triggered by theoretical considerations such as the pursuit of explanatory adequacy, this evolution turns out to have interesting consequences for natural language processing and hence for machine translation. Its modularity and declarative nature, as well as the most prominent role played by the lexicon and by lexical properties in the overall grammar, are all features which make GB increasingly attractive among computational linguists. There is no reason why interest in modern theoretical linguistics should not reach the world of MT, and clearly both MT and theoretical linguistics could only benefit from such a collaboration.

FIVE : PATRICK SHANN

MACHINE TRANSLATION:
A PROBLEM OF LINGUISTIC ENGINEERING
OR OF COGNITIVE MODELLING?

Ambiguity is one of the crucial problems in computational linguistics, particularly in machine translation. If the ambiguities are
not cleared up, one cannot expect to produce mechanical translations of reasonably good quality. The aim of this chapter is to
examine the claim of artificial intelligence that, in the field of knowledge representation and natural language understanding, A I offers
better solutions to the problem of meaning than linguistics does.

By using certain linguistic tools such as semantic markers and
selective restrictions ([600] Katz and Fodor 1963) or case frames
([582] Fillmore 1968), certain semantic problems can be solved in
MT. It is well known that a certain class of meaning problems
cannot be solved without considering knowledge of the world, a
domain that is traditionally excluded from linguistics. In opposition
to the linguistic approach that tries to draw a borderline between
purely linguistic semantic knowledge and general world knowledge,
research in A I assumes that there is no such line and that a semantic
theory of language must include general world knowledge that
influences the interpretation of sentences ([591] Hayes 1976).
Schank ([644] Lytinen and Schank 1982), for example, claims that
it is impossible to implement a working MT-system without the
existence of a theory of natural language understanding, which
includes a theory of knowledge representation. Natural language
understanding for human beings and therefore also for machines
involves access to a rich base of commonsense knowledge about
people, things and events as they happen in the real world.

This chapter will examine some experimental A I systems that
were developed for machine translation (the Yale project, Wilks'
preference semantics, SALAT and CON³TRA). We will con-

71

centrate on the different types of meaning representation and on how this knowledge is used for the solution of the difficult problems in MT. To explore the AI approach to MT, we will focus on certain types of ambiguities that are out of the range of purely linguistic techniques, and discuss the solutions proposed by AI.

The linguistic approach

The MT community normally agrees on the fact that semantic analysis of the input text is necessary to produce correct translation. One of the classic techniques originally developed by linguistics is the use of semantic features for lexical units, and selectional restrictions concerning the combination of the lexical items. This type of theory was first suggested by Katz and Fodor ([600] 1963) within the framework of generative grammar, and is used in MT-systems like SYSTRAN or TAUM AVIATION (see chapters 13 and 15).

The basic idea of this theory is simple. The different meanings of the lexical items are distinguished by the semantic features attached to them. These binary features range from the very general, for example [HUMAN], [ACTION], [PHYSOBJ] to the relatively specific, like [COMPARTMENT], [AGRICULTURAL PROCESS]. A dictionary entry for *ball* would contain the following features ([591] Hayes 1976):

> *ball*→concrete noun→(social activity)→(large)→(assembly)→[dance]
> *ball*→concrete noun→(physical object)→[sphere]
> *ball*→concrete noun→(physical object)→[cannonball]

For verbs, the dictionary contains a second type of information, the so-called selectional restrictions. These patterns describe the context expectations in terms of the syntactic and semantic features. They are called selectional restrictions because they select the appopriate sense of one word which can combine with the given sense of another word.

The simplest use of markers and selectional restrictions is, for example, to match the semantic markers that a verb expects for its subject and object against the markers of the nouns which are in the sentence. By this matching operation based on the identity of the features that are expected in the selectional restrictions and the features of the corresponding words, the meaningful reading of a sentence is detected, the word-sense ambiguities hopefully reduced and the semantically absurd combinations rejected.

Two examples from TAUM AVIATION will illustrate how this technique works. The first example ([260] Lehrberger 1981)

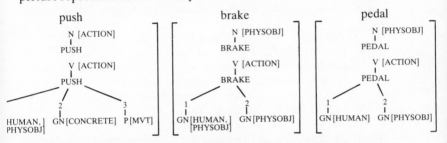

shows how syntactic ambiguities can be filtered with the help of selectional restrictions. The phrase *Cockpit left shelf* can have two readings: (1) The left shelf in the cockpit, or a second one, which is semantically absurd, (2) the cockpit has departed from the shelf. The interpretation of *leave* as the verb is ruled out by the fact that the verb does not accept as its subject the noun *cockpit* with the feature [COMPARTMENT]. Therefore the phrase is considered a complex noun group.

In the second example *Push brake pedal* ([215] Isabelle 1981), each of the three words can be either noun or verb. Again the matching process has to find the correct combination. The following picture represents the dictionary entries for each of the three items:

To parse this sentence the system has first to discover (1) that the verb is *push*; (2) that it has an implicit subject; and (3) that the object of the verb is the noun group *brake pedal*.

The example shows clearly that the relationship between the context pattern and the markers of the nouns is not direct and absolute correspondence as is suggested in the original proposal of Katz and Fodor (for more detailed criticism see [591] Hayes 1976). If the selectional restrictions are used in an absolute manner as Katz and Fodor propose, they risk eliminating a lot of metaphorical uses that should be treated by an adequate system. For this reason Wilks ([654] 1973) introduces the preference technique described in further detail below.

A second important problem is the choice of features. If they are to be useful, they need to be fairly specific. Theoretically there is no particular limit that indicates where semantic subcategorisation has to stop. The advocates of the sublanguage approach ([604] Kittredge and Lehrberger 1982) propose sets of semantic features specific to the linguistic particularities of specialised texts in restricted semantic domains. This approach resolves ambiguity problems in complex noun groups, which is a non-trivial problem, but its trade-off is that world knowledge of a restricted subject field is coded into the semantic features. The great risk of such sublanguage-semantics is to trap a system into a particular subject field, without the possibility of extension to other fields unless the whole of the semantics is redone.

A third problem is that the technique of markers and selectional restrictions is not powerful enough to solve certain types of ambiguities that have to be disambiguated to guarantee a correct translation. The following tentative list presents the problematic cases that a reasonable semantic component has to face. Most of the ambiguities can only be resolved with the help of world knowledge. It is for this class of meaning problems that A I claims to have solutions that can be interesting for MT.

(a) Word-sense ambiguity. German *Schloss* translates as *castle* or *lock*:

> *Er hat den Schluessel ins Schloss gesteckt.*
> (He has put the key into the lock)
> *Kommst du mit mir ins Schloss?*
> (Are you coming to the castle [with me]?)
> *Er hat den Schluessel im Schloss gelassen.*
> (a) He left the key in the lock
> (b) He left the key in the castle.

The third example cannot be solved with selectional restrictions.

English *to put* can be translated into German by *stellen, legen* or *setzen* according to the normal position of the object:

> *He puts the bottle on the table.*
> (Er *stellt* die Flasche auf den Tisch)
> *He puts the book on the table.*
> (Er *legt* das Buch auf den Tisch)

The normal position of objects is a matter of world knowledge.

(b) Prepositional ambiguity. A special case of word-sense ambiguity concerns prepositions, which are a particularly difficult problem for MT.

> *The committee reached its conclusions by long and protracted discussions.* (*nach* langwierigen Diskussionen)
> *The committee reached its conclusions by the end of the afternoon.* (*am* Ende des Nachmittags)

One preposition in the source language is normally translated by one of a wide range of different prepositions in the target language, depending on context. It is important to note that resolution of ambiguities of this sort can also resolve the structural dependence of prepositional phrases, another notorious problem.

(c) Syntactic ambiguity. The sentence *I saw the man on the hill with the telescope* has five possible readings. A phrase with one

more level of prepositional groups has 14 possible parsings and the following phrase, with 7 prepositional groups, has 429 parsings ([625] Winograd 1983):

> *The appearance   of the man   under the tree   with a broken*
> *branch   near the road   to the town   with the market.*

Complex noun groups represent another case of difficult syntactic ambiguity. According to the dependency relations that are computed, a phrase like a *small car factory* translates into *kleine Autofabrik* or *Fabrik fuer Kleinautos.*

This kind of problem requires knowledge of the context.

(d) Referential ambiguity of pronouns
   (The tree) in the garden (which has cones) is pretty.

The tree (in the garden which has pansies) is pretty.

The relative clause can be attached to either *tree* or *garden*. For semantic reasons it is clear that the relative clause *which has cones* would normally be attached to *tree*, just as the clause *which has pansies* is dependent on *garden*. This kind of problem becomes a translational problem as soon as one translates from a language which has no genders into a language which distinguishes between different genders. To translate the English sentence *The pebble hit the window and it broke* correctly into French, the reference of *it* has to be identified because the two possible nouns have different genders: *le caillou* is masculine (pronoun *il*) and *la fenêtre* is feminine (pronoun *elle*).

AI approaches

Although AI work generally does not concern itself with the problems of MT, the problems described in the last section are common to both, since any system depending on the analysis of natural language will encounter them. The difference between AI and standard MT philosophy is that AI has a strong emphasis on the role of world knowledge in understanding whereas MT is based on 'pure' linguistics. In the early seventies the work of Charniak ([634] 1973), Wilks ([654, 655] 1973, 1975) and Schank ([647] 1975) was directed towards the problems of word-sense ambiguities and the correct referring of pronouns. They tried to solve the questions by focusing on meaning representation and the use of inference rules for common-sense knowledge about the world in which we live. While Wilks suggests that a system with partial world knowledge is able to deal with a good number of ambiguities, Schank claims that only a system that 'understands' the input text in depth can be expected to produce good translations.

Our discussion of different types of systems will be organised around three questions:

(1) What type of world knowledge is used?

(2) What type of meaning representation do the different systems have and how is this representation used in the inference device?

(3) What ideas can be useful in an MT system?

Wilks developed his system of preference semantics in the context of research, focusing on problems like word-sense ambiguity, case ambiguity (prepositions) and difficult pronominal references, rather than aiming at the actual production of a large MT system. Within the AI emphasis of the early 1970s, it was built as a purely semantic system ([663] Wilks 1983) working with semantic primitives and case frames. The system parses texts into a semantic representation that can be used for different kinds of inferences. One of the basic ideas is 'preference', an alternative way of treating inference to that typified by logical deduction. The idea of preference is used in various ways throughout the system. It is based on the assumption that utterances are not either grammatical or wrong, but just more or less acceptable. Context patterns therefore are not treated as restrictions but as expectations which can be more or less fulfilled. Ambiguity resolution becomes then 'preferring' one reading to another.

The system contains about a hundred primitives, including deep cases (SUBJ, GOAL, CAUSE...), actions (MOVE, DROP, FLOW...), and other classes (THING, MAN, FOLK...). Each word has, in its lexical entry, one or more trees of primitives – formulae – according to the number of readings it has.

One meaning of *grasp* is coded in the following formula:

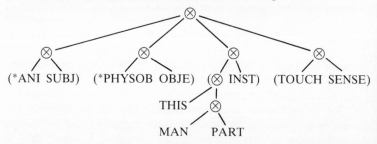

The main element of the formula, the head, is the rightmost branch, expressing the fundamental category to which the formula belongs. Each sub-formula at the same level specifies an act or a preferential case restriction of that act. The following representation reproduces the above formula in a more legible form:

Subformula	Case/Act	Value	Explanation
(*ANI SUBJ)	SUBJ	*ANI	the preferred agent is animate
(*PHYSOB OBJE)	OBJE	*PHYSOB	the preferred object is a physical object
((THIS(MAN PART))INST	INST	(THIS(MAN PART))	the preferred instrument is a human part, the hand
(TOUCH SENSE)	SENSE	TOUCH	the action is of physical contact

The semantic formulae offer more possibilities for a complex description of the word sense than do simple markers and selectional restrictions. The lexicographer can encode preference for the context without having to distinguish between linguistic and world knowledge. The formula may also have inference information attached: for example, that a particular verb may normally have a certain consequence or that the action may be done for a certain reason. Due to the fact that these subformulae express only preferences, the encoded consequences or goals do not follow in a logical sense if the predicate is used in a sentence.

One of the dangers of using semantic primitives is that the codings have a certain amount of vagueness. It is not known whether such a system can remain stable with big vocabularies containing several thousand lexical entries. Therefore, in later versions of the system, Wilks suggested the use of a thesaurus structure over the whole vocabulary to impose more consistency ([662] Wilks 1980).

The meaning of the input text is represented in a semantic block that is built out of a sequence of its fundamental units, the templates. Templates are patterns that consist of three primitives corresponding to an intuitive notion of a basic message that has an agent-action-object structure, for example [man have thing], [man cause thing]. This bare template structure is enriched by the semantic formula of the corresponding input words, whereby the head of the semantic formula has to correspond to one of the three template primitives. The power of this complex structure comes from the fact that the formula contains a lot of information about what is expected in the nodes of the other formulae belonging to the same template. If different possible templates can be assigned to the same piece of input fragment, a scoring mechanism computes how many preferences are fulfilled. The choice between the competing templates is made on the basis of a scale of acceptability rather than accepting or rejecting on the grounds of black or white grammaticality. By this approach an interpretation is accepted even if certain expectations are not fulfilled as long as no better alternative is available.

In order to construct a complete text representation, i.e. the semantic block, the templates have to be tied together by case ties and the anaphoras must be linked to their reference. This template binding is done by two different kinds of higher-level structures: the paraplates and the inference rules.

The paraplate mechanism shows an interesting way of resolving prepositional ambiguities and prepositional phrase attachments. The function of the paraplates is to connect two templates, one of which normally can be seen as a main clause and the other one representing a prepositional phrase. The paraplates assign the case tie relating template pairs, with the spacial-locative case (SLOCA) in the following example (where □ is a dummy agent):

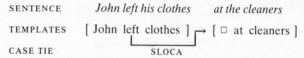

Paraplates are six-place structures describing possible template connections. On the left side of the arrow (see example on p.79) the primitives' triple corresponds to the first template, the right-hand side describes the primitives of the second template, the label under the arrow represents the case marker. Each paraplate is stored under some preposition and expresses one of its possible case ties. Paraplates have a filtering function insofar as all the nodes of the paraplate have to find a corresponding node in the template structures that they have to connect. For a given preposition the paraplates are partially ordered and the 'more preferred paraplates' simply means 'the paraplate applied earlier'. In the facing example we show the templates corresponding to a sentence and a stack of paraplates to explain how case ties get assigned on the basis of paraplate matching. Only the paraplate 2′ matches the template primitives given in brackets which correspond to the heads of the formulae. As a result of a successful match the case label mentioned under the arrow is assigned, and the semantic block is constructed by connecting the two templates and at the same time disambiguating the preposition. In our example the case DIREC-

SENTENCE   *He left Lugano   by the autostrada.*

TEMPLATES   [(*ANI) (MOVE) (WHERE POINT)]   [□ □ (WHERE LINE)]

PARAPLATES related to *by*

(1')   (*ANI) (MOVE) (WHERE POINT) $\xrightarrow[\text{(INST)}]{}$ □ □ (*REAL)

(2')   (*ANI) (MOVE) (WHERE POINT) $\xrightarrow[\text{(DIRE)}]{}$ □ □ (WHERE LINE)

(3')   (*ANI) (MOVE) (WHERE POINT) $\xrightarrow[\text{(DIRE)}]{}$ □ (*DO) (WHERE SIGN)

TION is assigned which allows the correct translation of the preposition.

Unlike a Schank-like system which calculates all the information
it has, inferences included, the semantic representation of Wilks'
system is of flexible depth. According to the 'laziness hypothesis',
the amount of information (or depth of meaning that is represented) depends on the amount of information needed for the
particular disambiguation to be done. To achieve this analysis at
flexible depth, the system uses two modes, the basic mode and the
extended mode. They are distinguished by the sort of knowledge
and mechanisms they use. In the basic mode of the system no real
world knowledge is involved beyond the conceptual knowledge of
word meanings that is encoded in the semantic formulae. As described above, its main instruments are a pattern matcher and the
calculation of the 'semantic density' of the matched combinations.
Certain pronoun ambiguities can be solved on the basis of this
preference calculation. An example: *Give the bananas to the mon-
keys although they are not ripe.* The basic mode can decide by
seeing, in the representation for the concept of hunger, that it
prefers to be applied to something animate, and that the concept of
ripeness prefers to be applied to something plantlike.

The extended mode of the system is only used if certain ambiguities or pronoun references remain. It is also divided into two
parts. The first, the extraction routine, tries to solve the problem by
deepening the representation with the information that is implicit
in the semantic formulae, that is, by creating new templates for
goals, consequences or results of actions that are not explicitly
mentioned in the text. The pronoun reference of the following
sentence can be found by simple extraction: *John drank the whisky
from his glass and it felt warm in his stomach.* From the semantic
formula of *drink* can be extracted that the result of the action is that
the whisky is [IN John]. The template with the pronoun (*it*) also
contains the case IN: [? it IN John + part]. The case identity allows

the inference that the referent of *it* is the whisky and not the glass.

It is only after failure of the extraction mode that the system uses common-sense inference rules. These rules code very general knowledge about the world, such as {If you strike somebody→he falls} or {If an animate subject MAKES that he gets an object→he likes the object}. These rules are not used to make strong deductions but both sides are matched against the templates in order to find the referent of the pronoun somewhere, by inference chaining in the semantic block.

The merit of Wilks' system is that he has shown ways to solve a number of crucial MT problems by using partial world knowledge on different levels. The idea of preference in combination with the calculation of semantic density in a structure containing rich semantic formulae seems to be quite promising for MT. Less convincing are his general common sense inference rules because it is in no way clear what sort of information has to be coded for a large-scale system.

The main interest of the Yale project, developed by Schank and his collaborators, is language comprehension and memory organisation, rather than MT. A series of programs have been developed focusing on the problem of understanding stories, particularly the text type of narratives. Most of the stories involve small day-to-day activities like visiting restaurants, car accidents, newspaper stories about terrorism or narratives about divorce, etc.

The Yale systems analyse the input texts into a deep semantic structure, which is a language-free conceptual representation. This semantic representation can be used for different purposes, such as abstracting, question-answering and translating. In a translation system the language-free representation functions as an interlingua that allows the direct generation of any target language. Two MT experiments have been carried out, one in connection with the SAM-system ([632] Carbonell et al. 1978), understanding and translating stories about restaurants, car accidents, and more recently a second system in the context of the MOPs-theory ([644] Lytinen and Schank 1982) that is able to translate some newspaper stories about terrorism.

Two points about representation will be discussed: (1) the use of primitives in the 'conceptual dependency' theory and (2) the high-level knowledge structures (scripts, MOPs (= Memory Organisation Packets), etc.).

The basic elements of the 'conceptual dependency' theory ([647] Schank 1975) are a small set of primitive acts. Associated with each act are a number of case-frames, including: actor, optional object, direction, an instrumental case. These case frames hold expectations of what conceptualisations should follow. The primitives aim to be very abstract so that they can be used to carry information required for the inference-making based on common-sense knowledge. Take a sentence like *A TV set was sold to Bill*. With a very abstract representation the inference that Bill now owns the TV set can be made independently of whether the word *bought, sold, got, took, gave, paid* or *acquired* was used. Schank's example below shows one of the major problems of conceptual dependency representation in the context of machine translation.

John sold a TV set to Bill

$$
\begin{array}{ccccc}
& P & O & & R \\
\text{John} & <=> & \text{ATRANS} & \leftarrow \text{OWNERSHIP: TV} & \leftarrow
\end{array}
\quad
\begin{array}{l}
\rightarrow \text{Bill} \\
\rule{0pt}{1.5em} \\
\rule{1em}{0pt}\text{John}
\end{array}
$$

$$
R \Downarrow \Uparrow R
$$

$$
\begin{array}{ccccc}
& P & O & & R \\
\text{Bill} & <=> & \text{ATRANS} & \leftarrow \text{OWNERSHIP: money} \leftarrow
\end{array}
\quad
\begin{array}{l}
\rightarrow \text{John} \\
\rule{0pt}{1.5em} \\
\rule{1em}{0pt}\text{Bill}
\end{array}
$$

Two instantiations of the primitive act **ATRANS** (non-physical transfer of something) are involved, each of which is seen as a result of the other. In one of these acts John transfers ownership of the TV set to Bill, in the second Bill transfers ownership of the money to John. One of the problems for translation with such a kind of representation is the fact that all surface information is eliminated. The system cannot make any difference between *bought, acquired, got* or any other synonym. Furthermore, information that is important to translation like active–passive, emphasis, register etc. is eliminated so that the Yale MT-systems cannot produce more than just meaning preserving paraphrases.

Under the influence of Minsky's proposal ([646] 1975) for higher-order knowledge structures, Schank developed similar high-level representation structures which he calls scripts (Minsky's term is frames). Minsky's basic idea is that people do not analyse new situations from scratch and then build new knowledge structures to describe those situations. Instead, they have stored in memory a large number of structures representing previous experience with objects, situations and events. To process a new experience, they evoke stored structures representing a stereotypical view of the event or object and then fill in the details of the current event. 'A frame is a data-structure representing a stereotypical situation, like a certain kind of living room, or going to a children's birthday party', writes Minsky ([646] 1975). 'Attached to each frame are several kinds of information. Some of this is information about how to use the frame. Some is about what one can expect to happen next,' The idea of frames has been applied to a variety of slot-filler structures. These representation structures can be seen as complex networks of nodes and relations with a lot of internal structure designed to make them useful in specific kinds of problem-solving tasks, as for example inferring unobserved facts about new situations or making explicit some information in a story that has not been mentioned but that is important for understanding.

Schank ([651] Schank and Abelson 1977) described his scripts as 'a predetermined causal chain of conceptualizations that describe a normal sequence of things in familiar situations'. The standard order of events in situations such as visiting a restaurant, shopping in a supermarket, for example, is formalised in a sequence of scenes that have stereotypical content. His original restaurant script looks as follows ([650] Schank and Abelson 1975).

```
script:    restaurant
roles:     customer, waitress, chef, cashier
reason:    to get food so as to go up in pleasure
               and down in hunger
scene 1:   entering
           PTRANS  self into restaurant
           ATTEND  eyes to where empty tables are
           MBUILD  where to sit
           PTRANS  self to table
           MOVE  sit down
scene 2:   ordering
           ATRANS  receive menu
           MTRANS  read menu
           MBUILD  decide what self wants
           MTRANS  order to waitress
scene 3:   eating
           ATRANS  receive food
           INGEST  food
scene 4:   exiting
           MTRANS  ask for check
           ATRANS  receive check
           ATRANS  tip to waitress
           PTRANS  self to cashier
           ATRANS  money to cashier
           PTRANS  self out of restaurant
```

Schank ([648, 649] 1980, 1982) has abandoned his first script theory for other types of high-level knowledge structures, for example MOPs. Basically MOPs have similar characteristics to those of the original scripts. Thus, they give possible orderings of 'scenes', the chunks that group together the primitive acts. One main difference in the new high-level organisation is that a single scene can be shared by several MOPs. This kind of development affects, essentially, only high-level memory organisation and so does not change the kind of problems that arise if this kind of approach to MT is adopted. We will therefore not go into any further details of Schank's new memory theory.

What are the advantages of high-level knowledge structures? According to Schank ([649] 1982), understanding a story, and therefore being able to translate it, falls into two main parts. First we infer the things we are not directly told and which are implicit information, and secondly we fit the input into an overall picture (a high-level knowledge structure) of what is going on. Frames, scripts or MOPs are very good knowledge sources for controlling the inference process and tying together texts in stereotype domains. Take the following story ([632] Carbonell et al. 1978):

> John went into a restaurant. He ordered a hamburger. When the hamburger came, he ate it.

Translating a highly ambiguous word like *came* per se is not easy. In this context the problem of translating, for example into French, is even more difficult because no word-for-word correspondence is available. A possible meaning preserving translation could be *quand le Hambourger fut servi*. To find this translation, knowledge of the topic and its implicit meaning are necessary: one has to know that in the context of restaurants food is normally served when a customer has ordered it. The restaurant script allows this kind of inference. Therefore the Yale group argues ([632] Carbonell et al. 1978 and [644] Lytinen and Schank 1982) that such ambiguity problems can *only* be tackled with the help of abstracting knowledge (conceptual dependency) and packaging knowledge (scripts or MOPs).

Frame-based systems might be very powerful indeed in limited and stereotyped domains. But in the case of texts without any limitation on semantic domains, a crucial problem with this type of approach is to know what frames will be needed for the analysis of a particular text. Take for example the following mini-story: *John dropped by at the supermarket for a quick snack on his way to book his plane ticket.* One could imagine several frames, like supermarket, restaurant, travel-agency, even if the choice is not clear. Another important problem is to know when the frames have to be activated and when de-activated.

On this point it is worth noting that the text types and subject fields treated by these systems are extremely narrow and have not been expanded much over the last ten years. A second negative aspect, related to the problem mentioned above, is that the stories treated are rarely longer than a few paragraphs. These arguments suggest that current frame-based systems offer little for building operational large-scale MT systems even if the approach is a very interesting area for long-term research.

The SALAT (System for Automatic Language Analysis and Translation) project ([199] Hauenschild et al. 1979) was developed between 1976 and 1980 by the Sonderforschungsbereich 99 in the University of Heidelberg and was, again, aimed at theoretical results rather than at an operational system. The goal was to investigate the possibility of using a data-base with different types of knowledge and an inference device in an MT system by combining translation theory, linguistic theory and artificial intelligence.

The system explores a different approach to the semantic problem. It incorporates formal, model-theoretic semantics and works with a common representation for texts and real-world knowledge without making use of special high-level structures like frames. The system is transfer-based, using a logico-semantic interlingua. The interlingua formalism is used to represent both the meaning of sentences subject to translation and world knowledge, including information about the situation. This representation formalism (ε-λ-context-free syntax for 'deep structures') was developed in order to facilitate the formulation of logico-semantic deduction rules in the inference device.

The design of SALAT is quite interesting. Apart from the text representation there is an independent data-base whose structure is shown in the diagram below:

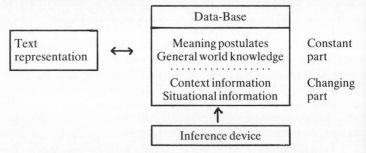

Several different kinds of information are stored in the data-base:
(1) contextual information (from the input sentence)
(2) information about the situation and the time of the

utterance (from the input sentence)

(3) general world knowledge

(4) meaning postulates (i.e. [everything ALPINE is something GEOGRAPHIC]).

Meaning postulates and general world knowledge are considered the constant part of the data-base whereas contextual and situational information are the changing parts. All four types of information are represented as expressions of the same logico-semantic ε-λ-context-free syntax.

The function of the data-base is not only to provide the necessary information for solving ambiguities by deduction with the help of general world knowledge, but also to define the semantics of the interlingua 'words' by a form of meaning postulates. SALAT follows the trend of MT development during the late 1970s, in that the representation of the meaning is done with other techniques than pure semantic primitives.

A disambiguation component is responsible for finding the correct reading if several deep structures have been produced by the analysis. The inference component, DEDUKT, uses information that is stored in the data-base, like context, situation, world knowledge and meaning postulates, as well as new information that is found by logical deduction.

The operation of DEDUKT can be illustrated by the following German sentence *Das alpine Gebiet ist klein* (The alpine region is small). German *Gebiet* has two French translations, *région* for its geographical reading, *domaine* for its scientific one. The data-base provides the information for the resolution of this ambiguity. Part of the explicit information is the meaning postulate already cited [everything ALPINE is something GEOGRAPHIC]. Applied to our sample sentence the system's inference rules interpret as follows:

(1) The above-mentioned meaning postulates, valid at all times and in all situations is also valid at the time B and situation A of the utterance of the sample sentence.

(2) If something ALPINE is GEOGRAPHIC and the *Gebiet* is ALPINE then the *Gebiet* is GEOGRAPHIC and translates therefore as *région*.

The weaknesses of SALAT are that the most interesting part of the system, the inference device DEDUKT, has not been fully implemented and tested over a large amount of data. Furthermore it seems that the example given above could be solved in an easier way by using a richer system of semantic subcategorisation. It could also be expected that the inference process would become very inefficient as soon as a large number of meaning postulates and facts about the world were integrated in the data-base, since the data-base has no further structure beyond being a set of logical clauses.

Nevertheless, the basic idea of having a special data-base for storing world knowledge is interesting. One could imagine it as a source of technical knowledge in specific semantic domains (for technical translation), a module in the system that can be easily enlarged or adapted to new subject fields. However, a major drawback of logical schemes is the lack of organisational principles for the facts constituting the knowledge base.

We have discussed different types of representation schemes used in A I, frame-type structures, Wilks' special devices (templates and paraplates) and the logical representation schemes developed in S A L A T. Each project made use of one of these representation devices. A different approach is taken by the CON³TRA (*Con*stance *Con*cept of *Con*text-Oriented *Tra*nslation) project ([640] Hauenschild et al. 1983, [165] Engelberg et al. 1984) in the sense that it does not construct a new representation scheme but takes as its basic idea the assumption that the translation problem can only be solved by the interaction of different knowledge sources. This research project, started in 1980 at the SFB 99 in Konstanz, combines ideas coming from linguistics, philosophy of language and artificial intelligence. Its objective is a formal reconstruction of text comprehension for translational purposes ('Uebersetzungsbezogene Kontexttheorie'). Research focuses on the specific problems of anaphora resolution and the reconstruction of the thematic text structure.

The basic assumption is that understanding is a heuristic process where different knowledge sources contribute to choosing the best hypothesis on the basis of accumulating evidence from various sides. Text comprehension is seen as a double question where on the one side the meaning gets constructed compositionally by its parts (point of view of logical semantics) and on the other side there is the basic problem that the parts by themselves are ambiguous. Their meaning in a particular text can only be determined with the help of context and background information (A I point of view). The more relations to context and world knowledge can be drawn the better the text comprehension is. A further hypothesis of the project is that at each stage of the analysis, partial results from the various sources have to be able to communicate, a method that has been used before by the Hearsay-II project ([639] Erman, Hayes-Roth et al. 1980).

The criteria for a successful selection of a pronoun's antecedent come from two different *types* of knowledge sources, pure linguistic knowledge on the one side, like morphological and configurational facts, the pronoun's closeness to the referent, the identity of its thematic role etc. and, on the other side, semantic criteria, like text consistency, situational and background information. The properties of the referent have to fit with the contextual information and general world knowledge. The model reflects this bipartition. The grammatical, lexical and world knowledge, which is needed for the analysis, is stored in the knowledge base. The source text itself is represented in two different ways, a multilevel linguistic representation containing pure linguistic data and a text model representing the content.

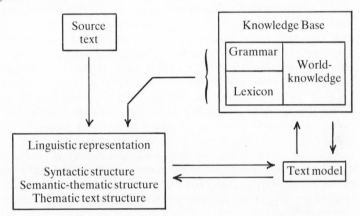

Each type of information has its own representation scheme. The linguistic part has three levels, the first giving morphological and configurational data, using a dependency grammar model. The second level represents the thematic roles and the third level the structure of the text themes. This last level shows the partial themes that the sentence deals with and how they relate to the overall text topics. All three levels of the linguistic representation follow the linear sentence structure and serve as basic information for transfer and generation during translation.

It is assumed that the linguistic levels by themselves are not good enough to disambiguate the sentences completely. The necessary complement for a successful analysis comes from the text model that is computed on the basis of and in interaction with the linguistic analysis and the knowledge base. The primary function of the text model – the semantic representation of the text – is to provide the necessary information for disambiguation. Therefore, text model and world knowledge (in the knowledge base) have the same formal structure. The representation scheme is a semantic network with complete nodes having frame-like structures. In the accumulation process of confirming a hypothesis from various sides, text coherence is considered the main factor and it is the text model's function to serve as a basis for its computation. Therefore, in the text model, in opposition to the linguistic representation, a referent is a single object, a node in the network, that contains all the properties and information concerning that referent. It is on this basis that the necessary consistency checks can be done.

From a static point of view, the text is represented twice, in a structure and an object-oriented way. Looking at comprehension as a dynamic model, the different levels of representation are important because they are considered independent components with their own structures. As anaphora resolution shows, each information source must be available at any moment and interact with the other components like an independent actor. During the interpretation of the text it must be possible to answer syntactic, semantic, thematic or encyclopaedic questions asked by some other module. Generally speaking, a successful resolution of all kinds of ambiguities is only possible if the different knowledge sources are not hierarchical structures but can interact at any moment. The components of CON³TRA are therefore constructed as procedures that can build and evaluate hypotheses with the help of the results of the other components. Each information source is like an independent actor that can interact with the others by sending and receiving messages.

As mentioned above, CON³TRA is a theoretical research project still going on. It will be interesting to see if its results will be incorporated into standard MT engineering techniques.

Conclusion

We have examined different ways of using world knowledge. This analysis has shown various types of knowledge representations that were developed for tackling natural language and MT problems.

Domain-specific technical knowledge is used in the TAUM projects where sub-language specific grammars integrate world knowledge into the formulation of well-formedness conditions. On the other side, Wilks claims that good results can be achieved with general common-sense knowledge encoded in the lexical units (formulae) or in general inference rules, an approach that is contested by the Yale researchers who work with high-level commonsense memory structures (frames) which they consider to be necessary for good translations.

The important question is what kind of knowledge is interesting for real production MT systems with very large vocabularies, and, of course, what kind of conclusions can be drawn from the AI experience in MT.

It was our goal to show that various knowledge sources are needed to produce high-quality machine translation. At the same time there is a serious consistency problem with big and complex dictionaries. An important consideration about large-scale systems is that the vocabulary coding is the most expensive part where a considerable number of lexicographers is needed. Therefore it is very important that the coding is not only controllable and manageable but also cost-effective.

Existing AI techniques are not ready for large-scale MT systems. Past experience in AI shows that in-depth coding of frame-type structures for large subject fields is not yet feasible. At the same time it is not necessarily interesting to calculate a very deep representation if the translation can be done by simpler mapping. Wilks' flexible depth approach follows a different way.

On the other hand, experience drawn from the development of operational systems shows that the linguistic engineering approach, which makes use of simple techniques such as semantic markers and perhaps semantic relations, produces reasonable results if the subject fields are restricted. The draw-back of the sub-language approach, where a system's grammar is dedicated to a specific text-type and subject field, is its non-extensibility. If a system is supposed to translate texts from different semantic domains, general knowledge should be carefully separated from domain-specific knowledge. Assuming that a good part of the ambiguities can be solved by specific technical knowledge about a domain, one should try to keep this knowledge separate in order to maintain the system's modularity. The ideal MT system would then have as its basis a general grammar (with a good syntax and a general semantic coding) to which different domain dependent modules could be attached. In the case of unsolved ambiguities, the system would use these specialised modules that contain the knowledge of the various technical domains, like agriculture, telecommunication, law, etc. Such a design would keep the whole system extensible and would facilitate the treatment of domain-specific problems in specialised modules. A first attempt in this direction has been made by Gerber ([183] 1984).

Despite the results of pure linguistic engineering further research is necessary in MT, and AI seems to be a fruitful research background for important problems such as the interaction of various knowledge sources that are needed for ambiguity resolution.

MT: A NONCONFORMIST'S VIEW
OF THE STATE OF THE ART

What I want to say, in a nutshell, is this. There are two different kinds of intellectual activity we call 'problem solving'. People who think about machine translation often seem to have a very wrong idea of what sort of thing *human* translation is. They think it is one sort of problem solving, but it is really the other sort. Once we realise what kind of thing humans are doing when they translate, we will modify our approach to *machine* translation. That does not mean that we will abandon MT as a hopeless dream. This is *not* going to be yet another paper on the theme 'MT is impossible'. In a way, just the opposite. Part of what I want to say is that MT workers are making their lives needlessly difficult, by tying themselves to a concept of translation which leads them to ignore all sorts of useful techniques that could help them.

We all know that the history of MT falls into two parts with a gap in the middle – the gap that followed publication of the ALPAC Report ([1] 1966). Before ALPAC, MT research worked with a fairly superficial model of language: people took it that human utterances or texts were little more than words arranged in a given order, so that, if you wanted to get a machine to translate, essentially all you needed to do was to get it to replace source-language words with target-language equivalents and change the order of words. And of course that did not work. When MT got going again after the ALPAC gap, the new MT researchers were much more sophisticated about natural language: they knew about Chomskyan deep structures, and the idea that there is much more going on in a sentence than meets the eye; and they realised that you cannot hope to *translate* a text properly until you *understand* it. So the watchword became: artificial intelligence. Of course (they said) first-generation MT was bound to fail when it tried to make machines translate word-for-word without going into the logic of what they were translating; but we will give our MT systems an AI component so that they *do* understand the texts they encounter, just as a human translator would. And if we do *that,* then there is no barrier in principle to perfectly accurate MT. Maybe in practice designing the AI component is going to be such a vast and difficult task that perfection is a pipe-dream or something for the twenty-first century, but perfection is what we are aiming at and perfection is the standard we will judge our imperfect results by.

Part of what I want to say is that this idea of skilled *humans* doing something that we can call perfectly accurate translation is a gross misconception of what sort of activity translating is. It is not just that human translations are always somewhat inaccurate: it is not clear what standard we are using when we identify a translation as accurate or inaccurate. I hasten to say that it is not only MT researchers who express this misleading concept of translation. The most extreme expression of it that I have encountered came from someone hostile to MT: P.J. Arthern ([64] 1979), responsible for translation for the Secretariat of the Council of the EEC, wrote in 1979 that MT would never be any use to him because 'the only quality we can accept is a 100% fidelity to the meaning of the original', and he did not envisage MT ever being that good. MT enthusiasts on the other hand say 'Yes, we will give you a 100% fidelity eventually, but it will take time'. To me this sort of exchange is weird. It assumes that there is something identifiable as 'a 100% faithful translation', which clever humans might manage to produce. I do not know what that is, or how you recognise it if you are lucky enough to encounter it.

Let me say, too, that I think the MT community is starting to become much more sensible about what they can reasonably aim at than they were a few years ago. I am very heartened by Peter Wheeler's discussion of SYSTRAN (chapter 13) with its recognition that what MT needs to do is to look for tricks that work more often than not and that, even when a mechanical translation is identical to what a human translator might have written, it is produced by a quite different method. And there is an excellent remark in Jonathan Slocum's chapter which I do not think one would have come across three or four years ago: 'there is no a priori reason why machine translations must be "perfect" when human translations are not expected to be so: it is sufficient . . . that they prove cost-effective' (chapter 17). Perhaps the tide is starting to turn. But, if I can help to bury once and for all the idea that somewhere out there is an algorithm that would give perfect human-like translating performance if only we could discover it, then this will not be any too soon.

What did I mean by two types of problem-solving activity? A clear case of one type would be: solving the kind of puzzle which begins by saying something like 'Andrew, Basil, and Charles are a painter, a violinist, and a lawyer. Their wives are called Millicent, Nettie, and Olga. The man who is married to Olga lives next-door to the violinist. The lawyer drives a Volvo. Basil never eats eggs . . .' – and after a lot more of this sort of thing it asks something like 'What is the name of the violinist's wife?' A clear example of the other type of problem-solving would be the work of a natural scientist trying to find an explanation for some puzzling observation: say, Newton's observation that an apple which becomes detached from its tree falls vertically to the ground.

What is the crucial difference between these two kinds of problem? It is *not* a difference of complexity – at least it is not intended to be. The Sunday-paper puzzles about violinist's wives' names can be very complicated indeed. And conversely Newton's theory of gravity was strikingly simple, once Newton had managed to formulate it – in fact simplicity is one of the hallmarks of science. Rather, I would identify the crucial difference by saying that one type of problem is *closed* or *bounded* while the other is *open-ended*. The problem about Andrew, Basil, and Charles is bounded in the first place because we know in advance what the range of possible answers are. If the question is 'What is the violinist's wife called?' then the answer can only be Millicent, Nettie, or Olga. And the problem is bounded also in a deeper way: only a fixed range of information is relevant for solving it, and all that information is explicitly set out in the printed statement of the puzzle. You are given a question, and you are given the information you need to deduce the unique correct answer; all you have to do is to make inferences from the premises until the answer emerges. (That is no small thing in itself, of course: as I have said, these puzzles can be very difficult.)

There is a point of view about science according to which scientific discovery is not very different. You notice something needing explanation; you assemble all the observations that are relevant, in terms of criteria of relevance that are taken as self-evident; and when you have assembled enough evidence then the answer to the problem is implicit, so all you need to do is to apply enough essentially mechanical intelligence to infer the answer from the heap of premises. People sometimes suggest that Francis Bacon thought science was like this. I believe that is actually unfair to

Bacon, and I am not sure that any of the great names really thought that science worked this way; but plenty of laymen do. It is fairly common in my experience for not-very-academic types to talk as if the word 'science' referred to the distillation of specially solid, specially certain kinds of knowledge from masses of observations in which that knowledge was somehow contained.

But, since the work of Sir Karl Popper in particular, sophisticated people have not taken that view of science very seriously. Popper has made us understand that scientific discovery is a far more open-ended thing than that. When we notice something that needs explanation, we are *not* at the same time supplied with a range of potential explanations. The scientist has to dream up a plausible explanation as best he can. Formulating a scientific hypothesis is a creative activity, like the work of a painter or composer. And it is not just the range of explanations for a phenomenon that is open-ended. The range of evidence bearing on the truth or falsity of any given explanation is open-ended too. Once a scientist has identified a problem and chosen a hypothesis about the answer to it, the ball is still in his court when it comes to deciding where to look for evidence to test his hypothesis.

The revolutionary consequence of Popper's model of science as an open-ended activity is that it becomes a quite uncertain one. To a Popperian, the theories of science are not specially solid and certain bits of knowledge: they are hazardous, unreliable, provisional beliefs. If we are not given a fixed set of possible solutions to a given problem, and we are not given a fixed range of evidence bearing on any particular hypothesis, then there is no knowing when someone might not dream up a new and better alternative to any of our current theories, or might not notice that some unimportant-looking phenomenon amounts to a refutation of a current belief. Newton's theory of gravity seemed for a couple of centuries to be as solidly and unshakably certain as the laws of logic themselves; and then along came Einstein and showed that Newton had been wrong all along. Anything we think we are sure of now may turn out tomorrow to be just as misguided as Newton's theory of gravity. Ultimately we do not know anything for sure. All our factual beliefs are guesswork – except when they relate to artificially bounded domains like the puzzle about the violinist's wife's name. In a case like that, once we have worked out that the violinist's wife is called Millicent we really can be sure that the question has been settled once for all, because the problem has been set up as a self-contained, closed system. But problems as bounded as that are very unusual in real life.

What has all this got to do with translation? Just this. Once we proceed beyond beginners' language lessons to real-life translation, translating a document is never a closed problem of the violinist's-wife's-name variety. It is an open-ended research project, like the task of working out why apples fall to the ground. Any translator is continually faced with questions that his best efforts do not allow him to resolve definitively – because, as Popper has taught us, definitive resolutions are not available to non-logical questions. Even the most careful piece of translation by the most skilled and responsible human translator is a tissue of guesswork, and in many cases different guesses would be equally defensible.

One consequence of this is that it makes very little sense to ask whether an MT system can produce 'correct' translations: as soon as we venture beyond the very simplest texts, no particular translation can be singled out as 'correct'. Another consequence is that any mechanical system which translates by applying an algorithmic process to a source-language text, however subtle the algorithm and whatever else its virtues, cannot be simulating the activities of a human translator: for a human, to translate is to engage in a Popperian research project, and Popperian research does not proceed by algorithm. It involves creatively dreaming up guesses and creatively inventing ways to test them, and two equally competent scientists (or translators), faced with identical data (or source texts), may produce different theories – or translations.

Let me quote some concrete examples. I did some research on the problem of resolving the reference of the English pronoun *it*. This is a typical problem that arises in MT; if you have to translate the word *it* into a gender language, you have to know what it is referring to. I got a sample of occurrences of *it* in context by trawling through the Lancaster–Oslo–Bergen Corpus, which is designed as a fair sample of real-life, unedited modern written English in machine-readable form; I extracted occurrences of *it* specifically from Category H of the Corpus, which is mainly government documents and the like – fairly sober kinds of prose, and a good match for the type of documents that practical MT projects like EUROTRA (chapter 19) are likely to have to deal with.

I began with 338 examples of *it*, among which I discarded cases where *it* was used non-referentially – cases such as, for instance:

> 1) *It* is easy for the Leader of the Opposition to suggest the idea of more and more new towns as a complete solution, . . . [H8.119]

where *it* is acting as a place-holder for the extraposed *for*-clause. Slightly more than half the examples were non-referential, so I was left with 156 cases of referential *it*. I then tried to work out just how sure I was in each case what the word *it* referred to. And the first thing I discovered was that in fully sixteen cases – ten per cent of the total – I really did not know. Consider, for instance, the following passages:

> (2) We recognise that there is some danger of the resources of the after-care service being strained by an excessive volume of 'voluntary' work, and that some method of selection to control *it* might become necessary. [H8.77]

> (3) The lower platen, which supports the leather, is raised hydraulically to bring *it* into contact with the rollers on the upper platen, . . . [H7.148]

> (4) The foregoing is an over-simplification of the pattern of Government, Government-aided and industrial food research in this country; *it* is uneven and thin in places, . . . [H9.88]

In (2), does *it* refer to *some danger,* or to *'voluntary' work*? Pragmatically speaking, it probably makes no odds in this case – but of course it might make a difference to the translation, if the target language had different genders for *danger* and *work*.

Likewise in (3), it might seem purely academic to ask whether *it* is the lower platen or the leather – if the leather is lying on the lower platen, then perhaps bringing the leather into contact with the rollers implies bringing the lower platen into contact. But I am speaking without any special expertise on this kind of machinery. Perhaps someone who knew what he was talking about might say: Of course *it* must be the leather – if the lower platen touched those rollers while the machine was running, God help anyone who was standing nearby. I do not know whether that is so – and offhand I would not even know where to find someone who did know.

(4) is a case where it is not just difficult to resolve *it* but it makes a real difference which way you resolve it. It could be *the foregoing,* or it could be *the pattern.* There is all the difference between saying that the *research* is patchy, and saying that one's *description* of the research is patchy; but I honestly cannot say for sure which the writer means (though I lean towards *the pattern* rather than *the foregoing*).

Of course it is no news that language is sometimes ambiguous, and that an MT system would have to make arbitrary choices in translating ambiguous passages because normally an ambiguity will not carry over into a target language. But what we usually mean when we call a passage ambiguous is that a competent, reasonably careful reader will notice alternative interpretations. These cases are not ambiguous in that sense. I have drawn your attention to the alternatives, so they probably seem obvious now; but I do not think that any of the ten per cent of passages which I ended up judging unresolvably ambiguous seemed at all problematic when I first read them. It was not until I forced myself into the unnatural discipline of asking 'Never mind the obvious referent: is there anything else the pronoun could refer to?' that I began to realise that there often were other possibilities which were sometimes just as plausible as the one I thought of first. Someone who reads a text normally behaves like a hack scientist, in Popperian terms: if he has one theory that makes sense of his observations he will not make life difficult by thinking of rival theories. But nothing guarantees that his first theory is actually the right one. It is a truism that language is often ambiguous; what I am saying is that ambiguity is far commoner than we usually realise, and that when we think some piece of language is unambiguous it is not because competent language-users have algorithms for resolving ambiguities but because we are good at not noticing ambiguities.

I do not think it is controversial to say that the main difficulty in MT lies in analysing the source-language text, rather than synthesising its target-language equivalent. There does seem to be a real danger that people may require MT systems to be better at the job of analysis than we humans are ourselves.

Even when a Popperian scientist does get to the stage of consciously assessing rival theories, nothing guarantees that the one he picks will be the correct one. Likewise with language a reader may choose the wrong interpretation even when he is aware of alternatives. In my research on the pronoun *it,* there was one case where I *know* I got the answer wrong. I know it, because I reached different conclusions about the same case on different occasions – I cannot have been right both times. This was the following example:

> (5) Sometimes an increase in the orders in hand is regarded as a wholly favourable feature, but this is only true to a limited extent. If the growth in the order book is not matched by parallel growth in engineering and manufacturing capacity *it* would adversely affect our delivery dates. [H27.82]

On one occasion when I read (5) I decided that *it* must be *the growth in the order book.* Then I looked at my notes and saw that a few weeks earlier I had been equally confident that *it* must refer to the whole proposition expressed by the *if*-clause. Reconstructing my thought-processes, I suppose that originally I laid weight on the fact that the verb of the *if*-clause is in the indicative; if the pronoun were intended to refer to the whole clause, a careful writer would have written *were* rather than *is.* But many writers ignore that grammatical subtlety. When I returned to the passage the second time, it did not occur to me to take the mood of the verb into account. That time, what I noticed was the semantic parallelism between the two sentences: if *it* means *the growth in the order book* then the first sentence is about the advantage of that growth and the second sentence about its disadvantage. This parallelism seemed to be an adequate reason to settle the question — until I discovered that I had settled it a different way for a different reason once before.

Now I realize that this discussion of my thought-processes may sound anecdotal and self-indulgent. But there is a serious general point I am using it to make. My attempt to resolve the *it* of passage (5) was a little scientific research project, and like other research projects it does not reach a definitive conclusion. We all know how scientific academic literature keeps returning to old issues when people think of new evidence or new arguments. Around 1960 the Chomskyans were busy proving beyond a shadow of doubt that context-free phrase-structure grammar was inadequate to describe natural languages, and we had to use transformations. Everybody was convinced. And then around 1980 Gerald Gazdar and his associates started to argue that transformations were unnecessary and that CF PSG is the correct formalism for describing natural languages, and people are becoming convinced of *that*. It is the same with translation. It does not make sense to demand that a translator produces a *correct* translation: that is like demanding that a research scientist produces only the correct theory of the phenomenon he is examining, and it is not given to humans to guarantee that achievement – all a scientist can do is produce *a* plausible-looking theory and defend it against alternatives. When a translator stops thinking about a translation problem and writes down a solution, it is not because he has settled the question beyond any possibility of re-opening it. It is because he judges that the cost of further research probably will not be repaid by a correspondingly valuable increase in the quality of his product. It is an entrepreneurial decision, not a logician's deduction. (Like any other entrepreneurial decision, it may be wrong – we cannot expect perfect decision-making by enterprisers, that is not what enterprise is all about.)

Entrepreneurial activity is fallible; and it is also creative. A commercial enterpriser has to dream up a new product or a new method of manufacture before he can decide whether to take the risk of investing in it. If the range of commercial possibilities were somehow given to us in advance, competitive free enterprise would have no purpose. Likewise with translation: the translator has to think up solutions to his problems, and he has to use his creativity to think of reasons why one solution is more plausible than another.

I shall discuss some more examples in a moment to show what I mean by that. But, first, let me contrast my point of view with someone who explicitly takes the opposite point of view. What got me interested in researching on the problem of resolving the pronoun *it* was an article by Jerry Hobbs called *Pronoun Resolution* ([642] Hobbs 1976). What Hobbs does in that article is to contrast two alternative systems for automatically finding the antecedents of anaphoric pronouns. In the first part of the article he defines an algorithm which works purely on the basis of the grammatical structure of the text, ignoring any considerations of the meaning or pragmatics of what is being said. It is quite a subtle algorithm (I shall not attempt to explain it here); and it works strikingly well. Hobbs tested it on sample texts drawn from several genres of prose. Even on the pronoun *it,* which is usually harder to resolve than the other anaphoric pronouns because there are more possible antecedents, Hobbs's grammatical algorithm gave 77 per cent correct results. And if the grammatical algorithm is supplemented with crude Katz–Fodor-type information about semantic collocations (for instance, if *it* is the object of the verb *eat* then do not identify it with an abstract noun), the success-rate rises to 83 per cent. Remember that when I tested my *own human* ability to resolve the pronoun *it* I only convinced myself that I knew the right answer in about 90 per cent of cases. And the example of passage (5) shows that sometimes when I was convinced that I knew the answer, I actually did not; so it is very possible that someone else might have scored my ability to resolve *it* at less than 90 per cent. Conversely, I can see various purely grammatical considerations that Hobbs's algorithm ignores but which might well be added in to improve its performance further. Comparing my human performance with the performance of Hobbs's grammatical algorithm, I am better, but Hobbs's algorithm comes in quite a creditable second.

You might think, then, that Hobbs would pat himself on the back for constructing this algorithm. Not a bit of it. Hobbs's name for this grammar-based algorithm is 'the naive algorithm'; and he only produced it as a negative example – an example of how not to resolve pronouns automatically. Hobbs calls his algorithm naive because it does no inferencing; it ignores the logic of the text it is examining and looks only at the grammar. To Hobbs it seems self-evident that a satisfactory automatic pronoun-resolution system must use inferencing. He gives two main reasons:

(1) the performance of the grammatical algorithm, while not bad, is well below 100 per cent;

(2) humans use logic rather than just grammar to resolve pronouns, so a mechanical system ought to do the same.

Point (1) is an argument only if *human* performance is 100 per cent or close to it; and we have seen that it simply is not. Hobbs almost seems to imply that if we were very clever we might construct an automatic system that was actually *better* than humans at understanding human language. But that is an utterly contradictory idea. Language means what humans take it to mean; in this domain there cannot be any question of machines outperforming humans.

As for the second point, I agree that humans commonly use inferencing rather than pure grammar to resolve pronouns. But when Hobbs discusses this I think it is he that is naive, not his algorithm. Hobbs actually describes in the second part of his article a semantic algorithm for pronoun resolution, which is what he thinks we ought to use instead of his so-called naive grammatical algorithm. He does not specify the semantic algorithm in rigorous detail, as he does in the case of the grammatical algorithm: he just sketches the general outline of a hypothetical future semantic algorithm. One of the things I find very odd in Hobbs's article is his willingness to take for granted that a procedure which he cannot state rigorously, when it *has* been stated rigorously will certainly outperform a procedure which is already defined in detail and produces very good results. This seems to be rather a common syndrome among the artificial intelligentsia; but surely it smacks more of blind faith than sober wisdom? More important, though, what Hobbs *does* say about his hypothetical semantic algorithm contains a clear mistake: he quite explicitly fails to take account of the provisional, Popperian character of human inferencing.

One of the ways Hobbs illustrates his semantic algorithm is by showing how it would deal with Winograd's example-sentence *They prohibited them from demonstrating because they feared violence.* How can a pronoun-resolution system identify the *they* in *they feared violence* with *they* rather than *them* in *they prohibited them*? Hobbs shows how an inferencing system will be able to construct a series of deductions that succeed in making sense of the word *because* just if *they* in the second clause is equated with *they* in the first clause.

But of course any kind of deductive system can only yield results if it is given some premises or axioms to deduce from. Hobbs takes it for granted that his pronoun-resolution system will be equipped with a data-base of factual general knowledge that the pronoun-resolution system can draw on. And he stresses that the bits of knowledge that are relevant for understanding his examples are standard, well-known facts. One of the crucial premises for understanding the Winograd example, for instance, is the fact that demonstrations cause violence, and Hobbs calls this 'A prominent fact about demonstrations, well-known to anyone aware in the 'sixties'.

Perhaps so. But it will not *in general* be true that the premises used in understanding language are well-known, established facts. Popperian thought is not purely deductive; it is *hypothetico*-deductive. We solve real-life intellectual problems not by deducing the solutions from a fixed set of axioms, but by *inventing* plausible premises and seeing if they help us reach satisfactory conclusions. Making deductions may be a reasonably mechanical sort of activity, but deciding on the premises to deduce from is often highly creative.

Turning back to my own research on resolving *it*: I have said that I often was not sure of the right answer, but another thing I found very striking was how often, when I *did* feel confident that I knew the right answer, I could justify that confidence only in terms of premises that I seemed to have invented for the occasion. Let us look at some examples. Each of the following passages is a case where Hobbs's purely grammatical algorithm produces an answer which I think is wrong; but the semantic reasoning that leads me to a different answer relies on premises which are far less straightforward than the premiss 'demonstrations cause violence':

(6) . . . gratitude to Sir Wilfred Fish, lately Honorary Director of the Department of Dental Science, for the outstanding services he has rendered to dental science in establishing the Department and guiding *it* over the past five years. [H26.152]

(7) That is the pattern we are now seeking to establish in legislation. Under the 'permissive' powers, however, in the worst cases when the Ministry was right and the M.P. was right the local authority could still dig its heels in and say that whatever the Ministry said *it* was not going to give a grant. [H16.24]

(8) Nothing in the Commercial Agreement as modified by the present Agreement shall –

   (a) require either Government to do anything contrary to any obligations to which *it* may be subject under the General Agreement on Tariffs and Trade; . . . [H14.50]

In (6), Hobbs's grammatical algorithm equates *it* with *dental science*, while I think the right answer is actually *the Department*.

But why do I as a human resolve the pronoun this way? Not because there is anything illogical about the idea of Sir Wilfred Fish guiding dental science – there is not. If I did resolve this particular ambiguity by reference to meaning, rather than by a grammatical algorithm somewhat different from Hobbs's, then I think I must have unconsciously argued something like this: It would be embarrassingly fulsome to describe a man on his retirement as having guided a whole subject practised all over the world, but quite fitting to describe him as having guided a particular Department; what is more, when two achievements are mentioned in close conjunction, like *establishing the Department* and *guiding it over the past five years,* then the effect would be ironic unless the two are on a comparable scale with one another – and a formal speech marking the retirement of a distinguished scholar is no occasion for waxing ironic about his career. But these premises are not exactly standard bits of general knowledge – I do not believe I had ever entertained them until the problem of understanding this passage forced me to invent hypotheses that would make it comprehensible.

Now turn to (7). Hobbs's grammatical algorithm gives *it* as *the Ministry*; I feel sure that *it* is *the local authority.* How did I reach that conclusion? If I can reconstruct my thinking I suggest it went something like this: In the first place it seems to me that etiquette between British local authorities and departments of national government implies that local authorities may flout ministries' wishes but they do not call them liars. *If* that premiss is true then *it* cannot be *the Ministry,* and I *think* the premiss is *probably* true; but I am not aware that the issue had ever arisen in my mind previously, and now the question has come up I do not feel sure of my ground – perhaps local authorities regularly accuse ministries of prevarication. I need more evidence. Well, now that I think about it, the phrase *dig its heels in* goes better with a scenario in which the local authority continues to deny a grant application despite pressure from above, than with a scenario where the local authority is calling the ministry a liar, because there is no suggestion that calling the ministry a liar would be a long-term operation needing heels to be dug in (though again I may be wrong there – perhaps it is only my ignorance of the relationships between local authorities and ministries that hides from me why *dig its heels in* is appropriate in connexion with the reading that I am taking to be the wrong one).

If my reading is seriously challenged, I can look for still more evidence to test it. For instance, it seems to me that a change in the law might allow national government to require a certain action from a local authority but not vice versa – though again I do not feel certain enough of the British Constitution to swear to this premiss, and again it had not consciously occurred to me previously.

There is no ultimate stopping-point to this process of looking for premisses that can be used to yield arguments for one or another reading, and testing them. How far you take the process in practice will depend on factors like how doubtful you feel about your original decision, and how much hangs on getting the answer right. If it is very important to get the right interpretation and you feel seriously unsure, you can go to all kinds of lengths to get evidence; for instance, you might try to contact the original writer and ask him what he meant. To think of mechanically simulating this process would surely be as inappropriate as trying to automate Sherlock Holmes?

Now look at (8), which is a very different case. Here the 'naive' grammatical algorithm equates *it* with *Nothing in the Commercial Agreement* rather than with *either Government,* and there is nothing straightforwardly illogical in that: 'something in a Commercial Agreement' might quite possibly be 'subject to obligations under a General Agreement'. I believe the main principle I used myself in concluding that *it* refers to *either Government* here was nothing to do with the semantics of the words involved; I think I assumed that, when a text contains an indented text within itself, this creates a presumption – a rebuttable presumption, but a prima facie presumption – that an anaphor in the indented part will not refer to an antecedent in the main text. Again I certainly do not believe that there is a piece of standard general knowledge, shared by competent human readers of English and available to be incorporated into an AI data-base, that says that anaphora is affected by indentation. This was a Popperian hypothesis which I formulated creatively because it helped me solve a particular problem.

To me it seems very clear in practice that the kind of problems which have to be solved in translating are often much more like the open-ended sort of problems encountered by research scientists than like the complex-but-bounded problems printed in the puzzle corner of a Sunday paper. But, if that is obvious to me, I am naturally led to wonder why it is not equally obvious to other computational linguists. Despite what I said earlier about recent writers on MT being less glib about the wonders of AI than they were a few years ago, it still seems to me that plenty of computational linguists remain faithful to the sort of assumptions I have been criticising in the case of Jerry Hobbs. And I think there are two reasons for this.

One is that people who work with computers find it hard to

appreciate the open-endedness of the translation task, because the problems they are used to are commonly bounded problems. I said at the start that in real life it is hard to find problems like the one about the violinist's wife's name; real intellectual problems are almost always open-ended. But one kind of real-life problem that really is bounded is the problem of designing a computer program. The range of possible answers to a programming problem is a perfectly closed system – it is defined by the syntax of the programming language. And all the elements of the programming language are perfectly rigorously specified, so if you know what your program is supposed to be doing then in principle purely deductive reasoning ought to give you a program which functions correctly.

It does not work like that in practice, of course. The reasoning involved is so complicated that programmers make mistakes. Programming *feels* like a very Popperian activity: you carefully write the best program you can, you expose it to the possibility of disconfirmation by running it, almost invariably it fails in one way or another, so you revise it and try again until some eventual version of the program seems to be error-free – for the time being. Programming has all the superficial appearance of Popperian learning through conjectures and refutations, without the essential open-endedness that makes the process of conjectures and refutations unavoidable in the sciences. For the programmer, trial and error is just a heuristically convenient alternative to the immense fag of ensuring that one's program works right first time. But because trial and error *is* so convenient for the programmer, even though it is not absolutely necessary for him, computer experts tend to overlook the reason why in other spheres of mental activity fallible trial and error really is totally unavoidable even in principle.

That is one reason why computational linguists fail to grasp the open-endedness of translation problems. But there is another reason, which to my mind is rather less excusable; and that is the strategy of concentrating on small and carefully-chosen samples of language. Like many other attitudes to natural language that I have been criticising, this one is surely something that the computational-linguistics community have picked up from the Chomskyans. Books about grammar emanating from the Chomskyan school commonly revolve round invented example-sentences about John being eager to please and the like, which often look fairly artificial and seem to be selected so as to contain only problems to which the writers know the answers. This may perhaps be defensible within the field of pure theoretical linguistics (though I feel suspicious about it even there). But in computational linguistics, which normally has at least half an eye on practical applications, surely it is quite unjustifiable. In our Unit for Computer Research on English Language at the University of Lancaster we take it as self-evident that techniques which work only for invented samples of language are not worth considering; all our research is based on 'raw', real-life English. (Thus, every example quoted in this paper is a genuine extract from a published source.)

But what seems axiomatic to us does not seem so to others. I was quite surprised, for instance, by the flavour of the papers given at the Inaugural Meeting of the European Chapter of the Association for Computational Linguistics, at Pisa in September 1983 ([564]). I have no wish to pillory any particular individuals, but let me quote a sample of typical bits of 'natural language' (so-called) that various speakers at Pisa analysed:

Whatever is linguistic is interesting.
A ticket was bought by every man.
The man with the telescope and the umbrella kicked the ball.
Hans bekommt von dieser Frau ein Buch.
John and Bill went to Pisa. They delivered a paper.
Maria é andata a Roma con Anna.
Are you going to travel this summer? Yes, to Sicily.

There was even a paper that argued that a parser ought to be robust enough to handle deviant sentences, and quoted as an example the sentence *Mary drove the car and John the truck*; if you are really fastidious about what counts as an English sentence, apparently this looks like a distortion of the 'proper' sentence *Mary drove the car and John drove the truck*. Out of thirty-odd papers at Pisa, I do not believe that more than two or three sounded as if their authors were actively interested in what I would call 'natural language' – because these neat, dapper little invented examples are utterly uncharacteristic of real-life 'natural language'.

Or at least, they are uncharacteristic of most genres of natural language. There is one genre where examples like the ones I have quoted are the norm: and that is language-teaching primers. Books with titles like *French Without Tears* are full of sentences like *The man with the telescope and the umbrella kicked the ball*. And there is a good reason for that: they represent a bounded kind of language – bounded by the vocabulary and grammar that the primer has introduced. If you write a language primer, you are not allowed to have a sentence in Lesson 3 including a construction that will not be discussed till Lesson 12: the pupils would not stand for it. But outside the schoolroom nobody limits himself to any particular bounded subset of language. It seems to me that the habit of always discussing this sort of example permits most computational linguists to ignore the open-endedness of natural language, and the open-endedness of the processes that humans use to comprehend it.

I realise that the computational-linguistics community would defend their habit by saying that they have to start with simple bits of language and get them right before they can go on to get more realistically complex language right. Chomsky again: we have to describe competence before we can describe performance. I do not believe it, I am afraid. My belief is that systems which process real-life language successfully will not be decomposable into a core system that deals with a dapper, polished little subset of language, together with peripheral components to handle 'performance deviations'; successful systems will be geared to real-life language through and through.

And talking of successful systems leads me on to the positive conclusion I promised at the beginning. So far I have been relentlessly negative; but I do not myself believe that what I have been saying implies a negative assessment of the prospects for MT. I have argued that computational linguists mistakenly think that understanding natural language is a complex-but-bounded problem, like the problem of designing a computer program, whereas it is really open-ended. Processes which *are* complex-but-bounded are good subjects for mechanical simulation; and it is a very reasonable AI axiom that, if you can make your artificial system achieve some intelligent goal using the same method as a human, it is silly to waste time designing a system which reaches the same goal by a different route. One practical consequence of the non-Popperian outlook of computational linguistics has been that language-processing techniques which seem unrelated to human language processing are ignored as irrelevant, even if they work. We have seen that in connexion with the Hobbs paper. Pronoun resolution by reference to grammar worked pretty well; but Hobbs believed that humans do not use grammar to do the job, so he quoted the grammatical algorithm only in order to dismiss it as naive. Adding Katz-Fodor semantic restrictions made the algorithm work even better; but Hobbs knew that real human language does not always obey Katz-Fodor rules (for instance, Katz and Fodor would say that the object of *eat* must be concrete, yet we often speak of people eating their words); so Hobbs felt that crude Katz-Fodor semantic restrictions could have no place in a serious automatic language-comprehension system.

Once we see human language-comprehension as a creative, fallible, non-algorithmic Popperian activity, then the idea of direct computer simulation no longer applies: we cannot program silicon chips to be imaginative, and anyway there is no special virtue in mimicking human behaviour exactly when humans do not understand language perfectly either. So what we ought to do is to look for any techniques that give good results – irrespective of whether we think humans use the same techniques, and irrespective of whether the techniques are 100 per cent accurate or merely better than the best previous techniques.

The payoff from this pragmatic approach to computational linguistics can be spectacular. For instance, one of the standard MT problems when the source language is English is the grammatical ambiguity of English words. Because English has so little inflection, a high proportion of words taken in isolation are ambiguous as between two or more parts of speech; this is one reason why MT researchers often think of English as one of the most difficult languages.

This particular problem is no longer a problem. At Lancaster we have cracked it. Our CLAWS system automatically assigns grammatical tags drawn from a 134-member tag-set to words of real-life written English text; tested on texts representing diverse genres of writing, including some downright ungrammatical writing, our system performs with monotonous regularity at a success-level of 96 to 97 per cent of words correctly tagged. 96 per cent is not 100 per cent; but I see no reason to believe that anyone could ever design an automatic tagging system that is 100 per cent accurate. What is much more relevant is that my colleagues and I have not managed to hear of any other computational linguists with systems that can do the job with 96 per cent accuracy, or anything like it. The other important point about our system is that it works in a very *un*natural way. Our tagging system 'knows' virtually nothing about the grammar of English. Almost the only thing it knows is the range of transition probabilities between adjacent pairs of tags; it does not know anything at all about the overall grammatical architecture of sentences (and still less does it know anything about what sentences *mean*). Now it hardly seems reasonable to imagine that *humans* resolve word-class ambiguities just by reference to immediate transition-probabilities. Surely, we do use meaning. But the point is that, unnatural or not, the system works, and that is what matters.

(Currently, we at Lancaster University are moving on to the larger problem of sentence-parsing using similar 'crudely unnatural' statistical techniques. This work is only in its preliminary stages, but already results are highly promising, with a first batch of real-life sentences parsed – correctly – in March 1984.)

I have formed the impression over the last two or three years that some computational linguists find the attitude I am expressing faintly disgusting – rather like purist academics shuddering with distaste at the messy compromises they encounter if they venture into the world of commerce or industry. Well, academics can preserve their illusions – provided they are careful never to step out of their ivory tower. And in the same way I expect computational linguists may be able to maintain the axiom that only 100 per cent performance is interesting, provided they stick to sentences about men with telescopes and umbrellas. But there is little virtue in living in a dream-world. If computational linguists can get people to fund them in order to analyse sentences about a ticket being bought by every man, that may be nice for them; but meanwhile there is serious work to be done. I see no reason to suppose that that work *cannot* be done. But it *will not* be done, as long as people dream that superfast electronic calculating machines can somehow be turned by clever programming into resourceful, risk-taking Popperian entrepreneurs of knowledge.

Humans are one type of imperfect translator. Computers cannot be the same as humans, but they might be another kind of translator; at least as imperfect as humans, but not necessarily much more so.

PART TWO : SOFTWARE

SOFTWARE BACKGROUND FOR
MACHINE TRANSLATION : A GLOSSARY

The aim of this chapter is to give a succinct account of some basic computational terms, notions or techniques mentioned in machine translation literature. It does not claim to be exhaustive or completely accurate. Through sketchy description and examples it should give an intuitive grasp of what the thing is or how it works. In each section, references to more specific descriptions will introduce the interested reader to the specialised literature.

In the following, terms that are part of this basic computational glossary are in italics.

Parsing

Parsing is a key component of all systems that deal with a language. A good introduction to techniques and problems of natural language parsing can be found in [602] King 1983 and [625] Winograd 1983.

A *parser* constructs, from a linear (unstructured) input and a *grammar,* a structure that represents a 'meaning' to be used by a subsequent process ([630] Barr 1981, 257). In the case of natural language parsing, the input is a sentence or a text (a sequence of words). The grammar is a computational representation of a linguistic theory of the language that is parsed. The structure produced is the grammatical analysis of the sentence.

In the framework of *formal language theory,* a grammar is a set of rewrite rules describing a particular language. For example, a grammar for the sentence 'John loves Mary' could be:

Rule 1: A sentence consists of a Noun Phrase followed by a Verb Phrase.
Rule 2: A Verb Phrase consists of a Verb followed by a Noun Phrase.

111

Rule 3: An instance of a Verb is the word 'loves'.
Rule 4: An instance of a Noun Phrase is the word 'John'.
Rule 5: An instance of a Noun Phrase is the word 'Mary'.

This grammar matches the intuitive knowledge that a sentence has a subject followed by a verb and its complement. In a more formal way this grammar is written:

R1: S→NP VP
R2: VP→V NP
R3: V→loves
R4: NP→John
R5: NP→Mary

The sentence 'John loves Mary' is parsed with this grammar to produce the *parse tree* or *derivation tree*:

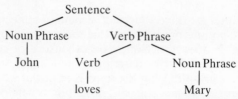

A strategy to parse this sentence could be the following:

(1) We assume that the input is a Sentence.

(2) Rule 1 predicts that the first part of a Sentence is a Noun Phrase (followed by a Verb Phrase).

(3) Then we concentrate on the first part and discover in Rule 4 that a Noun Phrase may consist of the single word 'John'. We are confirmed in our predictions by checking in the input that the first word of the sentence is 'John'.

(4) Then we apply the same strategy on the second part of the Sentence, repeating prediction and expansion. We find successively that a Verb Phrase is a Verb and a Noun Phrase (Rule 2), that (Rule 3) a Verb might be 'loves' (checked) and that (Rule 5) a Noun Phrase might be 'Mary' (checked).

This *parsing strategy* (or *enumeration order*) is called *goal driven* (because it first predicts a goal, and then tries to fulfil it) or *top down* (because the constituents of the parse tree are found from the top to the bottom).

Another strategy could be:

(1) To start from the words of the sentence, and try to find rules that apply to them: Rules 4, 3 and 5 say that 'John' is a Noun Phrase, 'loves' is a Verb and 'Mary' is a Noun Phrase.

(2) Then Rule 2 says that a Verb followed by a Noun Phrase is a Verb Phrase.

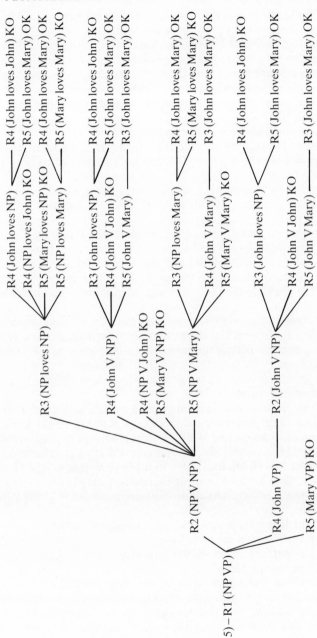

Figure 7.1. Choice tree, top-down parse.

(3) Finally with Rule 1, we can say that the first Noun Phrase followed by that Verb Phrase is a Sentence.

This strategy is called *data driven* (because the actual data present as input drive the parser) or *bottom-up* (because the constituents of the parse tree are found from the bottom to the top).

In the general case, at each step of both processes, many rules are applicable, and a choice must be made. In fact we can also draw the succession of choices as a tree. The choice tree for the top-down parse described above is shown in figure 7.1.

If we choose to apply one rule at a time on the previous result, the strategy is said to be *depth first* (because we go deeper in the choice tree).

If we choose to apply all applicable rules, the strategy is said to be *breadth first* (because a whole level of the choice tree is considered together).

In the depth first case, the parser may follow a branch of the choice tree that does not lead to the desired results. In that case, the parser should revise its choice and start again with another branch. This is called *backtracking* (because the parser follows its track (branches) back to the last point where a not-yet-tried branch was left).

There are various claims about the efficiency of those combined strategies. In general the depth first strategy finds the first result quicker, but for certain grammars can get stuck in an endless branch.

Production systems

Production systems have been used for a broad variety of applications, including psychological models (PRISM2 [539] Langley 1983), medical diagnosis (MYCIN [533] Davis 1977) and machine translation (Q-System [532] Colmerauer 1971). A concise description is given in [630] Barr 1981 and [665] Winston 1984. [557] Rosner 1983 provides a thorough analysis, [534] Davis 1975 gives an overview, and relevant algorithms can be found in [625] Winograd 1983.

A production system consists of three distinct parts:
(1) a production set (rule base)
(2) a context (database, short-term memory, buffer)
(3) an interpreter.

The *context* is a collection of information which represents the intermediate results during the process, the initial input data at the beginning, and the final result(s) at the end (if it ever ends!). For example, in a question answering system the initial data might be a question, and the result the answer. In a parsing system, the initial data might be a sentence and the result a parse tree.

The *production set* is a set of *production rules* which will be used to transform the initial input data into the final output. It is this part that contains the domain specific knowledge, and it is sometimes called a *knowledge base*. A production rule has a *left-hand side (lhs)* also called condition, pattern, or situation, and a *right-hand side (rhs)* also called action or assignment. In the very general case, the lhs describes the situation (an arbitrary condition over the current context) in which the rule should be applied. The rhs describes what actions have to be performed to modify the context. In many systems, the lhs is a *pattern*. If it matches a set of elements in the current context, then the rule is applicable. For example, in a simple *rewrite system* (closely related to formal grammars) the lhs is a string of symbols, which, if it is found in the current context, is replaced by the string of symbols of the rhs.

$$\begin{array}{cc} \text{lhs} & \text{rhs} \\ \text{abcde} \rightarrow \text{fghijk} \end{array}$$

In the Q-system ([532] Colmerauer 1971), this is still a string rewritten as another string, but the lhs is a more complicated pattern. It involves variables which will be bound if the rule applies and whose value can be used in the rhs:

$$A^* + B^* = NP(A^* + B^*).$$

The *interpreter* is the key of the system: it controls how the rules are applied. It is an iterative process:

(1) First, it evaluates in the current context the conditions of each rule of the production set and finds those that are applicable. If the lhs is a pattern, a *pattern matcher* tries to find in the context one or more sets of data that match the pattern.

(2) Then it applies one or more rules according to its *control strategy* (also called *conflict resolution* strategy). For example, if the application is *sequential* it applies only one rule, chosen in the set of applicable rules according to some criteria (first found, most specific lhs, etc.). If the application is *parallel* it applies simultaneously all rules of the set.

(3) Then the evaluation phase starts again. Thus the system evolves in a succession of *states*. The process stops when no rule is applicable or when a state fulfills some halting condition; for example, the context contains a complete parse tree or an answer to the question.

Another task of the interpreter is to decide how to perform the rhs. In the *replacement model* the action of a rule implies the removal or modification of all or part of the data over which the lhs has matched. Thus it prevents the infinite application of the same rule. In the *addition model* the rhs adds new data to the context, without altering the previous context. If this addition is not enough to prevent the application of the same rule the interpreter should have a means to do so (for example, deactivation of rules or marking data as in the Q-system).

On this basic theme, interesting variations are possible; for example, the set of production rules may itself be part of the context (the system can modify its behaviour according to the situation encountered), or the user may define his own sophisticated control (meta-knowledge).

Various computational terms

An *interface* is a device (hardware or software) whose only purpose is to connect two machines, processes or systems. Sometimes it is just the description of how two machines or processes communicate. For example, an American–European plug adaptor, a program that translates user input into machine-coded form.

A system is *interactive* if the user has a dialogue with it. The user inputs his data in a gradual way, and the system responds directly. The interaction usually involves a question and answer dialogue, with requests from both sides for help or commands. In a *batch* system, the user prepares all his input in advance, and after processing the system outputs all the results. If something fails the user has no direct control and has to start all over again.

A process that terminates for all possible input and produces each time the same results for the same input is called *deterministic*. Otherwise it is called *non-deterministic*. A weaker interpretation of the term non-deterministic means that at some stages of the process, a certain number of actions can be performed in any order (or in parallel), but the number is not known in advance (it is dependent on the input).

If a programmer does not know how to solve a problem in all cases, he can use a *heuristic,* expressing his intuition about the problem. The price to pay for this ignorance is that it cannot be proven that a heuristic process produces only correct solutions or that it produces all solutions. A heuristic is the programmer's rule of thumb.

If a process combines multiple partial results in all possible ways, and does that iteratively, the size of the database containing the partial results grows very quickly. This is called *combinatorial* or *exponential explosion*.

A BRIEF LOOK AT A
TYPICAL SOFTWARE ARCHITECTURE

At the most general level, from the user point of view, a translation system is a black box with one entry and one exit. It receives the source language text to be translated as input, and produces the target language equivalent text (translation). The concept of the black box is useful to express the fact that the user needs only to know how to invoke the translation procedure, but not how it operates. If we open this black box, we will be able to detail the system's content at a lower level, say for a MT specialist. At that second level the system consists of three parts: a translation process, a short-term memory, and a long-term memory.

Figure 8.1 diagrams this decomposition, and the arrows represent the flow of data inside the system. The entry, exit, source text and target text are still recognisable. For brevity's sake, long-term memory is now abbreviated to LTM, and translation process to TP.

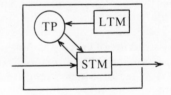

Figure 8.1.

Standard names have not been given to the various elements; on the contrary, such a decomposition is often found in documents. In chapter 7, this logical division can be seen in the paragraph on production systems where the production set is equivalent to our LTM, the context is equivalent to our STM, and the interpreter is equivalent to our TP.

117

Translation process

In figure 8.1, the TP system element is represented differently from the two others because of its fundamentally different character. The TP is a program in the computational sense of the word. Basically, this program uses its own indigeneous knowledge and knowledge saved in the LTM, and applies it to the STM. This STM transformation leads to the translation of the given text. This rough description has introduced two key terms, 'use' and 'apply'.

As can be seen from the arrow linking the TP and the LTM in figure 8.1, 'use' here means that information can only travel in one direction and, more importantly, that the LTM can not be modified. This notion of 'use' is very important and we will mention it again in the paragraph on LTM. The TP uses the information in the LTM and from it works out what actions are to be executed on the STM. These actions are rarely to be executed in all cases, and therefore often have conditions on them. These conditions must be fulfilled before any action can be executed. It is for this operation of condition fulfillment that the contents of the STM must be known. This explains the arrow from STM to TP. Executing the action modifies the STM, hence the arrow from TP to STM.

A very close link can be seen here between what we are describing and what chapter 7 describes in talking about production systems. Thus, the Q-system ([532] Colmerauer, 71), quoted here, is a system in which the TP interprets the system's rules. Outside machine translation software, for PROLOG, it is the interpreter which acts as the TP if the language is being interpreted. If, though, the language is being compiled it is the output from compilation, in the form of a programme, which does this.

Short-term memory

The STM concept hides a memory structure of varying complexity. We can not hope to give an overview of all STM structures, so here we will concentrate on the 'chart' structure which is recurrent in descriptions of this part of a system.

Generally, a translation process is divided into the three phases of analysis, transfer and generation (also called synthesis). Concentrating only on the problem of the analysis will be sufficient to clarify the memory structure of the STM.

Chart is the name given to a computer structure for representation of linguistic states and processes. Our aim is to link up the computational and linguistic features of this structure.

In King ([602] 1983), G. B. Varile gives the following general definition: 'A chart is a graph, that is a set of nodes and a set of arcs linking them. The arcs are ordered and oriented, and there is a unique first and last node. No loops may occur in a chart and each of its arcs bears some (possibly complex) information called a "labelling" which is not the same thing as a "label"; a label is part of the complex labelling.'

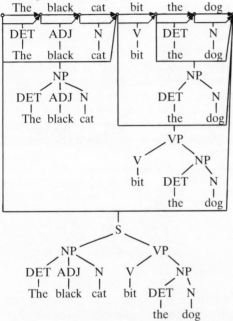

Figure 8.2. Representation of a sequence of states resulting in analysis of 'The black cat bit the dog'.

First, we can see that the orientation of the arcs determines the directionality of the chart, and that the first and last nodes determine start and end points. From this we can infer that any path not following the above three points would be invalid. Secondly, all valid paths represent different stages in text analysis.

Often, in drawing chart structures, different paths through the chart are shown as sharing common parts. This is merely a more compact way of representing the data. The compact structure can be transformed into an equivalent structure uniquely composed of disjoint paths.

The chart's capacity to record several paths can be interpreted in two different and non-contradictory ways:

(1) Constructive interpretation: the analysis algorithm is constructive in that it uses information which is given (text source) or created (partial analyses already done) to continue with analysis. This constructive approach is executed using the source text as initial data (bottom-up algorithm). This interpretation of the chart means that for each path there is a corresponding new element constructed in the analysis tree. The graphic representation does not always allow us to follow the steps of analysis chronologically but this constructive characteristic is still shown in figure 8.2.

(2) Selective interpretation: A critical and constant phenomenon in the treatment of natural languages is 'ambiguity'. To handle this phenomenon, one can also interpret a set of paths as a set of possible analyses of ambiguities. It is noticeable that all ambiguities appearing at a particular time are not necessarily still around at the end of analysis. This is the case in figure 8.3.

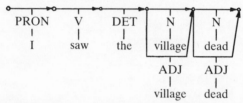

Figure 8.3.

It is quite obvious that in each particular case of a chart, the two interpretations can appear. Humans can easily distinguish between the two cases. The machine also manages this using a certain number of technical tricks. Without entering into too much detail, we can see that the paths where no arc has been used as a construction base must be interpreted according to the second possibility. This is shown in figure 8.4 (taken from Varile, op. cit.), where paths presented as double lines are formed from unused arcs only.

Long-term memory

As was stated before, this element of the system is basically fixed. The LTM shares the knowledge necessary for translation with the TP. From this point of view, it is difficult to discuss the contents of the LTM in detail. We can, however, put forward the following theses:

(1) the more general the TP (e.g. a simple interpreter of rules), the deeper the linguistic information held in the LTM;

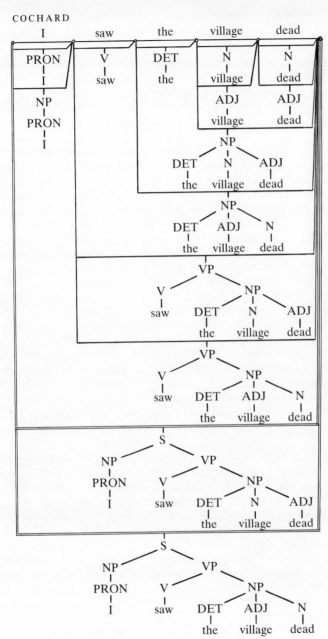

Figure 8.4.

(2) the more intelligent the TP (using linguistic knowledge), the more limited the LTM.

This needs to be kept in mind when understanding how information is split up.

Some examples will help to make this clear:

(1) The Q-system uses a general interpreter which interprets a file of coded rules. The LTM is formed by these coded rules, which are the result of a compilation.

(2) Prolog in compiled form creates a programme which can be executed. This programme is less general and has a certain amount of linguistic intelligence.

We find another aspect of LTM use when we look at efficiency in a translation system. Up to now, we have detailed the central mechanism which allows us to achieve an automated translation job. The mechanism described uses very general and powerful tools, but is also very heavy. For this reason, the global translation routine can be split into a set of subroutines built with tools particularly suitable to the problems to be solved. We do not intend to show the full range of specific treatments, all the more so as this question is discussed in considerable further detail later. We will restrict ourselves to a single case 1: the dictionaries.

The creation of monolingual and multilingual dictionaries is a constant in all translation systems. The most primitive realisation is to include dictionaries as a particular case of the production system. The general tool is used, with all the problems it raises. This weakness of the general tool leads to the development of a specific treatment for dictionaries. Let us consider this computational aspect and analyse then the role of dictionaries in the translation process.

The improvement of a system's translation capabilities generally implies (partly) an increase in the size of the dictionary. In practice this often leads to dictionaries of 500000 entries or more. So it is very important to have a means of controlling dictionary content and of efficiently accessing large amounts of data. Nowadays systems called Data Base Management Systems (DBMS) deal with those problems.

In such a brief overview it would be impossible to describe those DBMS. However, we will just mention that the DBMS provides tools to ease the input of a new entry, to control (to a certain extent) the integrity of the data, and to retrieve quickly some stored information.

From the linguistic point of view, we distinguish between two dictionary classes: the monolingual dictionary and the bilingual dictionary. The monolingual dictionary contains syntactic and semantic information that will be connected to a word (or part of it) before the text is parsed. Information is relevant only to part of a word if morphological analysis is performed. The bilingual dictionary contains information needed during the transfer process. A dictionary entry consists of a source-language word, including the syntactic and semantic attributes and of one or more corresponding target-language words, including attributes.

Conclusions

This chapter has tried, very briefly, to give the reader a first insight into what has become one of the classic software designs for a machine translation system. It has also pointed out the problem of generality in software tools versus computational efficiency of the system, which will be taken up again later.

NINE : ALAN K. MELBY

CREATING AN ENVIRONMENT
FOR THE TRANSLATOR

Since early design decisions generally have a significant effect on the final form of a system, all relevant criteria should be examined from the beginning. This chapter will describe a translator work station which provides an interface between a human translator and a variety of tools, including machine translation. The design of the translator work station will suggest some design criteria for machine translation systems.

A translator work station can be put together using existing hardware, operating systems, and programming languages. Existing wordprocessing and database software can be used with some modifications. Existing local area networking and telecommunications can be used advantageously. All the major elements are available – the question is what combination to use. The designer of a translator work station is a chef and thus is not guaranteed success by using the right ingredients – they must be combined effectively. The judge of the result must be the end-user, i.e. the translator, who tastes the product by using it to produce translations.

The design for the author's work station specifies three levels of aid thoroughly integrated. The first level requires no preparation of the text to be translated and consists of wordprocessing with terminology and telecommunications options. The second level requires the source text (i.e. the text to be translated) to be in machine-readable form but in exchange provides simple pre-processing of the text and facilitates terminology retrieval. The third level is used when a draft translation in machine-readable form, e.g. a machine translation, is available. Level three extends the translator's options, permitting selective post-editing in addition to the terminology and communications tools available at levels one and two.

According to this design, machine translation is a tool to be used at the discretion of a human translator. Even if machine translation eventually replaces human translation in some specific language pairs and subject-matter sublanguages; there will still be plenty of translation which needs the human touch. And it should be remembered that a system as refined and successful as METEO still routes some text of weather forecasts to human translators for review and correction. Now this is not intended to discourage research and development in machine translation. To the contrary, it is the author's belief that if translators begin using well-designed work stations, it will encourage machine translation by making it more usable.

We will now look at the three levels of aid in more detail. Then we will describe the use of the systems in various environments, including revision of human translation, individual translators versus teams, and co-operation with typists. It will be noted that some aspects of the design, namely integration and windowing, are special instances of a general trend in software design. Finally the implications of the work station for the design of machine translation systems will be discussed. An appendix will sketch the history of the translator work station described in this chapter and will give the background of the design team.

Level one

Level one is built on the foundation of wordprocessing software. A wordprocessing system is not specific to translation; it is a writer's aid. Of course, there may be a need to translate into several languages on the same work station, in which case multilingual wordprocessing software would be needed. But the system really first becomes a translator aid when wordprocessing and bilingual dictionary access are integrated, which means that the dictionary can be consulted without bringing down the wordprocessor and that when a translation is selected it can be transferred into the wordprocessor to avoid manually typing it. Other details are also important. For example, the access time to find a term once it has been typed in should not be more than two or three seconds, and access should be via normalised key so that a minor difference in capitalisation or hyphenation will not prevent retrieval of the desired term. Also, the format of the dictionary should be flexible, and the length of an entry should be variable. Of course, the translator should be able to add new terms and update existing entries in a local dictionary without leaving the wordprocessor and, if part of a team, should have read-only access to a dictionary shared with other translators. Such details can make the difference between the system being a help or a burden.

Besides wordprocessing and terminology access, level one provides integrated communications facilities. This means that without leaving the wordprocessor, the translator can open a window on the screen which emulates a terminal on a telecommunications network. This network could provide access to remote databases of documents, terminology banks, and electronic mail to other translators. Once a translation is completed, it should be transmitted electronically to a shared database of translations so that others can avoid retranslating the same document.

With level one, the translator can be handed a barely legible copy of a memo and sit down to begin translating it immediately.

Level two

If, on the other hand, the source text is available in machine-readable form, the system can take advantage of it and provide additional options. For example, the translator can request an immediate concordance of all occurrences of a given term. And the system can automatically look up all terms in the current segment of source text. Those terms with a standard translation can be displayed in a window which could be called a suggestion box. At first it may seem that the metaphor of a suggestion box is inappropriate since the translator is not putting anything into it. However, the intent of the metaphor is that the computer is in a subservient role, placing suggestions in the box for the translator's consideration.

Level two also provides the option of the translator pointing at a term in the source text and requesting dictionary look-up without having to type it in. The pointing could be accomplished by asking the system to number the words of the source text and then entering the number(s) of the desired word(s). Or the translator could use a mouse or other pointing device.

At all levels of aid, including one and two, there is a wordprocessing window and the option of a dictionary window. At level two and above, the word processing window may be accompanied by a window of source text and suggested translations of selected terms.

As always, the details of the recipe for level two are important. The exact arrangement on the screen or screens, the method of associating the suggested translation with its source text term, and the handling of single and multi-word terms will make a difference in the ease of using the system.

At level two, when a translation is completed, it can be transmitted to a shared data base along with its source text. Having both source and target texts in the data base, along with automatic segment matching routines, would allow a translator to request information on a term and get it from actual texts, bypassing the steps involved in adding new terms and translations to a term bank. Source and target text data bases also provide a basis for software which will assist in the translation of a revision of previously translated documents.

Level three

So far, there has been no mention of machine translation, but level three is, as the circus announcer would proclaim, last but not least. At level three, a translation is already available and is ready for post-editing. Being ready for post-editing means that the source text is divided into segments, the translation is divided into corresponding segments, and the source/target pairs are in machine-readable form. Accompanying each segment of source text and its translations is a grade which indicates whether there are any known or suspected problems in the translation. The grading system might include five levels from A (no indication of problems in the translation) to B (slight problem) down to E (severe problems or no translation at all). Clearly, the translation to be post-edited could be a machine translation, but it could also be a human translation done at level two on a translator work station and then revised at level three by the same or another translator. The grades are not assigned during level three processing. They are assigned during the draft translation phase by the human or machine producing the translation.

To use level three, the translator sets a tolerance level (A–E) once at the beginning of the session and then begins requesting segments of translation. For each segment, if the grade attached to the segment meets or exceeds the tolerance level, it will be displayed for consideration by the translator. If the translation is judged to be acceptable, the translator transfers it into the word-processing window where it can be post-edited as needed. If the translation is judged to be unacceptable, the translator simply ignores it and optionally switches to level two temporarily. If the grade attached to the segment is below the tolerance level, it is not even displayed, and the system automatically switches down to level two until the next segment.

Level three is extremely flexible because it does not mean giving up the tools available at levels one and two. Machine translation is known to be of uneven quality, and a translator who is forced to post-edit machine-translated text segment by segment may soon become frustrated. If, however, local and remote dictionaries and databases, as well as manual translation are always options available to the translator, frustration should be greatly reduced, and a machine translation which would be rejected by a translator using a strict post-edit approach may be accepted and useful to a translator using a work station with multiple integrated levels.

Work environments

The translator work station described above need not be implemented on a large computer, since the machine translation component is not in the work station itself. The machine translation system may be as near as another node in a local area network or as far as files can be transferred over telecommunications lines or by mailing diskettes. Indeed, the translator work station can be implemented on a modern microcomputer and should provide a standard interface for the translator when dealing with various data bases and machine translation systems. Let us look at a few possible work environments.

The simplest environment is an individual translator working at home. Here, the work station would be a microcomputer with a high-capacity diskdrive (for source and target text files), a printer, and a port for telecommunications. The translator could receive source texts through the phone lines or by mail. If the texts are in machine-readable form, the translator can use both level one and level two tools. Even if the source text is only available on paper, the level one options are ready as always. If the client requesting the translation has compatible work stations, a specific subject matter glossary can be sent to the translator on diskette along with the source text. Clearing-houses could also be organised to facilitate the exchange of glossaries on diskette. One could also have the source text translated by machine and send the source and target text to the translator to be revised using level three.

An important human factor is fatigue from working at a video display screen more than four or five hours a day. At levels two and three, the source text can be printed out on paper by segment with the associated machine translation and/or suggestion-box terms and blank space for the translation. Then the translator can either write in the translation or dictate it, referring to the segment number and suggestion numbers for efficiency and accuracy. Then the translator could enter the translation using level two/three or send a diskette and dictation tape back to the client who could arrange for a typist to enter the translation using the standard work station. In the last configuration mentioned the translator does not require constant access to a work station.

When a team of translators is working together on a set of related documents, the work stations could be tied together using a local area network, allowing the translators to share terminology files and printers. Sometimes translators work with typists, producing draft translations on dictation machines or with pencil and paper. In these cases, the translator may begin using a work station mainly for terminology and database access. If the typist uses a compatible work station, the draft translation could be returned to the translator on diskette. Then the translator could revise it using word-processing. If one translator produces a draft translation using level two, another translator could revise it using level three. If the organisation has a machine translation system, it could be a node on the network and translators could send texts to it for translation and then revise them using level three.

Thus, essentially the same user interface in the work station can be used in a variety of working environments.

Trends in computer system design

The work station described in this article uses integration and windows to create a productive working environment for the translator. During the past several years, while the work station design has been refined, the use of integration and windows has increased greatly from its beginnings at Xerox Research. This is encouraging and provides a supplementary argument for the soundness of the design.

A hardware issue, pixel density, is related to window design. A text window needs to be a full 80 characters wide and each character must be formed in a cell at least 8 pixels wide and 12 pixels high. Of course, non-Roman characters may require even more pixels per character.

Implications for the design of machine translation systems

Most machine translation systems are designed from the assumption that for every segment of source text there must be some kind of machine translation produced. In the author's experience, this leads to the spreading of linguistic development effort over a huge variety of problems, some of which occur rather infrequently. If instead, the system is designed from the assumption that only selected source segments will produce a translation but for those selected segments the translation should be very good, then the linguistic development effort will be concentrated on fewer, high-frequency phenomena. Also, most machine translation systems are designed to recover from problems in the linguistic processing and somehow produce a translation anyway. An alternative suggested by level three is to design a system to detect problems and grade their severity rather than continue at all costs.

Although much research is still needed to evaluate the effect this design will have on a machine translation system, it seems probable that such a system will have cleaner processing algorithms. And it is nearly certain that the work station will make machine translation more acceptable to translators and therefore more useful. However, the proof of the translator work station design is in the using of it. So let us get on with the using of translator work stations.

Appendix – a particular translator work station

The design described in this chapter has been produced as part of a long-term effort called the BYU–TAS project. There is also a team which is implementing a translator work station with this design. One of the team members (the author) is a computational linguist who participated in a machine translation project for ten years before turning to translator aids and becoming an accredited translator. Another is an accredited professional translator with considerable experience in terminology issues. A third member of the team is thoroughly experienced in multilingual word processing and systems analysis. An examination of teams developing machine translation systems sometimes reveals a lack of actual accredited translators. Translators should always be involved in developing a translator work station, and perhaps the lack of translators should be corrected in the development of future machine translation systems.

The current work station is the third major version. The first version, designed and programmed in 1981, was called the Suggestion Box translator aid system. It was written in PL/1, using a subroutine package of 3278-type screen handling routines, under the VM/CMS operating system on an IBM mainframe. A group of translation students who tested the system provided feedback showing that a multi-level system would be more desirable. User feedback also highlighted the negative influence of varying response time associated with a multi-user computer environment. So it was decided that the second version of our translator work station would be implemented on a micro-computer to allow consistent response time to wordprocessing and other personal commands, while still permitting access to other translators and to shared databases using various forms of networking.

The second version was programmed in 1982. A Z-80 based microcomputer using the CP/M operating system was selected. All three levels were programmed in a demonstration version which was again submitted for testing to a group of student translators. User feedback on the second version indicated that the text entry and dictionary look-up phase and the final revision wordprocessing phase needed to be integrated. The users also requested on-screen display of accented characters in their true form. It became evident that the typical 8-bit, 64K, CP/M microcomputer was not suited to the needs of a translator work station.

The third version is based on the current generation of 16-bit microcomputers compatible with the IBM personal computer. The standard IBM monochrome display adapter provides a 256-character set (instead of the standard 96-character set) and displays 25 lines of 80 columns with each cell containing 9 rows of 14 pixels. The MS-DOS operating system allows over 500K for a single user program and provides an interrupt structure which makes it feasible for one program, such as a word processor, to invoke another program, such as a dictionary lookup program, and regain control later with both programs remaining resident. The current (third) version was tested by a group of student translators and a team of professional translators in 1983, and an improved version three system was further tested and made commercially available in 1984.

Many improvements are being made to the current version of the work station, such as investigation of alternative displays with more than 25 lines and 80 columns, but it appears that a minimal level of performance has been achieved which justifies the use of this work station by many translators while further research and development are being conducted. Another factor which facilitates the use of such a work station is the growing availability and lowering cost of IBM PC-compatible microcomputers, such as the Apricot.

Of course, low cost versions of the bit-mapped display machines originally developed at Xerox Research provide an enticing hardware/software environment for a future version four translator work station. And emerging computer controlled laser digital video disk technology suggests a solution to dictionary storage capacity problems for personal work stations.

Conclusion

The BYU–TAS project is committed to the notion of a translator work station which provides wordprocessing, personal terminology files, communication with other translators, access to various data bases, and a link to machine translation systems.

Research on such a translator work station should be done in parallel with research in machine translation by other projects.

RESEARCH AND DEVELOPMENT ON MT AND RELATED TECHNIQUES AT GRENOBLE UNIVERSITY

The Groupe d'Etude pour la Traduction Automatique (GETA) has been studying the problem of automation of the translation process since 1972, pursuing the work done by CETA since 1962.

Philosophy of automated translation (CAT)
We feel that a more appropriate designation for research on machine translation would be research on *automation of the translation process*. Although it may be very interesting to investigate the subtleties involved in the translation of a set of test sentences, and how to emulate them by machine, it is even more promising to attack the real problem of automating (totally or partially) the translation of *documents*. Hence, it would be better to envisage MT systems as Computer Aided Translation (CAT) systems. As a consequence, they should offer functional aids in a working translation environment.

We distinguish four essential phases in the processing of a document in a translation environment: *acquisition* of the document; *rough translation,* possibly done in parallel by several translators; *revision,* sometimes done in several passes. For technical documents, a technical revision by a (possibly monolingual) specialist in the field is often required; *output* of the final document, including figures, charts, etc.

A document may be created in a translation environment (as in the EEC), or sent to it in its final form. As soon as some automation is envisaged, machine aids are currently available for putting the document in machine-readable form (text-processing systems, possibly coupled with OCRs).

Strictly speaking, the creation of a document is not a function of the translation process. But, if this creation can be linguistically controlled by some linguistic process, automation of the rough translation becomes a lot easier. The TITUS system illustrates this point (see chapter 12).

CAT techniques centre around the total or partial automation of the rough translation process. Two main approaches have been tried. In the first, translation is done by a program, in batch or interactive mode. GETA, TAUM, SFB–100, METAL all follow the batch line. BYU (ITS), ALPS and WEIDNER have tried the interactive approach. Such systems are generally referred to as Machine Translation systems, or as Human Aided Machine Translation (HAMT) systems.

The second approach is generally called Machine Aided Human Translation (MAHT). Here, emphasis is on the automation of the translator's office, with specialised text-processing systems, fast access to on-line terminological data banks, spelling checkers, etc.

Within the HAMT approach, two strategies are possible. First, one can try to *define* some subset of a given natural language as a formal language. Then, an analyser is built. If a given unit of translation is 'legal', it will be translated. If not, it will be rejected. Hence, the automatic system is a 'partial system', because it translates only a percentage of the input. TAUM–METEO (see chapter 15), or the first CETA systems (before 1970) are good examples of this strategy.

The second strategy, followed in all current GETA systems, is to build a 'total system', which will always attempt to translate all the input, even if it is partially ill-formed with regard to the implemented linguistic model.

The revision of a document is usually done with the help of text-processing systems. But the automation of the revision function itself has not yet been attempted. It seems that the level of understanding and of general knowledge required to perform even a 'linguistic' revision is higher than the one required for translation. This is even more true in the case of 'technical' revision.

Machine aids for the output of the final document are widely used, and are not an object of study for research groups in MT.

There are two other functions which should be automated in a CAT environment. First, the management of a large data base of documents, together with a record of the actions performed on them (modifications, translations, revisions, etc.). In other words, a translation environment should interface nicely with a textual database system.

Second, in the case of HAMT, creation, debugging, maintenance and evolution of the 'linguistic software', abbreviated here as *'lingware' ('linguiciel')*, require a 'programming environment', that is, a specialised database centred around one or several SLLP (Specialised Languages for Linguistic Programming).

At GETA, we have developed an integrated programming system for HAMT, ARIANE-78. Only the batch approach has been implemented for the rough translation phase. But the system also supports a subenvironment for MAHT (THAM in French), used for the revision of the rough translations as well as for purely human translation (and revision).

Contrary to general practice, the translation systems developed at GETA have been designed to be multilingual. Hence, translation must be composed of three logical steps:

(1) *monolingual analysis;*

(2) *bilingual transfer;*

(3) *monolingual synthesis* (also called 'generation').

Thus, a given analyser may be used to translate from one 'source language' into several 'target languages', and the same synthesiser may be used to translate from several source languages into the same target language. The same division is used in modern 'multi-target' compilers for programming languages.

Everybody agrees that a very good translation requires a very deep understanding of the text. But this is not achieved, even by good human translators, in particular in technical fields, or there would be no need for revision in the first place!

Hence, the objective of MT systems is rather to produce *good enough* rough translations, that is, to give some very crude translations which may be revised with less than twice the effort needed to revise an average human translation of the corresponding text.

By using such expressions as 'very good', 'good enough', or 'medium', we implicitly suppose the existence of some 'hierarchy' of understanding. In actual fact, understanding cannot be defined in an absolute way, but only with reference to some domain. The hierarchy of understanding is organised around a hierarchy of *levels of interpretation*. We distinguish between *linguistic levels* and *extra-linguistic levels*.

The *linguistic levels* are:

(a) *morphology:* this is the level of the analysis of words or idioms in terms of morphemes, lexical units, potentialities of derivation, semantic features, valencies, etc.

(b) *syntax-1:* at this level, *syntactic classes,* such as noun, verb, etc., are associated with words, and *syntagmatic classes,* such as nominal phrase or verbal phrase, with groups of words. This gives a 'bracketing' of the text (or several in the case of ambiguity), often represented as a tree giving the 'constituent structure'.

(c) *syntax-2:* this is the level of representation in terms of *syntactic functions,* such as subject, object, attribute, or (equivalently) of *dependency relations.*

(d) *logico-semantics-1:* at this level, the *logical relations* between parts of the text are identified. They are sometimes also called *inner cases.* In the GETA systems, they are named ARGO (logical subject, or 'argument 0'), ARG1 (logical object), etc.

(e) *logico-semantics-2:* the *semantic relations,* such as consequence, cause, concession, measure, localisation, etc., are essential to translate correctly the 'circumstants', as opposed to the 'arguments' of a predicate. On circumstantials, semantic relations are also sometimes called *outer cases.* Of course, they may also be attached to the arguments (for example, a logical object may be interpreted as a patient). But they are less indispensable (for translation), because the semantic relations of the arguments of a predicate are often very difficult to compute, and because, even if they are computed, a good translation may often be obtained simply by using directly the lexical unit of the predicate, plus restrictions on the (semantic features of the) arguments.

This list of levels is not exhaustive. The implemented lingwares also use the representation of a sentence's actualisation features (surface tense or abstract time, aspect, modality, determination), type of statement (declarative, interrogative, exclamative, imperative, negative) or emphasis (theme/rheme, intensification). But we consider that all of them are relative to a knowledge encoded in a formal system of a linguistic nature. Hence, we characterise this type of understanding as *implicit understanding.*

The *extra-linguistic levels* are:

(a) *expertise:* here, we refer to some static knowledge about a particular subject matter, consisting in a collection of facts, rules and procedures. This level has also been called 'static semantics', in contrast to the 'feature semantics', incorporated in the linguistic knowledge, and to the 'dynamic knowledge', which is discussed below.

(b) *pragmatics:* this level is taken to be the highest level of understanding. At this level, understanding a document means creating a representation of the facts, events, suppositions, scenarios, etc., described by the text. This presupposes the ability to learn facts and structures, to reason by analogy, and to abstract. In short, pragmatics is related to the most ambitious themes of AI. Until now, only very small illustrative computer models have been presented.

Understanding at some extralinguistic level may be called *explicit understanding*. Typical applications where it is needed include expert systems, which should be able to explain their actions. However, for translation purposes, implicit understanding is often sufficient, *at the translator's level*. At the level of a technical revisor, explicit understanding is required.

Second generation CAT systems rely only on 'implicit' understanding. In the past, and in some current systems still, the previously mentioned levels are mutually exclusive. By this, we mean that a given text will have separate representations for each of the defined levels. This usually leads to a sequential strategy of processing, with all its drawbacks.

At analysis, for example, it is often difficult to compute the semantic relations for all parts of the unit of translation, especially if the size of this unit is large (one or several paragraphs). In the sequential approach, one is then forced to refuse the unit, or else to translate the complete unit at the previous level.

This is why GETA uses *multilevel interface structures* to represent all the computed levels on the same graph (a 'decorated tree'). Detailed examples of this kind of structure have been given elsewhere, and are also included in chapter 16 by J. Ph. Guilbaud.

In short, such structures are in effect *generators* of representations at different levels, and also *factorise* various types of ambiguities.

Incidentally, this type of structure was first proposed by B. Vauquois in 1974, during sessions of the Leibniz group, which led to the launching of the Eurotra project by the EEC. Since then, it has been refined and tested, on a variety of applications, including Russian–French, Portuguese–English, English–Malay, English–French, English–Chinese, English–Japanese, French–English and German–French.

In translation, by human or by machine, specialisation is indispensable in order to obtain good quality. A literary translator is usually at a loss to translate a computer manual. This specialisation follows two lines: first, specialisation to a certain *typology* of texts; and second, to a certain extent *domain*. In CAT systems, as for humans, there is a core knowledge, plus knowledge specific to the application. We may say, in a first approximation, that grammars incorporate the typological specialisation, and dictionaries the domain specialisation.

This is why modularity is essential in the construction of CAT systems. The same core should be the base of several versions, tailored to different *sublanguages*. It is interesting to note that R. Kittredge, the first linguist to attack the study of sublanguages, and particularly technical ones, from a theoretical point of view, has long been active in the TAUM project.

One might argue that specialisation to a certain sublanguage amounts in fact to the incorporation of some extralinguistic knowledge in a CAT system. But the *form* of this knowledge is not what is required in order to qualify as expertise, because it is expressed by some *combinatorics of classes* ('combinatoire de classes', to translate one of Vauquois' favourite expressions). Rather, we may say that, as in Plato's cave, the real world is 'reflected' in the structure of texts and in their peculiarities. In particular, 'in-house' writing habits correspond to some sociological conditions governing document creation.

Let us now recall the main principles which have guided the implementation of CAT systems at Grenoble. In principle, there are many ways to implement lexical and grammatical knowledge. In SYSTRAN and other first-generation systems, one uses the assembler or macro assembler level. In 'first and a half' generation systems, the implementation language may be some high-level programming language, such as FORTRAN (SUSY I), COBOL (Saskatoon 1970), PL/I (ITS), COMSKEE (SUSY II), etc. The drawbacks are evident. In particular, either linguists are burdened with ancillary tasks, such as implementing data and control structures, or they require the help of some 'slave' computer scientists to translate their wishes into working programs, with the result that their desires, incorrectly formulated, are also incorrectly translated.

Nowadays, certain groups are trying to use (also directly) very high-level general programming languages, such as SETL (Novosibirsk), LISP (NTT) or PROLOG.

In second-generation systems, and in projected third-generation systems, emphasis is placed on the use of Specialised Languages for Linguistic Programming (SLLP), which offer built-in data and control structures, with an underlying powerful mechanism.

This is the case in all 'rule systems', based on (extended) CF-grammars (CETA, METAL, SFB99, ETL-Lingol), adjunction grammars (LSP, LADL), Q-systems (TAUM), ATNs (BBN, TAUM), (extended non-deterministic) finite state transducer (GETA–ATEF), tree-transducers (Friedman, Petrick, GETA–

ROBRA, SF99-TRANSFO). The built-in data structures are usually particularly classes of graphs or hypergraphs, such as decorated trees, Q-graphs, 'charts' (MIND), etc.

Choosing one or more implementation languages for SLLP is another matter. The highest and most efficient level should be selected. There is an inherent conflict in this dual goal. Hence, compromises are made, sometimes by using several implementation languages. For example, ARIANE-78 is implemented in ASM360 (macro-assembler) and PL360 for the compilers and interpreters of the SLLP, PASCAL and EXEC2 for the other tools, the management of the database and the interactive interface ('monitor').

As in other fields of AI, the declarative and procedural approaches are in competition. The declarative approach leads to rule systems with an underlying 'combinatorial' algorithm, which produces a set or a subset of 'solutions' in some fixed way. It is best exemplified by analysers built on (extended) CF-grammars, or by Q-systems. The main advantage is the relative ease of programming. But it is almost impossible to implement powerful heuristics, because, in essence, the computations of the various solutions are independent.

The procedural approach has been followed in the more recent second-generation systems ('second and a half'?). For example, the TAUM-AVIATION system uses REZO, a Q-graph transducer based on the ATN model (chapter 15). In ARIANE-78, ATEF and ROBRA give even more possibilities of heuristic programming. This added power, however, requires more programming skill.

SLLP are designed to be easy to use by linguists and terminologists who have almost no computer science background. Hence, they must be integrated in some 'user-friendly' environment. In ARIANE-78, this environment is implemented as a specialised database of lingware files (grammars, dictionaries, procedures, formats, variables) and of corpuses of texts (source, translated, revised, plus intermediate results and possibly 'hors-textes' – figures, etc.). A conversational monitor interfaces the database with the users (in French or in English).

Subenvironments are defined to permit the preparation, testing, debugging and maintenance of the lingware, to manipulate the texts, to check the spelling of a list of corpuses (or of individual texts), to produce mass translations, and to revise the translations. The database system ensures the coherence and integrity of all the applications and texts in a given user space (since the system is multilingual, it is perfectly possible to have several translation systems in the same user space, sharing one or several analysers or synthesisers).

It is interesting to note that the needs of linguists led to the creation of such a 'programming system' before this type of system became a main theme of R&D in software engineering.

An interesting comparison can be made between this sort of CAT system and compiler–compiler systems for programming languages. The various SLLP are in effect tools used to build morphological analysers, structural analysers, transfers and generators.

ATEF, for instance, may be compared with LEX, used to write lexical analysers for programming languages. Of course, the richness of the information contained in each word, and the inherent ambiguity of language, make such a tool more complex: it is necessary to handle large dictionaries (as opposed to small sets of reserved words and of identifiers with no a priori content) and to offer advanced control structures, such as non-deterministic programming with or without heuristic functions.

Development and research in software

This part of the work is progressing in three directions. First, some work is being done on ARIANE–78, in order to improve it and to augment its practical usability and friendliness. Second, in the framework of the French CAT project, there is the specification and development of an industrial system, which, in its most complete version, should ultimately run on a database machine. Finally, research is being done on a new generation of CAT software.

Let us briefly recall the structure of the translation process under version 4 of ARIANE–78, which has been presented in detail elsewhere (see figure 10.1).

All phases are mandatory. The lexical information for analysis is essentially contained in the AM dictionaries, although it is possible to handle special cases (of small classes of words) directly in the AS grammar.

A priority scheme is used in TL to choose, for a given lexical unit, the equivalent(s) given by the highest-ranking dictionary. The priorities may vary from one execution to the other, for example according to the domain being handled. Some dictionaries may even be ignored for a given translation.

The structure of version 5 of ARIANE–78 offers new facilities for lexical processing. EXPANS is a new LSPL based on TRANSF. Unlike TRANSF, it may use a limited context in the

Logical phase	Physical phase	+/-	Language (SLLP)	Files for ling. data: Type (number)
ANALYSIS	Morpho-logical AM	+	ATEF	Variables (2) Formats (3) Grammar (1) Dictionaries (1 to 7)
	Structural analysis AS	+	ROBRA	Variables (1) Formats (1) Procedures (1) Grammars (1 to 7)
TRANSFER	Lexical transfer TL	+	EXPANS	Variables (1) Formats (1) Procedures (1) Dictionaries (1 to 7)
	Structural transfer TS	+	ROBRA	as AS
GENERATION	Structural generation GS	+	ROBRA	as AS
	Morpho-logical generation GM	+	SYGMOR	Variables (1) Conditions (1) Formats (1) Grammar (1) Dictionaries (1 to 7)

Figure 10.1

input tree to create the output tree. In the assignment parts of rules, it is possible to use local 'decoration variables' and backtracking. This improvement has also been added to ROBRA.

Essentially, the optional phases AX and AY make it possible to incorporate the lexical knowledge in a more modular fashion. For example, AM may be used to associate lemmas and strictly morphological information with the morphs. Then, AX may be used to relate the lexical units to the lemmas, and to give them syntactico-semantic information, such as valencies, argument frames and semantic restrictions on the arguments. AY could then include information about the syntactico-semantic behaviour of idioms whose components are separable or inflectional, such as compound predicates or German verb-particle constructions.

The overall design of a particular translation system may be more modular. For instance, it is possible to construct a structural analyser as a sequence of transformational systems using the same decoration type.

Other optional EXPANS phases have been included. The most important is GX, in which it will be possible to index all properties of target lexical units not necessary at transfer time – that is, almost all. Until now, all lexical target information (not including the strictly morphological information) had to be included in *all* TL dictionaries for all language pairs with the same target language (see figure 10.2).

Work is also being done on supporting tools. ATLAS, for example, is an environment to aid in the indexing of 'coded' dictionaries. It is built around a language in which it is easy to write 'indexing charts' which are in effect special types of labelled loop-free graphs. Internal nodes contain questions, exiting arcs are labelled with the possible answers, and terminal nodes contain the codes to be assigned. The language allows for three types of comments. Among them, SCRIPT/VS commands are used to produce formatted outputs of the source, using IBM's SCRIPT formatting program. The system includes a facility for generating the output in a graphic form, and hence for directly producing the indexing manuals on paper. There is an interpreter, coupled with a screen-handling routine, used for interactive indexing. The system generates appropriate menus while the answers of the user direct it to traverse the indexing chart towards a final answer (e.g. a code, or a recipe to choose among several competing normal forms for a lexical unit). The answer may be directly transferred to the appropriate area in the 'dictionary window' (four lines of the dictionary under construction). During such an interactive session, the incorporated syntactic editor may be accessed to change the chart, and to produce the correspondingly altered version of the source. The system is written in PASCAL, with a small routine in assembler for screen-handling.

THAM is an environment for Machine Aided Human Translation (Traduction Humaine Aidée par la Machine). It may be used alone (for translation or revision), or under ARIANE–78 (for revision of machine translations). It is completely programmed in XEDIT/EXEC2, the programming language for the powerful XEDIT screen-oriented editor. Basic ideas for THAM come from various sources such as IBM's DTAF system (only used in-house on a limited scale) or A. Melby's TWS, presented at COLING–82 (cf. [549] Melby 1982 and chapter 9). The user may access several 'rings', each containing a maximum of four files:

Logical phase	Physical phase	+/-	Language (SLLP)	Files for ling. data: Type (number)
ANALYSIS	Morpho-logical AM	+	ATEF	Variables (2) Formats (3) Grammar (1) Dictionaries (1 to 7)
	Lexical complement AX	-	EXPANS	Variables (1) Formats (1) Procedures (1) Dictionaries (1 to 7)
	Lexical complement AY	-	EXPANS	Variables (1) Formats (1) Procedures (1) Dictionaries (1 to 7)
	Structural analysis AS	+	ROBRA	Variables (1) Formats (1) Procedures (1) Grammars (1 to 7)
TRANSFER	Lexical transfer TL	+	EXPANS	as AX
	Lexical TX complement	-	EXPANS	as AX
	Structural transfer . TS	+	ROBRA	as AS
	Lexical TY complement	-	EXPANS	as AX
GENERATION	Lexical GX complement	-	EXPANS	as AX
	Structural generation GS	+	ROBRA	as AS
	Lexical GY complement	-	EXPANS	as AX
	Morpho-logical generation GM	+	SYGMOR	Variables (1) Conditions (1) Formats (1) Grammar (1) Dictionaries (1 to 7)

Figure 10.2

S (source text), T (translated text), R (revised), and D ('natural' dictionary). The dictionary has a very loose format, with the only requirement being that the keys are sorted in ascending order. No access method is compiled, in order to allow for dynamic changes at any time. Normal addition of an item under THAM inserts it, of course, at its proper place. Several useful functions have been designed and implemented, for the comfort of typical users. For instance, one can delimit a string on any window and copy it in the revised text. Another powerful function enables the user to delimit any number of non-overlapping strings (in R or D) and to define any permutation on them. Also, a 'suggestion mode' has been designed. The layout of the windows is also controlled by a very simple THAM command. Initial experiments have shown this to be quite useful.

VISULEX is a tool used to visualise a part or the totality of the lexical database of any ARIANE–78 application in a synthetic and 'natural' way. It is evident from the above diagrams that the lexical database may reside on around fifty files, for a given language pair. In a given dictionary entry, one typically finds a lexical unit, with several codes defined in other files (formats, procedures), referring to one or more lexical units, with the same type of information. VISULEX presents the lexical database with access by the lexical units. According to the 'part' chosen, it produces the source information (coming from AM, AX, AY) and/or the target information, for each equivalent (this information comes from TL, TX, TY, GX, GY, GM). It offers two levels of detail. At the first level, the information is presented by using only the *comments of the codes,* found in the various ARIANE–78 files. At the second level, a parallel listing is produced, with the codes themselves, and their symbolic definition. For the Russian–French application, VISULEX output gives two listings of around 150000 lines each. This makes it a lot easier to detect indexing errors, at all levels. This is a first step towards improved 'lexical knowledge processing'.

A portable and enhanced version of ARIANE–78 is being developed as part of the effort of the French national CAI project to achieve a significant degree of 'technological transfer' from research laboratories to industry, the ultimate aim being to develop commercial systems based on the second-generation CAT technology developed at Grenoble University. A similar project has been launched in Japan, under the scientific guidance of Kyoto University (Prof. Nagao).

Emphasis is on portability and maintainability of the basic software. Although there are very few problems with the current implementation of ARIANE–78, it can only run on IBM or IBM compatible machines, under VMSP. The specialised databases for texts and linguistic files are implemented using the CMS file system, and the monitor makes heavy use of the EXEC language (CMS command processor, comparable to the Shell of Multics or Unix).

ARIANE–78 is very useful for building translation systems and experimenting with them on fairly large corpuses. But, in order to integrate this technology in the working environment of translation services, it seems advisable to rely on general database systems rather than on ad hoc implementations using some particular file system.

This is why the new version should ultimately run on a database machine. Hence, the function of human or machine translation could be incorporated, in a typical application, in a technical documentation database.

Of course, this organisation will also allow for the handling of the lingware (grammars, dictionaries) in the same manner as before (see also below), that is, through the database standard mechanisms.

The final implementation should run under Unix on such a machine. The monitor and the compiler will probably be written in a language such as LISP or PROLOG. The interpreters of the SLLP may be written in a combination of LISP/PROLOG and PASCAL/C, depending on the final technical specifications.

The database aspect should be interfaced in such a way that it is possible, if necessary, to execute translations on another, more powerful processor. Intermediate versions could for instance offer ARIANE–78's core system on an IBM processor linked to the database machine acting as supervisor.

An important aspect will be adequate processing of heterogeneous documents, containing text as well as formulas, charts, diagrams, etc.: the use of a 'multimedia' database machine may prove decisive, together with high definition bit-map screens.

Finally, research aimed at producing a new generation of MT software is progressing.

Second-generation CAT technology is now well understood and has been fairly well tested. This technology now appears to be ready for industrial development. But this is not to say that the ultimate has been achieved. On the contrary, as work progresses, we more and more realise the depth of our ignorance about the fundamental phenomena of thought and language.

What Bar-Hillel wrote twenty years ago still remains true. He said in effect that *explicit understanding* is mandatory, in order to achieve really high quality in machine translation, if the domain of discourse is even moderately large.

Hence, it may seem that a complete understanding of the functioning of the human brain would be required before one could imagine really better CAT systems.

Recent trends in AI seem to invalidate this line of reasoning, however. Significant advances have been achieved by designing so-called *expert systems,* which try to incorporate expert human knowledge about a given area. This knowledge is necessarily partial, and given to the system in a form which nobody pretends is similar to the hypothetical representations in the human brain. Rather, one tries to approximate some 'natural' expression of this knowledge.

Hence, one may say that, from the point of view of software design, the central (research) problem in CAT is to find principles around which to build *expert systems for translation.* Another type of problem is the search for still better data and control structures to support linguistic representations and processes, even at the second generation level.

Seeing that linguistic expertise is already quite well represented and handled in current ('closed') systems, we are orienting our research towards the possibility of *adding* extralinguistic knowledge (knowledge about some technical or scientific field, for instance) to existing CAT systems.

Also, because current systems are based on transducers rather than on analysers, it is perfectly possible that the result of analysis or of transfer (the 'structural descriptors') are partially incorrect and need correction. Knowledge about the types of errors made by linguistic systems may be called metalinguistic.

In his recent thesis, R. Gerber ([183] 1984) attempted to design such a system, and to propose an initial implementation. The expertise to be incorporated in this system includes *linguistic, metalinguistic* and *extralinguistic* knowledge. In R. Gerber's thesis, the system is constructed by combining a 'closed' system, based only on linguistic knowledge (an application written in ARIANE–78), and one or two 'open' systems, called 'expert corrector systems'. The first is inserted at the junction between analysis and transfer, and the second between transfer and generation. Figure 10.3 may be useful.

The decorated tree structures used in ARIANE–78 are very useful, but have some limitations. For example, structural ambiguities are represented by coding them in the tree structure, in a manner analogous with the use of and/or trees. In his thesis and at

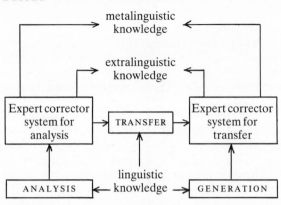

Figure 10.3

COLING–82, N. Verastegui ([562] 1982) proposed a new data structure, called 'sliced multitree', which seems to factorise the representation of multiple tree structures, while still allowing for the use of tree transformational systems. In fact, he has designed a new SLLP, STAR–PALE, derived from ROBRA, and outlined its implementation. This idea may be combined with the use of generalised Q-graphs (or 'charts') for representing ambiguities either by branchings in the graph, by the structure of the multitrees, or at the level of the decorations. In his thesis and at COLING–82, Clemente ([531] 1982) presented a graph rewriting system for 'E-graphs' (Q-graphs with decorated trees – not multitrees). The difference with Q-systems lies in the fact that Clemente's algorithmic model is a substitution model rather than an addition model, and that the schemas may denote subgraphs with branches rather than fixed-length paths. These ideas have yet to be tested in the context of the development of real CAT systems.

Last, but not least, it seems very important to implement the various SLLP by using parallel algorithms, because of the anticipated development of multiprocessor parallel architectures based on a very large number of identical processors. Currently, even SLLP such as ROBRA, which rely on a model of parallel rewriting in decorated tree structures, are implemented with sequential algorithms designed for monoprocessor von Neumann machines.

In his thesis, N. Verastegui also proposed a truly parallel implementation of STAR–PALE, using processors (for each 'schema element') working on the same multitree and communicating by exchanging messages through the structure itself.

Development and research in linguistics

For more than three years now, GETA has been developing and testing a Russian–French system on a large corpus of scientific and technical abstracts. In 1983, this involved the processing (input, machine translation and human revision) of around fifty abstracts per month, for a total of about 60000 words. The revision has not been done by specialists in the domain, but by members of the team, whose aim has been to detect the errors of the MT system and to devise a strategy for maintenance and further development of the system. As a matter of fact, it seems that maintaining and upgrading the lingware, even in an admittedly limited second generation CAT system, requires quite a lot of expertise.

The current Russian–French system contains around 9000 lexical units. Individual translation errors are usually due to coding errors in the dictionaries. This is, however only an 'a posteriori' remedy. It is far more difficult to verify and correct such dictionaries 'a priori' in a systematic way. The VISULEX system is now being used for this purpose.

Another problem is how to increase the size of the lexical database. The 'a posteriori' method consists of indexing the lexical units found in the texts and absent in the dictionaries, in their descending order of frequency. Here, the a posteriori method seems rather more effective than the a priori method, provided one begins with a reasonably sized core.

As a matter of fact, we have tried to systematically index the terms of a given domain, by simply taking a 'normal' dictionary and indexing it. On the one hand, many words indexed in this manner never appear in the texts we translate, and, on the other hand, such dictionaries very often lack terms which appear frequently in recent, and hence interesting, texts!

Furthermore, there are very few reliable sources of multilingual technical terminology, especially for this language pair.

Grammatical maintenance and development is very much like maintaining and upgrading classical software, because of the procedural character of the SLLP. Here, the control structures of the various SLLP are very important, because they force the linguists to program in a modular way, and make it possible to construct powerful debugging aids such as local parametrised tracing or step-by-step execution.

Hence, if the source of a translation error is thought to come from one of the structural phases, it is quite easy to trace it to a particular grammar, a particular rule, and even to a particular procedure or elementary condition. If the fault is a simple bug, that is, bad programming of an otherwise valid idea, correcting it is very easy.

It must be said, however, that a significant proportion of the detected errors (maybe 20 to 30 per cent) derives from flaws in the linguistic model itself. Very often, we are limited by a lack of linguistic knowledge. For instance, in translating from Russian into French, it is very difficult to analyse the determiner system, and, even when it can be correctly calculated, to generate the correct article (or none), because nobody has yet given a complete and correct explanation of the use of the article in French.

The usual academic way of doing grammatical maintenance is to attack the most interesting (and most difficult) errors first. But, from the point of view of an operational system, another approach is more efficient. We prefer to determine which errors make human revision more difficult, and, beginning with the most damaging, to correct the simplest of these errors first.

Another area of R&D work focuses on the presentation of the machine translations. First, we try to produce a layout similar to the input. This is possible, because we handle large units of translation, and because the text-formatting commands are not erased, but passed to the linguistic processor, which usually handles them as special kinds of punctuation.

Also, the analogy with compiler–compiler systems leads us to try to produce the equivalent of *error messages* in the output. The most simple form is to present multiple equivalents for a given term, in case the linguistic processor has been unable to completely solve the ambiguity. In other cases, where the analyser produces marks of doubt about the correctness of its output, the generator may produce appropriate messages in the output text, in order to direct the reader's or the reviser's attention to that particular portion of the text.

In the long run, dictionaries turn out to be the costliest components of CAT systems. Hence, we are working towards the reconciliation of 'natural' and 'coded' dictionaries, and towards the construction of automated verification and indexing tools.

Natural dictionaries are usually accessed by the lemmas (normal forms). Coded dictionaries of CAT systems, on the other hand, are accessed by morphs or by lexical units. Moreover, the information the two types of dictionaries contain is not the same. Natural dictionaries usually give some morphological information, and very poor syntactic information, along with a lot of examples of use and such miscellaneous information as pronunciation or etymology.

For example, a natural dictionary will rarely give all the syntactic constructions in which a verb may appear, with corresponding semantic restrictions on its arguments. Even the exhaustive classification of verbs produced by a university project like LADL (M. Gross) doesn't give the *semantic relations* associated with the arguments of the verbs, in their various syntactic constructions.

On the other hand, coded dictionaries in second-generation CAT systems have no use for examples (or only within comments). But in order to produce deep level analyses and paraphrases in generation, coded dictionaries must provide semantic restrictions on the arguments of predicates (or semipredicates), and the interpretation of the various argument slots at the semantic level, if possible.

Finally, it is highly desirable to maintain some degree of coherency between the coded dictionaries of a CAT system and the natural dictionaries which constitute their source, for documentation purposes, and also because these computerised natural dictionaries should be made accessible to the revisers. It is of course quite puzzling at revision to find that a term has been given the wrong equivalent by the automatic system, although the (natural) dictionary contains the correct one!

N. Nedobejkine and Ch. Boitet ([526] 1982) proposed an initial solution to the problem of defining an integrated structure for natural and coded dictionaries. This proposal was presented in 1983 at an ATALA meeting in Paris.

The central idea here is to start from the structure of modern dictionaries, which give access by the lemmas, but use the notion of lexical unit. Each item may be considered as a tree structure. Starting from the top, selections of a 'local' nature (on the syntactico-semantic behaviour in a phrase or in a sentence) give access to the 'constructions'. Then, more 'global' constraints lead to 'word senses'.

At each node, codes of one or more formalised models may be grafted on. Hence, it is in principle possible to index *directly* in this structure, and then to design programs to construct the coded dictionaries in the formats expected by the various SLLP.

Up to this level, the information is monolingual and usable for analysis as well as for generation. If this language is source in one or more language pairs, each word sense may be further refined, for each target language, and lead to equivalents expressed as *constructions* of the target language, with all other information contained in the dictionary constructed in a similar way for the target language.

Hence, we have not tried to define a unique 'many-to-many'

dictionary, but rather a set of 'one-to-many' dictionaries, in particular because we don't know of any dictionary of the first type developed for the basic core of several natural languages, which would include relatively unambiguous technical terms (usually noun phrases) in a given domain as well as highly ambiguous and difficult verbs.

This kind of work seems to be a particular case of what researchers in AI, and particularly in expert systems, are nowadays calling 'knowledge processing'. As we have just seen, a major problem is of course to find a good way of *representing* this knowledge.

For us, there is another problem, perhaps even more important. Because of the cost of building machine dictionaries, we need some way to *transform* and *transport* lexical knowledge from one CAT system to another.

This is obviously a problem of translation. Hence, we consider this type of 'integrated structure' as a possible *lexical interface structure*. Research will soon begin on the possibility of using classical or advanced database systems to store this lexical knowledge and to implement the various tools required for addition and verification.

In the case of ARIANE–78 lingwares, VISULEX can be considered as a first version of a tool for incorporating knowledge coming from coded dictionaries, and ATLAS may be seen as one type of interactive tool to be developed for interactive verification and indexing.

To give an example, suppose the interface structure requires a certain type of information which is not present in some coded dictionary given as input. The information contained in this dictionary and required by the interface structure will first be added to the database. Then, the system should be able to require the missing information and to acquire it, in an off-line or on-line manner.

Just as in current software engineering, we have long felt the need for some level of 'static' (algebraic) specification of the functions to be realised by algorithms expressed in procedural programming languages. There have been several proposals for specification tools in some classical areas.

For example, researchers have been able to specify a PASCAL compiler fully by describing the correspondence between the programs and their abstract trees, and between the abstract trees and the corresponding target code. From this specification, it has been possible to automatically produce a compiler.

In the case of CAT systems, the problem is more difficult,

because there is no a priori correct grammar of the language, and because natural language is inherently ambiguous. Hence, any usable specification must specify a *relation* (not a function) between strings and trees, or trees and trees: many trees may correspond to one string, and, conversely, many strings may correspond to one tree.

Working with B. Vauquois in this direction, S. Chappuy presented in her thesis ([528] 1983) a formalism of *static grammars,* presented in *charts* expressing the relation between strings of terminal elements (usually decorations expressing the result of some morphological analysis) and multilevel structural descriptors. This formalism is currently being used for all new linguistic developments at GETA.

Of course, this is not a completely new idea. For example, M. Kay proposed the formalism of *unification grammars* for quite the same purpose. But his formalism is more algebraic and less geometric in nature, and we prefer to use a specification in terms of the kind of structures we are accustomed to manipulating.

Before trying to automatically generate analysers or generators from their static specification, which may prove to be difficult and may lead to very inefficient programs (written in ARIANE–78's SLLPs), we want to study the formal properties of this new formalism, and its relations with other formalisms proposed in the literature. We also want to consider the possibility of designing and building a system that would contain a syntactic editor of such static grammar charts, verification tools for ensuring the coherency of a set of charts, and facilities for graphic output of the geometrical parts of the charts.

The next step will be to augment a set of charts with an explicit representation of some strategy, in the form of metarules, in order to generate and test analysers and/or generators which we hope could be reasonably efficient.

Hence, lingware engineering may be compared with modern software engineering, because it requires the design and implementation of complete programming systems, uses specification tools, and leads to research in automatic program generation.

If this automatic program generation proves to be possible, we might then converge with still another line of software engineering, in a very interesting way. It will then be feasible to design a *syntactico-semantic structural metaeditor,* that uses a static grammar given as parameter, in order to guide an author who is writing a document, in much the same manner as metaeditors like MENTOR are used for programming languages.

This could offer an attractive alternative to interactive CAT systems like ITS, which require a specialist to assist the system during the translation process. As a matter of fact, this principle is a sophisticated variant of the 'controlled syntax' idea, like that implemented in the TITUS system.

Its essential advantage is to guarantee the correctness of the intermediate structure, without the need for a large domain-specific knowledge base. It may be added that, in many cases, the documents being written are in effect contributing some *new* knowledge to the domain of discourse, and hence *cannot* be already present in the computerised knowledge base, even where one exists.

Conclusion

Although it seems to be widely believed in AI circles that current work on CAT systems revolves around old ideas of the 1960s, we hope to have shown that, on the contrary, current research and development is very much in line with modern work in software engineering and AI in general, and presents some very interesting and difficult particularities, because of its object of study.

Finally, let us not forget to mention that part of this current research effort is being done in the framework of the EUROTRA project.

Acknowledgements

Many thanks go to E. Macklovitch, who has helped me with the intricacies of English!

MACHINE TRANSLATION AND
SOFTWARE TOOLS

In this chapter we discuss the desirable properties of a software environment for MT developments, starting from the position that successful MT depends on a coherent theory of translation. We maintain that a good development environment should not just provide for the construction of instances of MT systems within some preconceived (and probably weak) theoretical framework, but should also offer tools for rapid implementation and evaluation of a variety of experimental theories. After a discussion of the properties of theories of language and translation as candidates for consideration for MT, we go on to describe a prototype software system which is designed to facilitate the kind of theoretical experimentation advocated.

Basic premisses
Let us consider first of all some basic assumptions:
 (1) an MT system is more than just an explicit list of pairs (or n-tuples, in the multilingual case) of texts;
 (2) an MT system is of finite size;
 (3) the size of the set of distinct texts to be translated is unknown a priori.
Note that these assumptions are reasonable not just for fully automatic high quality MT, but also for more modest goals such as human aided machine translation and machine aided human translation. Note also that (2) is just as much an observation on the amount of information about the translation process which we can hope to supply to the machine as it is a comment on the potential size of the physical hardware resource available on the computer itself.
 If these assumptions hold, then, in order to do MT at all, we need

154

to be able to provide a finite characterisation of an effectively infinite set of pairs of texts.

Thus far, we hope, what we have said is relatively uncontroversial – indeed much of it is an extension to translation, and especially MT, of what has been a commonplace in linguistics for some decades. What is perhaps less acceptable is our contention that such a finite characterisation can only be achieved by stating the appropriate generalisations about languages and the relations between them. Once these generalisations are made explicit, what we have is a theory, however tentative, of translation.

We are accustomed to hearing many objections to this position. One such objection runs: 'MT is a practical enterprise, not a theoretical one; theory always breaks down when we have to deal with real data'.

To this we would counter that if a theory breaks down in the face of 'real data', either we have a wrong theory or we have come across a special case. If we have a special case, a good theory will tell us why it is a special case, and hence how to deal with it; if the theory is wrong, we should be prepared to discard it and build another. If there is no explicit theory, every case is by definition a special case: we have no way of integrating the new information gleaned from the failure into what we already know. The option of discarding or modifying the theory is simply not available if there is avowedly 'no theory in the system'. In reality theory-free systems actually do have theories embedded in them, but cannot be modified either because the inventor is unable to make the theory explicit or because the relevant theoretical statements are buried in a mass of program code.

A related, but subtly different, objection goes like this: 'Linguistic theories are not about real language but about an abstraction which never occurs in real life'.

This is a specific complaint about the relevance of existing linguistic theories to MT, and as such it had some force in the past, although the trend in theoretical linguistics these days is towards more realistic theories of language (cf. chapters 3 and 4). The objection is not, however, as it stands, an objection to a theory of MT, only to the uncritical adoption of an inappropriate theory within a domain for which it is unsuitable. With this we are wholly in accord, and we shall point out later that a very important feature of an MT development system is the flexibility to allow testing out of new theories.

The most radical opposition of all, however, takes the position: 'There is no possibility of ever having a theory of MT: the domain is just too chaotic, and we would be better off just getting down to the job'.

We cannot take this position very seriously: if it is the case that there are no general statements which need to be made about the way MT works, then it is hardly reasonable to expect to write a program to do the job, *unless the general principles are already built into the programming tools*. We shall argue that part of the purpose of MT software design is precisely to facilitate the construction of problem-oriented tools, so that the majority of users can concentrate on setting what they know about language within a well-defined framework.

On the adequacy of theories

We have argued that for a complex task such as MT we need a theory, if only so that we can understand what the MT system does and – even more important – convey that understanding to others who will have to develop and maintain it. We suggest that any but the most trivial program in reality contains some theory of its domain of application, although in general in MT that theory is not made explicit. At the same time we recognise that there is currently no adequate theory of translation which can be embodied in an MT system – indeed we have yet to determine what an adequate theory would be like. We consider briefly the question of adequacy of a theory in a computational context, beginning with an illustrative example. Consider the assertions:

(4) (a) $S \rightarrow NP\ VP \mid SUBJ(\$\$) = \$1, \$3 = \$2$
    (b) $VP \rightarrow V\ NP \mid \$\$ = \$1, OBJ(\$\$ =) = \$2$
(5) (a) $NP(U^*,/,U1^*) + VP(V(V^*),W^*) ==$
                $S(NP(U^*,/,SUBJ,U1^*),V(V^*),W^*).$
    (b) $V(U^*,/,U1^*) + NP(V^*,/,V1^*) ==$
                $VP(V(U^*,/,GOV,U1^*),NP(V^*,/,OBJ,V1^*)).$

The notation used in (4) is an annotated phrase structure notation borrowed from the compiler generator Yacc ([537] Johnson 1975): annotations are separated from the rest of the production by a bar ($\mid$), $\$\$$ references the left-hand symbol, and $\$i$ references the *i*th right-hand symbol. The formalism of (5) is based on Q-systems ([532] Colmerauer 1971), a language for representing transductions from strings of trees to strings of trees: in this notation, parentheses, comma and plus are metasymbols used for representing tree structure, the double equals is the rewrite symbol, and symbols ending with the asterisk are all variables ranging (in this case) over strings of trees and local to a production.

The intended effect of both (4) and (5) is comparable. Both

assert that an NP followed by a VP is a valid S, a V followed by an NP is a valid VP, that the feature and relationships of the VP transfer to the S, that the V is the head of the structure, the first NP is the subject of the S–VP–V and the second is the object.

The first thing we observe is that the complexity of (5) is much greater than that of (4), although it is not evident that (5) is more informative. Rather (4) is more compact than (5) because it states only those linguistic facts which are not predicted by the theory. The global theoretical statement, that particular synctatic relations are derivable from constituent structure, is implicit in (4) but has to be made explicit in (5).

Second, we could also, using a different interpretation of tree structure, have written (5a) as

(6) $NP(U^*,/,U1^*) + VP(V(V^*),W^*) ==$
$$V(NP(U^*,/,SUBJ,U1^*),V^*,W^*).$$

Is (6) a 'better' statement of the linguistic facts than (5)? Is either better than (4)? Such questions are unanswerable outside a theoretical framework which gives an interpretation to the syntax of a rule.

The framework of (5) and (6) is extremely weak. It makes no commitment on questions of representation beyond a limitation to sequences of trees. It forces the rule writer not only to state linguistic facts but also to build representations explicitly. Its syntax is therefore effectively uninterpretable; it means what we want it to mean, and it is our responsibility as users to be consistent in our interpretations.

On the other hand, a strong theory will make claims both about representation and about the building of that representation. Suppose we restate (4), by applying a simple syntactic transformation, as

(7) (a) S→   NP   VP
           ↑SUBJ=↓   ↑=↓

     (b) VP→   V   NP
           ↑=↓   ↑OBJ=↓

now recognisable as assertions in the formalism of Lexical Functional Grammar (LFG) ([599] Bresnan and Kaplan 1982). As suggested above, LFG uses a fairly complex representational theory containing both constituent structure and functional representation. However, the formalism of (7) requires no indication whatsoever about the algorithm for building phrase structure trees or the method used to compute functional representations. As a result, the rules are simpler, and the rule-writer is not required to restate the same underlying principles over and over again.

Although LFG is a much richer theory than the simple fragment given here suggests, we do not wish to propose that LFG is the appropriate theoretical framework for MT. Whatever its merits as a linguistic theory it is clearly far from being a theory of translation, which is what we need for MT. It does, however, provide a good illustration of the way in which a strong theory can contribute to a clearer perception of the problem and a more rational division of tasks.

An environment for theoretical experimentation

An MT system is necessarily larger and complex. Most of the size and complexity comes from the fact that in order to translate we appeal to a vast fund of linguistic information (much has been made of the need for MT to have extensive world knowledge, but strangely, outside MT circles, there has been very little comment on the enormity of the linguistic knowledge required). To make MT work, we need to find ways of organising that linguistic knowledge so that it can be accessed and interpreted both by the machine and by the humans who will maintain and augment the system.

Our contention has been that the more global knowledge of a domain human and machine have in common, the less onerous will be the task of stating and organising the specific details of the domain. The fundamental job of MT software design ought then to be that of building the appropriate principles into the machine, and providing the user with a language for stating the necessary specific details in a natural way. In other words, we have to design and implement a problem-oriented language, within a habitable environment.

Introduction to the software system

The first half of this chapter presents the production of MT software as a paradoxical undertaking: the result has to be problem oriented, but at the same time, the design must reflect the fact that the machine translation problem domain is currently without an adequate definition. When faced with such a paradox, there are essentially three options available to the software designer:

(a) He may impose problem orientation on the system by constraining it by whatever means are available. Although this yields a problem-oriented system, the result is likely to be oriented towards the wrong problem – a discovery is only made after the release of the system to end users, when it is too late to make the necessary changes.

(b) He may sacrifice the goal of problem orientation by constructing a system which is theoretically speaking, *weak* – i.e. which is so general as not to exclude any possible behaviour that might be required by the user. Here, the objection is essentially the same as that which the user would raise were he to be presented with a Turing machine for performing desk calculations. That is, the device requires the user to get involved with details which have nothing to do with the problem at all.

(c) He may accept the difficulties inherent to the paradox, and try to construct an architecture aimed at reducing them in a principled way.

It will come as no surprise that we reject (a) and (b). Our interpretation of the third possibility involves a distinction between the end product of system design, which is an *implementation* of some sort, and the *framework* from which the implementation emerges. Although both arise naturally in some form or other during system design, the amount of attention given to each may vary.

When the problem domain is well defined, the criteria of success for the implementation are by definition predetermined, so that if the product satisfies the criteria, the system designer will have discharged his responsibilities. In such cases, it is perfectly correct for the designer to concentrate as much attention as possible on producing the highest quality implementation, i.e. one that is optimised as far as usage, and space/time occupancy are concerned. Here, the framework not only takes second place, but is often never consciously externalised. As an example, the task of writing a compiler for a defined programming language might fall into this category.

However, when the problem domain is not well defined to start out with, it is foolish to concentrate all one's efforts into the implementation, not only for the reasons explained in (a) above, but also because the provision of an optimised implementation takes a great deal of time which might better be spent on finding out what it is that is supposed to be implemented.

The scheme to be described below embodies this philosophy by setting up a framework within which released software is consciously ascribed an experimental status. This is not to be seen as an excuse to supply poor quality implementations. Rather, it implies that the framework must be solid enough to

(a) generate prototype implementations that are adequate for the performance of experiments by those that are in the best position to generate critical feedback – namely, the user community, and

(b) accommodate the feedback in concrete terms by supporting the generation of counter implementations. Of course, we do not believe that this process continues ad infinitum. If we do our job poperly, and the feedback is well founded, we can assume that there will be a tendency to converge on a product or a series of products that meet the original requirement of being appropriately oriented towards the appropriate problem domain. At this point, a third desideratum would be

(c) to provide enough information about the experimental implementation to enable an optimised version to be constructed in a more conventional, probably industrial mode of software development.

However, the success of the whole scheme depends very much on the ability of the framework to fulfil its intended purpose, i.e. offering the maximum support in the face of user inspired criticism. We take this goal as our starting point for transforming the framework from a vague idea into something concretely realisable.

Modules

From the external point of view, an experimental implementation has the form of a compiler for user programs, that produces a source-text to target-text translation program, together with an interpreter that runs the program generated by the compiler.

The abstract meaning of a user program can thus be thought of as a relation of some kind between text representations. The job of an experimental implementation is thus to compute that relation, which is obviously a complex one. Part of the job of software design is to find a way of breaking the computation down into parts between which the relations are well understood. There are two good reasons for this. The first is that the parts will of necessity be smaller, less complex, and easier to construct than the whole. The other is that this strategy accords with the old-fashioned but sound principle of modularity whereby the changes in the behaviour of the whole that follow from a change to one or more of the parts will be predictable, orderly, and therefore controllable.

A well-known technique for handling the meaning of a complex language, which applies as much in the domain of formal semantics as compiler design, is to choose a simpler language and then translate expressions from the first language to the second ([558] Scott 1973; [556] Richards and Whitby-Strevens 1980). There is nothing to stop this process being iterated any number of times, although the whole exercise would be rather pointless if the last language in the chain did not have some special status. In formal semantic terms, that language must have a semantic theory which maps expressions into extra-linguistic entities which may be abstract (e.g. individuals, predicates, functions, etc.), or physical (i.e. actual pieces of the world). In computer language terms, the same function is performed by an *interpreter,* that is, a device which 'executes' expressions in the language directly. In this sense, every computer is an interpreter of its own instruction set.

We have embodied this technique into the architecture of an experimental implementation by breaking it up into parts that we will subsequently refer to as *modules.* These are characterised by an input language, which specifies the class of programs for the module, an output language which specifies the class of meanings of those programs, and a mapping between the two languages. Each module thus expresses a theory of meaning about the input language by implementing a particular mapping between input and output, and is essentially a *translator.* The underlying shape of a generalised experimental implementation consists of a collection of such modules arranged in a chain. This chain, although apparently hierarchical, behaves functionally as though some sections of the chain are compressed together. This happens when the mapping effected by each can be composed in an algebraic sense, as discussed further below.

However, for the output of the whole chain to mean something, both in the theoretical sense of having a meaning, and in the practical sense of producing behaviour, the last item in the chain must be an interpreter. This is also characterised by an input language, but there is no output language. Instead, there is a space of *computations,* and the mapping is between a sentence in the input language and a *particular* computation.

Specification languages

The framework we propose must respond to criticism. Our first observation is that criticism, in the sense intended above, will be directed towards experimental implementations. For this to be possible, the framework as a whole must be able to support *descriptions* of experimental implementations, which is to say, *descriptions of modules,* and the relationships between them. It seems natural, therefore, to think of input to the framework in terms of a *language* of some kind, in which to each module there corresponds a sentence (or perhaps a set of equivalent sentences) in that language. For reasons of clarity, we will henceforth refer to such a sentence as a *specification,* and to the language as a *specification language.*

Now in principle, any language can serve as a specification language, although obviously, some will be more suitable than others. There are various different dimensions along which they can be compared, the most important of which are probably *formality,* and *expressive power.*

At the informal end of the formality spectrum comes English, diagrams, and a wide variety of quasi-mathematical notations. Specifications in English, although common, are far from satisfactory for the fairly obvious reason that there is nothing in English to prevent vague, incomplete, and possibly inconsistent specifications from being expressed. A further disadvantage is that there is no well-defined relation between the specification and the implementation that can be exploited to speed up delivery of the goods.

We will say that a specification language is formal if it has a precisely defined syntax and semantics, and for the time being we will take this to mean that if a specification has a meaning, then that is both determinable and unambiguous. Another way of saying this is that the relation between a specification and the thing it specifies – in our case, a module, is formally describable. Formal specification languages have many advantages. For instance, formality in the language encourages rigour in the specification and provides a medium for intercommunication during the design process. Other advantages of such languages are discussed in [540] Liskov and Berzins 1979. However, from our own point of view, the most important aspect is that if the S-relation is formally describable in the sense intended, then each module, and therefore, each legitimate specification can be regarded as denoting an equivalence class of *implementations.*

Moreover, each specification is a formal object in its own right that can be studied using formal techniques (in correctness proofs for example), and manipulated according to formally defined operations. It can also be represented inside a computer. These last two aspects are particularly relevant to our needs. For if a specification stands for a class of implementations in a formal sense, then there exists an algorithm that will generate at least one member of the class. So a formal specification is, in principle at least, *runnable,* in the sense that it can be made to construct a chosen representation of an implementation.

Whether it is *actually* runnable clearly depends on whether a generation algorithm has been implemented. When it has been, there is of course no guarantee that the implementation is optimal, but then, as we have already established, this goal is secondary to our needs. When working in an experimental framework, the advantages of being able to generate an implementation automatically far outweigh the minor inconvenience of having to put up with its inefficiency.

As far as the framework is concerned, we require that all modules be generable in exactly this sense, and henceforth we will refer to a program which generates a module from a specification as a *module generator,* or more briefly, just 'generator'.

Above we argued in favour of describing each module of the system using a formal specification language. However, our definition of formality is extremely general, and includes *any* language which has a specified syntax and semantics. This lets in a large number of languages that we would not normally consider suitable for the purpose of module specification – for example, PASCAL ([536] Hoare and Wirth 1973). Clearly, another dimension of classification that we need to consider is *expressive power.* This is closely linked to the notion of problem orientation, so that the arguments to be adduced here are really no different to those that would be used in favour of the choice of any problem-oriented language with respect to a given problem domain. In a nutshell, this can be reduced to two requirements:

(a) Understandability: the specification should be more perspicuous, and easier to reason about, than the implementation. In particular, the terms of the specification should be similar, if not the same as, the terms of the problem domain, permitting a succinct and natural expression of the problem.

(b) Precision: the specification should say no more and no less about the implementation than it has to. From the declarative standpoint, we might say that the specification should describe *what* the implementation is supposed to realise without saying *how* to realise it.

PASCAL fails to meet both (a) and (b) in so far as a specification *in* PASCAL would actually *be* an implementation. The reason is that the model of computation presupposed by PASCAL as reflected in the constructs available in the language – the knowledge of computation employed by the virtual PASCAL machine if you like – is essentially at a *low level*. In other words, to 'specify' a computation, a PASCAL programmer has to explicitly *tell* the virtual machine what to do. To win the kind of advantages described in (a) and (b), the virtual machine that interprets the specification language must 'know' rather a lot about the problem domain over which the implementation is supposed to operate. For only then is it possible for the specifier to delegate to it the job of filling in the gory details of a computation by a simple act of omission, as reflected in an appropriate choice of syntax for the specification language itself.

So generators, and their associated specification languages, are theory-sensitive in an exactly parallel sense to that described earlier. And just as we would not wish to impose artificial constraints, or artificial generality upon the machine translation domain by predefining the theory (or lack of it), so we would not pretend to claim that there is just one specification language. For the choice of specification language must depend on the problem domain of each module, and there is no reason to believe that these operate over the same domain.

Optimisation and industrial specification

Although the main aim of the experimental approach described is to generate, by successive approximation, an implementation of an adequately problem-oriented MT system, we recognise that flexibility in the framework tends to conflict with efficiency of released software. As pointed out above, an equally important aspect of our work consists of providing for the construction of an optimised version of a given experimental implementation or, as is probably more likely, some piece of it.

In order to optimise a 'piece' of an experimental implementation (and this includes the degenerate case where a piece is equal to the whole), one has to be able to define both *what* the piece is, and also to know how it will interface with whatever exists already. Now, the notion of such a piece follows directly from the chain architecture of an experimental implementation, which is in places *compositional* in nature. Where this is so, if m1 and m2 are two adjacent modules, then we can define m3 to be functionally equivalent to m1 and m2 by equating it with their product:

$$m3 = m1 \circ m2$$

This permits us to define modules of arbitrary length for optimisation, and know exactly where to put them when completed.

Second, and perhaps more importantly, the task of optimisation requires an adequate specification, and the *implementation* of a module does not amount to such, any more than a giraffe is adequate specification for an animal with a neck. However, by arranging for all modules to be *generated,* we have in a practical sense already provided an adequate specification in the form of input to the generator. All we require of the optimiser is that he understand the specification *language,* and that he employ his expert knowledge of *programming* to generate the optimised module implementation.

An experimental implementation within the framework

The description of an experimental implementation above is agnostic about the number of modules that constitute a chain. However, the current experiment is essentially a three-level device: the outer level being the user interface, the intermediate or procedural level providing control, and the inner providing abstract data definition facilities. The interpreter is essentially a procedural device with powerful database management facilities.

As mentioned earlier, the function of the whole is essentially a compiler which maps user programs to the 'object code' of an abstract database machine. The function of the outermost module is analogous to the syntax-analysis phase of compilation, just as that of the second and third is to perform code generation.

The aim of the syntax-analysis phase of compilation is to map sentences in a syntactically rich, user-oriented language to sentences in a language which is syntactically constrained. This mapping is an entirely syntactic matter, in the sense that the explicit semantic content of the output is equal to that of the input.

The generator at this level should therefore be oriented towards the generation of syntax analysers. Because it would be unreasonable to impose constraints in the one place where the end user can and should have a say in the kind of syntax best suited to his needs, the generator must be equipped to construct modules whose input language may range over a wide class. At the same time, it need not be oriented toward the description of code generation.

To date, the user level generator consists of Yacc/Lex in combination with the Unix C compiler. Although this has enabled us to construct a working user interface rapidly, we feel that it is not ideal for two reasons. First, the class of languages for which the combination is capable of generating compilers is $LALR(1)$ – and this imposes a number of annoying restrictions. Second, the description of code generation employed by YACC/LEX is in fact C code, and this is far too general for the task to hand. As pointed out above, the nature of the mapping performed by modules at this level is syntactic, and given that the syntax of the output language is simple, the proper place for knowledge about *building* output sentences should be inside the generator itself, not in the specification. We are currently building a generator whose specification language overcomes both of these difficulties.

The outer-level module accepts user descriptions of linguistic processes, containing (a) patterns describing configurations of linguistic objects and (b) relationships with other processes. These are mapped into S-expressions ([547] McCarthy 1960), which has been chosen as the internal language of the whole compiler.

This contrasts with the requirements at intermediate levels, whose aim is to perform the code generation phase of compilation. Because code generation embodies the semantics of the input language, the primary function of a module here is to make it explicit. At the same time, the module should have to do as little in the way of syntactic manipulation as possible. We ensure this by defining input and output languages to be the same. Making the semantics explicit now amounts to providing information about what in the input language are uninterpreted symbols. Here, there are two possibilities. One can either do macro-generation, whereby all occurrences of those symbols are replaced by their definitions, or else one may collect the definitions up and pass them on for subsequent interpretation. We have chosen for the time being to adopt the latter solution, mainly because it makes for an extremely simple generator once the form of definitions has been chosen.

The procedural-level generator's specification language consists of FP ([523] Backus 1978) definitions, since this is ideally suited to the description of procedural constructs in a clean, formally elegant and theoretically sound fashion. It is assumed that the module will (a) construct Lisp definitions for each symbol and (b) append them to the input.

The procedural-level module defines the meaning of control constructs in the user language. These concern the execution of user defined processes, which are currently defined as being under iteration, in parallel, or in sequence. The meanings of these constructs are incorporated as defined above.

The aim of the inner-level generator is to define data abstractions that will be used in user language pattern descriptions. The specification language is PROLOG, and the technique for incorporating a specification into the module is similar to that for the procedural level. That is, we construct a LISP definition that will implement its meaning when run.

The inner-level module defines the meaning of data configurations expressed in the user language. Currently, this is oriented towards patterns which specify sequences of trees, the nodes of which are described in terms of generalised lists of properties and values.

The interpreter is currently the LISP system, augmented with primitive functions that permit communication with a PROLOG database. The LISP system interprets definitions, some of which are passed on to PROLOG directly. Although this arrangement yields an extremely flexible interpreter, we are, for obvious reasons, not entirely happy to use PROLOG as the underlying model for data. For one thing, it contains several constructs that are both unintuitive and undesirable, such as the cut operator. For another, it is too general for our purposes. We are therefore engaged currently in the development of a special-purpose database definition and manipulation system which will be incorporated into the next experimental implementation.

Acknowledgements
The authors would like to thank the European Community for funding the Eurotra project, out of which this research has emerged, and our colleagues Nino Varile, Steven Krauwer, and Dominique Petitpierre for invaluable comments and criticisms.

PART THREE : MT SYSTEMS

A BRIEF SURVEY OF SOME
CURRENT OPERATIONAL SYSTEMS

Later chapters of this section will give quite detailed descriptions of some of the systems currently available or under development with a view to operational use in the relatively near future. Most of those systems are intended to be fully automatic, with comparatively little post- or pre-editing, and with no interaction with the user. This chapter will take a brief look at some other systems, which do not follow this pattern. In the description of each system, emphasis is put on one particular characteristic of that system, in an attempt to give the reader a feeling for the range of possibilities available. Sometimes, it has been difficult to decide which characteristic to associate with which system, since often two or more systems share a common interest in some particular aspect. Thus, the reader should not take discussion of, say, the text processing environment of one system, as implying that other systems do not also make use of a text processing environment. Furthermore, no attempt has been made to provide an exhaustive list of systems, nor is there any intended implication that systems not mentioned here are not worthy of attention. As just one example of this latter point, the SPANAM system, used for Spanish–English translation by the Pan American Health Organisation is reliably reported to produce results of a quality sufficient to reduce manpower requirements by some fifty per cent and is apparently heavily used by PAHO. However, no attempt is made to describe it here because of its obvious family resemblance both to the Georgetown system described in the earlier chapters on history, and to SYSTRAN, which is described in detail in an independent chapter.

A further word of caution should be given with respect to the commercial systems discussed here. For natural reasons, the companies developing these systems are reluctant to divulge detailed technical information. Therefore much of what is said here is based on deduction from literature which was primarily produced for commercial purposes rather than for a technical audience, and from demonstrations of the systems. The accuracy of the information is therefore not guaranteed.

CULT

The first system considered here, the Chinese University Language Translator (CULT), is primarily interesting in that it concentrates very heavily on pre-editing as a means of producing relatively high quality translation. Although the literature is not very explicit, it seems that the pre-editor introduces into the text syntactic markers defining, for example, sentence boundaries and the boundaries of prepositional phrases, semantic markers indicating choices between different word senses, markers indicating where definite and indefinite articles should be inserted and so on ([283] Loh et al. 1978).

It is also interesting to note that Loh remarks, in the same article, that given the degree of pre-editing (about 5 per cent of the input text is pre-edited either manually or automatically, [280] Loh 1976), many of the problems encountered are problems of terminology rather than of translation proper.

After pre-coding of the text and character coding (needed since the source language is Chinese), translation follows a very conventional pattern. Dictionary look-up is done according to a longest match principle, syntactic categories and groups are established, the constituents found regrouped according to English surface order, and finally English equivalents determined.

As can be seen from the above, it is relatively unlikely that CULT could be extended without major work to deal with other language pairs. By its nature, it is essentially bi-lingual in design. However, CULT has been successfully used to translate the Acta Mathematica Sinica into English since 1975, and the Acta Physica Sinica since 1976.

ALPS

Our second system is based on the notion of interaction between the user and the system at several points during the translation process. ALPS does not pretend to be a fully automatic MT system: rather, it provides the translator with a set of translators' aids, i.e. software tools, to automate many of the tasks encountered in everyday translation experience.

The ideas of ALPS are based on a decade of experience in MT

at the Brigham Young University. The first attempts at fully automatic machine translation using junction grammar were abandoned in favour of an interactive approach, recognising the potential contribution of the computer during the translation process. The private company Alps was formed on 1 January 1980 in order to adapt the tools originally developed at BYU for smaller and more cost-effective equipment running on a powerful dedicated work station. The work station benefits from advances in office automation in that it can receive text from another computer, from magnetic tape or via optical character recognition and output the resultant translation to a variety of output devices such as printers or typesetting equipment without retyping the text. This contributes greatly to the cost-effectiveness of the system. The available language pairs are English into French, German, Spanish and Italian, as well as French into English. Further pairs, for example English into Arabic, are under development.

ALPS views active translator participation in the translating process not as 'settling for less', but as a gain in flexibility and in stylistic freedom. The central tool of the system is a menu-driven wordprocessing system coupled to the on-line dictionary. The system works at any of the levels described in the table below:

Human translation	Multilingual wordprocessor
Machine Assisted Human Translation (MAHT)	Selective dictionary look-up (SDL) Automatic dictionary look-up (ADL)
Human Assisted Machine Translation (HAMT)	Computer translation system (CTS)

At one end of the range, the computer can be used as a multilingual wordprocessor, allowing the translator to compose his translation on the screen. The wordprocessor is designed for translators, providing, amongst other features, multi-lingual character set support for virtually every language written in the roman alphabet, and simultaneous display of source text, translation and on-line dictionary information. At the other end of the scale, the computer can produce first-draft translation with interactive guidance from the translator. The raw translation can be post-edited in the same post-editing environment described above.

The machine-assisted human translation (MAHT) version of

ALPS consists basically of different types of on-line dictionaries, the Selective Dictionary Look-up (SDL) and the Automatic Dictionary Look-up (ADL). Both types of dictionary are user-defined and provide help for technical terminology. Each specialised dictionary contains words and phrases related to a particular field, like aviation, metallurgy, etc.

SDL can be considered as a computerised version of a translator's personal card file. The system includes facilities to consult and update SDL-dictionaries on-line in a special window during the translation. The dictionary entries can be incorporated into the translation text by a simple key stroke.

ADL provides automatic word-by-word dictionary look-up for user-defined technical terminology. Morphological processing allows all technical terms in the source text to be looked up sentence by sentence in their base form. When there is more than one translation for a word, phrase or idiom, the computer asks the translator interactively to select the proper one for the context of that particular sentence. For example, in the translation of the sentence: 'The required chip is a dual-input tri-state NAND gate' the translator would be asked to choose the proper translation for the word 'chip'. Depending on whether it was a translation into French, Spanish or German, he would be given some choices like these:

French chip	Spanish chip	German chip
1. *fiche*	1. *micropastilla*	1. *Chip*
2. *micro-plaquette*	2. *microplaquetta*	2. *Mikrobaustein*
3. *micro-chip*	3. *recorte*	3. *Stanzabfall*
	4. *ficha*	4. *Schuppe*

The translator could then select one of the translations suggested by the computer by simply typing the number of the translation he wanted. Or, if he did not like any of them, he could type in his own translation.

The amount of interaction can be cut down by two facilities. For each translated text a document dictionary is created into which the translator can merge the specialised dictionaries he wants to use plus all the terms for the particular document he needs. A Key Word in Context (KWIC) listing helps to decide which translations are needed for a particular document. This environment helps to ensure consistency in the terminology for the entire document, particularly if several translators are working on the same text.

The highest level of the ALPS system is the Computer Translation System (CTS), which produces first-draft translations. It includes therefore not only the same dictionaries as ADL but also a general vocabulary and syntactic processing. The CTS program handles most aspects of inflection, word-order and agreement. In the case of ambiguities, the system interacts with the translator.

CTS first examines the text sentence by sentence and produces a list with words it did not find in the dictionaries. The next interaction is ambiguity resolution on the lexical and grammatical levels. If the program encounters difficult or ambiguous grammatical structures, it comes to the translator for help. For example, in the sentence: 'Did you believe the boy's broken promise?' it would turn to the translator because there are two possible interpretations, the s in 'boy's' could be a possessive marker or a contracted form of 'has'. The translator is responsible for the correct interpretation. Since the quality of the grammatical processing depends on the amount of information the user can add in the lexical entries, the reader can deduce the status of the linguistic processing from the information that is requested when new dictionary entries are created. The ALPS documentation describes it as follows. 'Since CTS includes grammatical processing, some grammatical information is needed in the dictionaries. For the most part, it is sufficient to provide the category (part of speech), such as noun, verb, adjective, etc., using short codes. For example, some of the translation equivalents in the dictionary entry for the word 'back' could look like this:

French
back
dos[n
soutenir[v
arrière[aj

Occasionally, additional information (such as gender on nouns) is required. In any case, all grammatical information can be given in a few simple characters which take only moments to include in a dictionary entry.' This quotation shows that the system relies totally on its user interacting with it during the translation process, for semantic information as well as for information about the type of construction. The system assumes that the target language items have the same type of construction as those in the source: in all other cases the user has to intervene. The lexicographer cannot record information about structural changes caused by particular lexical items. In other words, interaction with the translator is being used to take the place of powerful linguistic processing.

WEIDNER

Our next system can be viewed most easily as a sophisticated aid to document preparation. It is significant that one of the first (and largest) companies to employ WEIDNER saw it as an effective way of streamlining its electronic publishing system.

WEIDNER was established around 1978. The language pairs offered include English, Spanish, French, German, Portuguese, Japanese, Arabic. Around 1981, the pair English–Arabic was sold to OMNITRANS with a view to translating parts of the *Encyclopaedia Britannica* into Arabic. Given the flexible nature of the system, centred mainly on the dictionary, new language pairs can be added quite easily. The WEIDNER system is used by MITEL in Canada, by Bravice (a translation service bureau in Japan) and by OMNITRANS. At the time of writing, twenty-five systems had been sold. The system runs on a PDP–11, and has recently been implemented on an IBM personal computer.

The actual translation component of the system, as we shall see in more detail shortly, comprises only a few simple dictionary look-up routines together with a few surface syntactic manipulations. Yet, the system is popular with translators, partly of course because there is not a wide range of choice at this level of MT, but mostly, we would argue, because it provides the translator with an advanced information processing tool far superior to any tool traditionally used in this profession. Indeed, the translator market was ripe for commercial exploitation: for example, at ASLIB conferences of recent years, it has become increasingly evident that translators have become very conscious of the capabilities of word-processing as a professional aid.

The WEIDNER system does not, however, offer just a super-wordprocessor. The core of the system, as with ALPS, is organised around a dictionary. The simple fact of employing a machine dictionary in some information task enhances speed of processing and ensures a high degree of consistency. This is well known, and such considerations have prompted the now widespread use of term-banks of all shapes and sizes. This is important, since, although MT may often be thought of as a purely linguistic problem, at a practical level, an MT system has to face up to the severe problems of terminological accuracy. A human translator normally spends up to 60 per cent of his time doing terminological research, which may in the end result in his coining a term. Existing termbanks tend to be large information retrieval systems housed on dedicated mainframes. They are therefore not immediately accessible to other than in-house users. Moreover, they are mostly confined to large organisations. Despite the existence of public information networks, e.g. Euronet, termbanks are not yet available for consultation by the translating public at large. Eurodicautom alone is connected to Euronet, and offers only limited possibilities to potential users whose main interests do not coincide with those of the EEC translation services.

This short disquisition on termbanks may appear to take us some way from our subject. But the point is this: the existence of an up-to-date, accurate machine dictionary, mounted on a smallish computer owned by a middle- or small-sized company, or operated by a translation bureau, can help to increase productivity and the consistency and accuracy of translation by a substantial amount. Systems such as WEIDNER should be seen in this light, that is, not as small-scale, cheap, flexible MT systems, the descendants of large-scale fully automatic MT systems, but rather as scaled down termbanks, with restricted technical terminologies, embedded in some sophisticated wordprocessing or text-processing environment.

It is perhaps worth repeating here, that while both WEIDNER and ALPS may on the surface appear to be direct descendants of the Brigham Young MT efforts, neither system claims to offer any substantial degree of machine translation. What is offered is a multilingual terminological data manipulation system, together with sophisticated wordprocessing devices.

However, WEIDNER is nonetheless more than a termbank. An attempt is made at producing a rough translation equivalent. As is usual, the user constructs dictionary entries for the system. The dictionary up-date module asks questions of the user which imply that the lexical entry is rather differently constituted to that of a termbank record. As the system asks for information regarding inflection, or syntactic behaviour, we may feel confident in expecting it to employ this information as any normal MT system would. The questions asked are quite detailed, and concern the syntactic and inflectional behaviour of individual lexemes or so-called idioms. However, analysis of these questions, and of the eventual output of the system, before post-editing, reveals that the linguistic conception underlying the system is localistic, dealing with lexemes in some very restricted context.

As regards the most intractable problem of MT, that of *ambiguity,* there is no indication that procedures exist to handle any ambiguities other than the simpler types of lexical ambiguity, i.e. homographs. There is apparently no strong analysis component, and indeed all available WEIDNER documentation and user comments make it clear that the most complex part of the system is the dictionary, and that badly constructed dictionary entries will result in bad translations.

One problem is worth mentioning in passing here: the dictionary update system is menu-driven, and, although this is at first attractive to an unexperienced user, it has the unfortunate effect of obliging the user to select some feature that is 'most like' the feature under consideration. Such an inflexible approach cannot of course produce high-quality translation as a result. In MT systems in general, it is accepted that 'dictionary' information is dependent on the linguistic strategy adopted. Given that the only truly linguistic part of the system, the 'linear compaction' module, is supplied in the form of a black box, users cannot change the system to accommodate different syntactic constructions, etc. Without any knowledge of the linguistic rules employed, and being constrained to use a menu-driven dictionary up-date module, the user can only make guesses at the eventual behaviour of the system when he gives some 'most like' response.

Thus, although much of the user's time and effort is expended on creating dictionary entries, in many cases the system will not produce an adequate translation. Typical questions asked of the user concern noun agreement, or the ordering of nouns and adjectives. Weidner documentation talks about analysis, transfer and synthesis routines, as well as morphological processing, and makes particular mention of 'cosmetic routines' which ensure correct agreement and ordering. However, raw output shows many cases in which such processing has failed, and user comment emphasises that relatively trivial matters (such as agreement) often have to be seen to by the post-editor.

Another consequence of the menu-driven system and of the localistic processing adopted, is that once a string is declared as an idiom it will always be treated as such, despite the fact that each element of the idiom must also be entered as a separate lexical unit.

One user has drawn attention to the need to provide information about what WEIDNER calls the 'agency' of words. The available menu here is as follows:

Noun: human, group, body part, animal, inanimate, concrete, abstract.

Verb: none, direction, location.

Although this selection of features appears rather restricted, obviously some effort is being made to capture more 'semantic' information. However, the same user comments: 'Weidner has yet to satisfactorily explain these categories and the reasons for them to us, since most of them have no direct effect on the translation.' Of course, without knowledge of the linguistic routines, one cannot know in any detail how (or even if) all the information asked for during dictionary coding is used. However, comments such as that above, together with inspection of the raw output produced, suggest that even the introduction of semantic information is of only limited utility.

Analysis is carried out in local mode, and is concerned only with the noun phrase or verb phrase level. It consists of three distinct independent stages: idiom search, homograph disambiguation and structural analysis. The 'linear compaction' routine apparently recognises strings such as idioms that act as one linguistic unit and compacts them into a single unit flagged with syntactic and semantic information. It is claimed that examination of local context leads to homograph resolution. A straight left-to-right parse is then performed to build up a phrase structure tree. There is no indication that an attempt is made to carry parallel analyses. The main problem with the local approach is that the wider context often necessary for achieving a correct analysis is not available when irrevocable decisions must be made.

Transfer is a simple direct affair and has few points of interest. We note only that, with a dictionary-based localised system, if a word or a phrase is flagged in the dictionary for re-ordering, it will always be treated thus. Also, phrases compacted locally in analysis are merely expanded again in transfer, hence preserving the 'local' interpretation.

The translation process is rounded off by a synthesis component which is meant to ensure correct agreement, inflection, etc.

As for results, let us quote one WEIDNER user: 'In another case we receive the translation *"ce brochure est conçu vous pour montrer opérer commensement la console"* for "this booklet is designed to show the operation of the console". This is certainly not very good French, but it is not all that bad either. It is now just a matter of using the word-processor functions of the terminal to change *"ce"* to *"cette"*, to reverse the words *"vous, pour"*, to correct the word *"commensement"* and to reverse *"operer comment"*. This is about as good a translation as can be expected of the WEIDNER system.'

As can be seen, there is little evidence of correct agreement and no attempt at re-ordering to suit the target language structure. The translator has quite heavy post-editing to do on the translation produced, even when the correct target lexical items are obtained.

The same user complains that the system still has trouble in disambiguating homographs, so that the word 'light' in 'the light box' will be translated by *'lumière'* or *'allumer'* rather than *'légère'*. So despite the detailed questions regarding homographs during dictionary up-date, there are still cases resisting satisfactory resolution.

To summarise all this, WEIDNER relies heavily on dictionary information, is confined to treating linguistic phenomena on a purely local basis, but nonetheless provides its users with a way of achieving consistent terminology plus an increase in productivity via the powerful text-processing environment in which it is embedded.

Significantly, users of the system see any improvement in quality as being merely a function of the size of the dictionary. They are apparently content to have to carry out extensive revision of raw output, provided that the lexical items have been supplied in some basic translated form. The fact that strange idiomatic distortions may take place does not seem to cause users problems. Where the system scores is in providing a cheap means for translators to have access to up-to-date, consistent, accurate terminology and in providing powerful wordprocessing facilities.

LOGOS

Like the previous two systems our next system, LOGOS, makes heavy use of dictionary information. ALPS was picked out because of its emphasis on several different levels of help to the user, and for its heavy use of interaction with the user. With the WEIDNER system, we concentrated on the use of the system as a way of ensuring consistent terminology and on the use of a powerful word-processing environment. The chief interest for us with LOGOS is its concentration on providing the user with powerful and easy to use tools for up-dating and extending the dictionary, plus the fact that LOGOS seems to use rather more sophisticated linguistic techniques than those found in the majority of current operational systems.

The system described here was first made public in 1982, although LOGOS is by no means a newcomer to the field of MT. Logos Corporation was founded in 1969 by Bernard E. Scott, who is its president and principal linguist. Research work, supported by the American government and by private sources, had begun in 1964 and culminated in 1971 in an English–Vietnamese system for the American air-force. Other projects dealt with English, Vietnamese, Russian, French, Spanish and Farsi. The most recent project is a German–English system. English–German has been announced for the near future.

Like WEIDNER, LOGOS is embedded into a sophisticated wordprocessing system. It runs on a Wang OS 140/145 (Office Information System), giving the user full access to the Wang's wordprocessing facilities. This includes not only the preparation of the files to be translated as well as the post-editing and formatting of the documents, but also access to the dictionaries for entering new words and modifying translations. The actual translation component receives the files in batch, and processes them in background mode. Translation speed for the Wang configuration with 256k is reported to be approximately 1000 words per hour (corresponding figures for the smaller configurations are not given).

Before turning to the dictionary component, let us first take a look at the overall system design. The system seems to be table-driven, with the tables relevant to source and target kept in separate data files. Thus, at least in principle, it should be possible to add new language pairs rather easily. However, since how these tables are organised and the way they are linked to the program as a whole remains a company secret, it is difficult to estimate the exact implications of adding new language pairs.

The basic parsing strategy is a bottom-up, left-right analysis, with one sentence being analysed at a time. Once a sentence has been analysed, even if not all constituents have been accounted for, each word or phrase is translated according to priority markers, probably based on simple feature marking.

The LOGOS documentation puts a lot of emphasis on the use of semantics, but we shall see later that the semantic features seem only to be used as a simple device to assign appropriate syntactic category in parsing, and as an aid to choosing correct translation equivalents. Semantic features on nouns seem to be checked only with respect to the governing verb and not across major constituents, with the result that, for example, relative pronouns are often mistranslated and possessive adjectives are not disambiguated.

LOGOS describes the type of algorithm employed as a mixture between an interlingual and a transfer grammar. It bases its claim to be interlingual on the fact that each string is first translated into SAL (semantic abstraction language) before parsing begins. In other words, a set of semantic features is attached to each lexical unit, and this information is subsequently used during parsing. The interlingual thesis is further justified in the documentation by 'the ability of a single-source language analysis to allow the generation of multiple target language translations'. However, the dictionary, at least the part which can be accessed by the user, clearly shows how heavily the system depends on the controlled and explicit transfer of individual lexical units, especially when a complete parse of the sentence has not been achieved and the system falls back on word-to-word translation. It seems then that the basic algorithm can better be described as a transfer approach aside from the case frames and valency patterns established for each verb – on which we assume the analysis is based – the quality of the output obtained suggests a lack of any very complex linguistic analysis. For example, a complex nominal phrase often remains unidentified and the output sentence merely translates each of the individual words, the target language result reflecting source language word order.

Let us now turn to the dictionary component (ALEX, automatic lexicographer). Before the text is sent for actual translation, an automatic scanner signals words not found in the dictionary, and the user may add new entries. ALEX also allows 'alternate' translations to be added according to different semantic domains and even allows translations to be changed according to the user's preference, via a company identity number which serves as a signal to the system to choose one translation rather than another.

Only nouns, adjectives or adverbs may be entered or modified. The user does not have access to verbs nor to other words 'of structural importance or high frequency', although suggestions for such items can be fed into the company via a customer representative. In spite of these restrictions, a great deal of information on how the system works can be gleaned from the lengthy user manual for entering words, especially by looking at the semantic information the user is prompted for as he is led through the menu.

Words, with their appropriate translations, are entered in canonical form according to word classes. The user may choose up to five subject matter codes from a list of 246 (order is significant). For each entry the user also adds semantic features, which the system prompts for either explicitly or by asking for 'semantic resemblances'.

For each new entry, based on the word class given by the user, 'alternate word classes' are generated. For example, when a descriptive adjective is entered, the form and translation of the corresponding adverb will be asked for; a participial adjective will lead to questions about a verb. The alternate word classes allow the user to specify different target expressions, thus blocking the automatic generation of incorrect forms. For example, for the adjective 'good', the user would specify that the corresponding adverb was 'well', thus blocking generation of 'goodly'. In some cases, where the integrity of the lexical data base is at risk, the user is blocked from making certain entries. An example of this is participal adjectives where the transfer would be different from the transfer equivalent of the verb the adjective is derived from, e.g., *'reizend'* for 'charming' but *'reizen'* for the verb 'to irritate'.

The extensive subclassification of nouns gives an indication of the relative complexity of the semantic information used in the system. The broad categories are

     (1) noun derived from verb and retaining its verbal sense,
     (2) abstract noun,
     (3) mass noun,
     (4) information-type noun,
     (5) measure-type noun,
     (6) human noun,
     (7) human organisation,
     (8) animate noun (non-human),
     (9) place noun,
     (10) time noun,
     (11) concrete noun,
     (12) aspective noun.

Each of the categories is further sub-divided into more specific classes, e.g. measure can be standard units of measure, continuous measurable concept or discrete measurable concept. The pattern given in the user's manual, 'measurable concept + number + unit of measure', could be an indication of how this information is used to build constituents in the analysis phase.

In order to guide the user, who is not expected to be a trained linguist, nor to have any understanding of how the system translates, a fixed number of choices are presented with an associated example and type of definition. A great deal of effort has been invested into making this component explicit, fail-safe and user-friendly. Nevertheless, as has already been mentioned when discussing the WEIDNER system, menu-driven techniques have their dangers. This point can be made more concrete here via the following example, taken from the LOGOS system.

    Select meaning of entry closest to *Stoff* in your word ——————— *STOFF*
    A. Stuff or material, i.e. mass.
    B. Topic or material, i.e. data.
    C. Substance, i.e. property of a thing, as in chemistry, physics or general.

——————————*Sauerstoff* (oxygen)
——————————*Unterhaltungsstoff* (material)
——————————*Hauptnaehrstoff* (chief nutritive substance)

As can be seen from this example, the choice the user is confronted with is not always obvious. Here, the problem is the difference between 'mass' and 'data'. The general problem, of course, is that of ensuring consistency between existing entries, the linguistic rules, and user-defined entries.

Since LOGOS takes German as its source language, it has to cope with the problem of compound words. No matter how large the dictionary, there will always be compounds not contained therein. The program and the dictionary together deal with this in a way which often provides intelligible, if not elegant output. If no target expression is found for a given compound, it is segmented into a 'head noun' and one or more 'half nouns'. The head noun is defined as 'the last noun in the compound string that is a thing'. A half noun is any German noun occurring in any position other than the final position in the compound. Half nouns may be coded with alternate transfers, providing the half noun form (which may include the infix *'s'* or *'ene'*) is not identical with an inflected form of the noun. Thus *'Gelegenheits'* could be coded with the translation 'occasional' instead of the default 'opportunity'. Each part of the compound is then translated separately. This algorithm is by no means infallible, and indeed the separation algorithm sometimes seems to miss words which probably are in the dictionary, but it does go some way to dealing with the problem of unknown compounds.

The last aspect of the dictionary we shall consider is the entry of nouns derived from verbs. As has already been mentioned, the user is not allowed to access verbal entries. However, it seems plausible that the information asked for when words derived from verbs are entered gives solid clues about the ways verbs are represented and of their role in the system.

When a noun is entered which derives from a verb and retains its verbal sense, the user is asked whether the word governs (1) a case, (2) one or more prepositions or (3) neither case nor preposition. (The same information is asked for if an adjective or an adverb derived from a verb is entered). The manual instructs the user on the difference between strong associations (valency bindings) and weak associations of the type adverbial, genitival and agentive. The example cited is the sentence 'The search by the children of the closet for dad's slipper', where the verb 'search' is described as having two 'semantic hooks' plus an agent, 'the children'.

Word	Object No.1 the thing sought	Object No.2 the thing searched
Verb		
search	'for' the x	the y
Noun		
search	'for' the x	'of' the y

The explanation continues: 'This example is more important than the "house on the hill" example above, because it compares two genuine semantic relationships of different strength. "Of" with the noun "search" is strong because the object of "of" is the object of "search" when "search" is a verb.

"Of", nevertheless, is decidedly weaker than "for". "For" stays with "search" throughout the metamorphoses of "search" between the noun and verb states. "Of" is only present when the verb takes on the garb of a noun. It is an example of a weak semantic association. . . .'

The quotation has been much abbreviated, however enough has been said to suggest confirmation of the hypothesis alluded to earlier, that the underlying linguistics is a simplified case system, where each verb has a fixed number of arguments whose slots are filled in accordance with tests on semantic features and syntactic information provided by the word classes of the lexical units.

This somewhat lengthy discussion of the automatic lexicographer has served, we hope, to provide also some understanding of the strategy embodied in the LOGOS system. It has also illustrated how heavily the system relies on lexical transfer and how well developed the dictionary component is. The primary means of achieving improvements in LOGOS seems to be through user interaction with the dictionary.

TITUS

The last system we shall discuss in this chapter is quite different from all those examined so far, both because of its function as only one part of a larger system, and because of its reliance on restrictions on the input to the system in order to produce correct translations. Let us consider these two points in order.

Machine translation is often thought of as a stand-alone process, where the primary aim is to translate texts. However, we should keep in mind that the possible implications and applications of MT are extensive, and that an MT system can be conceived of as one small, but necessary component of some larger information system. In Europe especially it is becoming a *sine qua non* of any modern information system that it is multilingual. A database which stores vital information on some aspect of science and technology can only be fully exploited if users can query the database in their own language (or at least in more than one language). The information explosion also makes it impossible to keep up with primary sources even within a single discipline. Hence, in recent years, greater attention has been paid to abstracts as a means of providing a concise version of the salient points of a document.

Abstracts are commonly written in some standard fashion that may be more or less controlled, depending on a variety of factors. We may therefore consider them as examples of restricted syntax documents, and note that some of the most successful MT systems owe their high performance to just this feature, i.e. a constrained syntax for input texts results in high quality MT. The best known of these systems are the SYSTRAN version used by Rank-Xerox, and TAUM–METEO.

It is therefore not surprising, given this introductory background, that someone had the idea of conceiving an MT system, or rather a series of systems, based on abstracts. The system is called TITUS, was developed by the textile industry and is operated by the Institut Textile de France.

We should immediately point out, however, that TITUS is not just an MT system. Indeed, it is true to say that MT is not even its primary purpose. It was set up in 1969 (TITUS 1) to satisfy a growing need within the textile industry for multilingual access to abstracts. The successive versions of TITUS are therefore as much fullscale abstract-indexing search and retrieval systems as they are MT systems.

What the textile industry required, then, was multilingual access to abstracts. The design adopted was not one of translating all abstracts as they arrived, indexing and storing them for later retrieval. Rather – and here the system was somewhat innovatory, given that the design decisions were taken in the early 1970s – each abstract was to be translated only once, then converted into an interlingual representation. This representation could then be stored in a form much reduced compared to the original abstract, and used for several purposes. The original input could be reconstituted with accuracy of meaning; equivalent abstracts could be generated in any of the languages of the system (French, English, German, Spanish); the contents of the representation could be searched, and the results (sets of abstracts) output in the language of the user's choice. Thus we have a multilingual information storage and retrieval system, based on interlingual representation of abstracts, where the input may be in any of the languages of the system, as may the output.

Before going into details concerning TITUS, let us dwell for a moment on the consequences of employing an interlingual approach to MT. As is well known, there was high enthusiasm for the interlingual approach amongst early workers on MT. The advantages over a transfer-based approach, for example, are that, with a (good) interlingua, the number of modules to be written is sharply reduced, and that the addition of a new language is achieved with little effort, namely just the construction of one analysis and one synthesis module. Early research workers were optimistic about finding a universal interlingua. Proposals were made to use, for example, Esperanto, whilst other proposals involved using one of the languages of the system as a switching language, or using systems based on semantic primitives, where the representation would be devoid of all surface syntactic information, hence giving a 'pure' meaning representation. However, these early attempts met with failure. There were just too many problems involved in finding one universal interlingua which would represent 'pure' meaning.

Even today, there are no serious claims made in this respect. Any interlingual systems that are around are either severely constrained, or are derived mainly from one language using a few semantic primitives which are then imposed on other languages for translation purposes. More and more, we will find systems being developed that will consist of several interlinguas, one for each specific group of languages, or indeed for each text type or subject field.

Where then does TITUS fit into this picture? We have stated that it is interlingua-based, and that it operates on abstracts which are normally written in a standard fashion. Herein lies the strength of the system: abstracts for input are in fact re-written by a human analyst in a severely constrained manner, using a restricted syntax (controlled syntax) and also a limited vocabulary. In TITUS II, this work was done completely manually. TITUS IV, version A, accepts input sentence by sentence in an interactive mode, and will prompt the analyst to resolve any ambiguities. It effectively allows a slightly less restrictive syntax.

In TITUS II, abstractors had to draft abstracts in LDC (*Langage Documentaire Canonique*), a restricted language from the point of view of syntax and vocabulary. TITUS II vocabulary aimed for a perfect correspondence of 'concepts' as opposed to word-for-word equivalence of terms, since the concept may be expressed in a different language by a simple word or by a whole phrase. This expression of a concept is called UL (*unité lexicale*).

In TITUS IV, version A, analysis is still performed by TITUS staff, but LDC has been abandoned for the interactive approach mentioned earlier. Documents received at the ITF, written in any one of the four languages, are interactively input in accordance with the rules of a new restricted language (a modified form of LDC).

Any divergence from these rules, or any ambiguities, are resolved by the operator. Abstracts are entered sentence by sentence, each one being validated before entry of the next. Syntactic constraints are severe in TITUS IV, in the first version, A: for example, only simple sentences (no subordination) are allowed. In the next version, for some pairs of languages (for example, French–English) problems like relativisation and subordination will be tackled. (A version allowing for subordination has already been demonstrated).

The basic model of the sentence according to TITUS IV is shown in figure 12.1. Only the first GN (subject) is compulsory. All the rest is optional. The GNC1, GNC2 and GNC3 are optional, with up to three allowed. The verbal group (GV) is optional, and contains a verb in active or passive voice. GVC indicates the complement of the verbal group, and is also optional. Constraints concerning the verb in TITUS IV are strict. For example, verbs may only be written in the third person, and the English progressive is not allowed. The overall organisation of the system is shown in figure 12.2.

Now let us turn to the lexicon. The lexical component of TITUS is made up of:

(a) a set of *fixed multilingual elements,* namely, prepositions, conjunctions, determiners, auxiliaries, adverbs of quantity and of negation which are not in the lexicon itself but in the software;

(b) variable elements listed in the multilingual lexicon. The basic element is still the *unité lexicale,* which may be:

UL substantives
UL adjectivals
UL verbals
UL adverbials.

With the ULs, there may be 'scope notes', which are used to indicate polysemy, and thus to trigger interaction with the operator where necessary. For instance, *contrôle* has more than two equivalents in English (and is also a homograph). In these cases, the system asks the operator which meaning is required and the problem is solved interactively. The same interaction solves polysemy

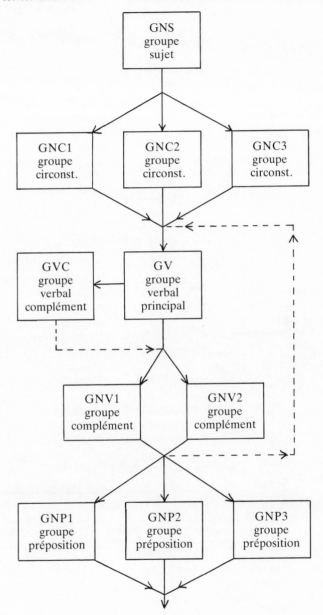

Figure 12.1. TITUS IV: the basic model of a sentence

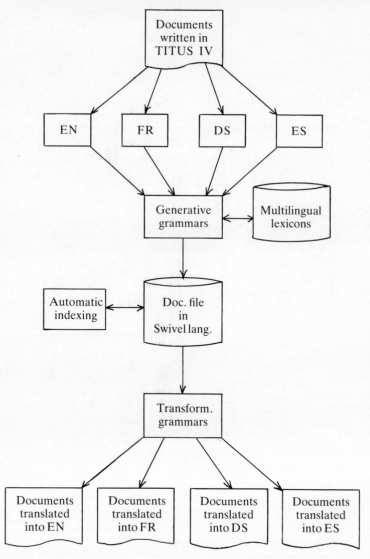

Figure 12.2. The overall organisation of TITUS IV

problems which might also confuse noun/verb or adjective/adverb distinctions.

As we have seen, TITUS is an interlingua-based system, using interaction during input to check the controlled syntax and vocabulary on which the system relies for correct translation. However, it should not be forgotten that the primary purpose of the system is not to produce first-rate MT, but rather to provide an information service.

Conclusion
This chapter has tried to give a fairly brief overview of some of the systems currently operational, picking out specific characteristics of individual systems in order to emphasise their use and importance in modern MT systems.

Acknowledgement
The author would like to thank John McNaught, CCL, UMIST for his helpful advice concerning termbanks.

SYSTRAN

The inclusion of SYSTRAN in this context reflects a radical change in attitude to MT in general and to this system in particular. It is not so very long since SYSTRAN's lack of academic pretension would cause it to be ignored. How, after all, could one take seriously a system which, if it did not produce the apocryphal sentence about steak and vodka, certainly did translate *La Cour de Justice envisage la création d'un cinquième poste d'avocat général* as 'the yard of justice is considering the creation of a fifth general avocado station'?

The fact that SYSTRAN is included indicates, perhaps, a growing understanding that what is really needed to bring MT to operational fruition is not so much theoretical investigations into the fundamental structure of language, important though such investigations are in an academic context, as the design and production of systems which will actually produce a translation, however flawed, which can be handed to a real customer in a real working situation. This I suggest would seem to explain the rapid rise to commercial acceptance of systems such as WEIDNER, ALPS and LOGOS, systems whose whole philosophy is to produce something, anything, on which a customer can work.

It is probably unnecessary to go into the pre-history of SYSTRAN, as other chapters are covering the topic. Suffice it to say that P. Toma worked in La Jolla, in the 1960s, at his dream of machine translation, and finally achieved his breakthrough when SYSTRAN was bought by the American air-force to translate from Russian to English. He then turned his attention to English–French, with an eye to the Canadian market, and it was this system which the Commission of the European Communities purchased in 1975, after first ordering a study of all the systems available worldwide, subsequently published as 'Handbook of machine translation and machine-aided translation' ([114] Herbert Bruderer 1977).

Having bought English–French, a system of some 28000 lines of programming wth a dictionary of about 7000 entries, and having started development work on it, the Commission next purchased French–English in 1977 and English–Italian in 1978. Today our SYSTRAN system amounts to some 100000 lines of programming, and the dictionaries for each language-pair amount to about 70000 single-word entries and 35000 multi-word expressions.

Following several years of continuous development, these three language-pairs went into production in 1981, and a translator receiving a job to do in any of the three language sections concerned is at liberty to request a SYSTRAN raw translation of it. At present, the production operation is centralised; in the future it would be in the hands of the translators themselves. For the present, the source-language text is input on a Wang wordprocessor, batched up with other texts for the same language-pair, and sent down a dial-up line to a bureau-service IBM 370 operating on MVS. It may be worth mentioning here that SYSTRAN requires about one megabyte of central memory and about 35 megabytes of virtual memory, and that we transmit to the computer at 1200 baud. We could go faster, up to 9600 baud, but find that we do not need to.

Since we do not use a dedicated computer, there is now a pause of some minutes, typically four or five, and then the translation comes back up the telephone line. Actual estimates of how fast SYSTRAN translates seem to vary from week to week and with whom you ask, but an order of magnitude figure would be 150000 words per hour. In any case, in our sort of operation – it is no doubt different for a setup such as the USAF translating millions of words for information scanning only – such figures are of purely academic interest. All that concerns my production team is that they should come back and have a look at the terminal about a quarter of an hour after sending off the text, while all that concerns the translator – who has plenty of other work to be getting on with in the interim – is that the SYSTRAN version will arrive sometime during the morning or afternoon in which the input was finished.

The raw translation is then passed to the translator, either on paper via a high-speed matrix printer or direct on the wordprocessor screen, for post-editing. The polished text is reprinted and leaves the translation service to gladden the heart of some bureaucrat somewhere in the labyrinthine Jean Monnet building which is the headquarters of the Commission in Luxembourg.

Our English–German system is just about ready to go into production in the same format, while French–German should be operational in about mid-1984.

However, the story does not stop with a satisfied requester. For the linguists on the SYSTRAN project, it is only just beginning. We automatically receive back the translator's corrected version and use it as the basis for our work of incorporating improvements into the system.

To understand the sort of improvements which we make, let us look at the way SYSTRAN works. Underlying the whole process is the dictionary, or rather a pyramid of dictionaries of increasing complexity. The first is the Stem, which is basically a simple word-to-word dictionary, with the source word on one side, rudimentary syntactic and semantic information in the middle, the target translation in the next column and target syntactic information on the right-hand side.

And so SYSTRAN starts by looking up all the words in a dictionary. Here, it seems, we have one of the fundamental differences between machine translation, at least as SYSTRAN embodies it, and 'real' translation. For while some machines do actually reproduce a human activity, where translation is concerned the best a computer can do is to simulate the *effect* of the activity. Even when the result is perfect, and is exactly what a human translator would have written himself, it is achieved by a very different route to that taken by the traditional translator.

Setting about its task in a manner quite unlike a real translator, therefore, SYSTRAN first looks everything up in the Stem. The first stage of lookup is handled by the subroutine LOADTXT, which scans for all the words which will need a Lexical routine, by comparing them against LEXTBL. When one is found, LOADTXT writes the offset address of the routine into a couple of bytes of the word, for use later on by the program LEXICAL. We shall return to this later.

LOADTXT also executes the lookup of the High-Frequency dictionary, which contains the rigorously unambiguous high-frequency words such as 'and', 'but', letters or numbers acting as enumerators, punctuation – included because analysis may be helped by knowing that there is a punctuation mark present – and fixed idioms such as *à cet effet* or *à ce propos*.

HF words are loaded with all the grammar information stored for them in the HF dictionary, then taken out of the list of text words, in the interests of speeding up the Main Dictionary Lookup.

Also at this stage Idiom Replaces are recognised. An Idiom Replace in SYSTRAN parlance is a multi-word expression which is easier to handle if it is treated as one word. Given a complex idiom such as *dans le même ordre d'idées*, for example, SYSTRAN could waste its time analysing *dans*, and that this is modified by the adjective *même* (having first disambiguated *même* to be sure that it isn't an adverb), and that *ordre* in its turn governs *de*, but how much simpler to have a procedure which transforms these six words into a single one, namely '*dans . le . même . ordre . d'idées*' and enters this single word in the Stem dictionary as an adverb, with some translation such as 'similarly'.

This device is also very useful for dealing with standardised phrases – a lot of our bureaucratic work is standardised — where the target version is equally standard but syntactically quite different from the source side. An example is the phrase *Les entreprises devront soumissioner conjointement ou solidairement au sein du groupement qu'elles pourront former,* which is transformed into the massive single word '*Les.entreprises.devront.soumissioner.conjointement.ou.solidairement.au.sein.du.groupement.qu'elles.pourront.former',* and entered in the Stem with the quite different translation: 'Consortia must have joint or several liability.' Here is a case where the system is indeed genuinely aping the translation process, because a Commission translator, seeing that standard phrase after a couple of times, does not translate it either, he simply rattles off the equally standard English phrase.

At the point we have reached here, LOADTXT is busy recognising such groups of words and transforming them into the dummy new word which, in its turn, is put back in the list of Low Frequency words for the further Stem lookup.

Now follows Main Dictionary Lookup, or MDL. In systems with highly inflecting source languages, such as French, German or Russian, morphological variants are handled by tables of endings allowing MDL to recognise all possible forms of verbs, adjectives and nouns which may occur.

It is important to understand that at this stage of the look-up the longest possible entry has priority. The word *durent,* for example, is not a stem, so SYSTRAN drops a letter off and tries again. '*Duren*' isn't a stem either, drop another letter and try again. And again. '*Dur*' looks at first as if it could be an adjective, but '*ent*' is not an acceptable adjective ending. Thus the present tense verb is taken. What has not been taken, because the system stops as soon as it is satisfied, is the possibility of the verb being the past historic of *devoir.* This possibility has to be covered by more complex dictionary coding, which we will come to later.

Furthermore, where two entries are of the same length, such as *dure,* third person singular present verb or feminine singular adjective, the one whose part of speech code is lower will be taken by default, and the alternatives handled in the homograph routines.

Also at this stage a dummy acute accent is added in most cases where a letter 'e' occurs in a Not-Found-Word (NFW), and the word thus created is rechecked against the Stem. This feature was designed to cover the case of capitalised French words, in which the accents do not appear. It could be extended to other letters and other accents, but the e-acute is by far the commonest problem, and the solution has given very worthwhile results.

MDL goes right through the words in the text, adding the grammatical information from the Stem dictionary where it can and dumping the words it can't find into the NFW list.

Treatment of NFWs breaks down into several steps, the first being the treatment of hyphenated words. The first stage is to strip out the hyphen and look again, as it is quite possible that 'carpark' is in the dictionary but 'car-park' isn't. This feature was originally designed for use with English-source texts, but has proved to be equally useful with French.

The outline then splits up the remaining hyphenated words into their component parts and looks up the separate halves in the Stem dictionary. The two halves are translated as the two different words, still hyphenated together, but take the part of speech, gender and number of the rightmost part. If one or both of the words are still not found, they remain in their source-language form like any other NFW.

Words which are hyphenated because they are questions, incidentally, are dealt with in two stages. Pronouns attached to a verb by a hyphen, with or without the euphonic '*t*', are first stripped off the verb, and dealt with as a separate Stem entry. Question Routine to handle the actual syntactic processing of the questions comes later, after the homograph resolution stage.

The words still remaining are compared in NFWRTN itself against a table of endings, and when a match is found, part of speech codes, number, gender, person or tense are given to presumed nouns, adjectives or verbs. This basic information will help parsing even though the words will have no meaning attached to them and appear in the TL text in their SL form.

However, when translating from English to French or vice-versa, the SL form may in fact be identical to the TL form, with the result that it does not matter to the post-editor that he or she may be looking at an unfound French word in an English text – if the letters are all the same, for post-editing purposes it is an English word! And vice versa.

As an added refinement of this characteristic of French and English, certain standard endings on NFWs are transformed into corresponding standard suffixes in the target language. SYS-TRAN French–English, for example, not finding the word 'radiologue', would strip off the ending, add a corresponding English ending to give 'radiologist', and hope for the best! Being an optimistic little creature, SYSTRAN would also add some semantic information – in the case of '-ogue' to '-ogist' it would be the codes for 'human, profession'.

In this particular case, the subroutine would have worked brilliantly, the post-editing translator wouldn't even notice the difference, and the linguists who devised the scheme would grin smugly.

There would be less enthusiasm, however, if it had tried the same dodge on the word *pirogue,* a canoe, and decided that it was the human profession 'pirogist'. But as ever with SYSTRAN, if it gets it right more often than not, it's worth doing!

Similarly, NFWRTN will detect NFWs starting with figures and ending with the sequence *'ième',* or even a single *'e',* recognise the word as an ordinal numeral and add the appropriate target-language ending. (Handling of numbers in any form is also part of the NFW routine.)

Even when these various attempts at 'guessing' the identity of a NFW have been made, the word still carries the information that it was not actually found, which will be useful in resolving certain later ambiguities: for example, if it is recognised that the antecedent of a pronoun is a NFW, and various other checks are positive, it is quite likely that the antecedent is in fact a proper name, thus resolving the 'who/which' 'he/it' problem.

Other NFWs, to which none of these operations can be made to apply, receive a null part of speech.

When all the words of the text have been compared against the Stem, they are sorted back into their original sequence, now with the dictionary information added to them.

Up to this point tests have all been based on character matching, no other information of any sort being yet available.

So that analysis proper can start, the program GETSENTN takes the sentences one by one and puts each one in turn in the analysis area, together with all the information from the HF and LF dictionaries including the all-important homograph area attached to each word.

Analysis itself starts with LS LOOKUP. LS (Limited Semantics) expressions are noun groups such as 'Institut Dalle Molle', 'Palazzo dei Congressi', or 'Villa Heleneum'. It is LS lookup that allows us to have *assiette de vol* rendered as 'attitude', *assiette d'approche* as 'approach attitude', but *assiette d'impôts* as 'taxation basis'. If SYSTRAN had just used the Stem, they would all have come out as some sort of dish!

For each word which was marked in MDL as potentially being the principal word of an LS, the program scans all the appropriate LSs, looking for a character match. Where two expressions overlap, the longest match will be chosen, which is one of the things a beginner dictionary coder tends to forget.

The next stage is the program HOMOR. At dictionary look-up, each homographic word in the text was marked as such, and HOMOR now looks from left to right down the sentence for all the words having the homograph notification byte not equal to zero. When one is found, HOMOR calls the appropriate subroutine (its number being equal to the value found in the identification byte).

The HMRTN thus called resolves the homography according to syntactic context, based on all the syntactic information drawn from the dictionary (nothing else being yet available).

The SYSTRAN English-source systems have over eighty different homograph routines, since over fifty per cent of all words in English are homographic. The French–English system has almost seventy, as the homograph problem in French is slightly less – although considerably more than we expected to find before we started work on our French–English system. These routines vary in length and complexity, and in the range of their applicability – the longest is of the order of 150 lines of programming, the shortest being the delightfully concise 'Bump to the end of the sentence and if the last word is a question mark, conclude that the homograph under investigation is an interrogative adverb and not a subordinate conjunction' – but their structure is always the same: hundreds or even thousands of questions about the syntactic context of the word to be resolved. These questions, and the conclusions to be drawn from the answers, are written in a macro-language specific to SYSTRAN.

The routines work by looking both to the left and to the right. Looking leftwards is safer, on the assumption that all preceding words have already been correctly resolved as to part of speech (not, in fact, a safe assumption) but looking rightwards, while more demanding, tends to give better results.

The routine thus looks from the Current Word to the one on its immediate right. If that word is not a homograph, it then becomes possible to determine what parts of speech the Current Word *may* be, on the basis of syntactically admissible combinations, or rather, strictly speaking, what it *can't* be, on the basis of syntactically impossible combinations. Gradually the various combinations which remain possible are whittled away, until all being well only one is left.

If the Current Word is 'la', for example, and the next word is unambiguously a verb, such as *chantons,* then *la* must be an object pronoun, as the combination 'article – verb' is impossible. Conversely, if Current Word + 1 is unambiguously a noun, then *la* must be an article.

If Current Word + 1 is itself a homograph, however, such as *porte,* then SYSTRAN cannot yet decide, and has to look further. The same sort of tests are carried out on Current Word + 2, which we hope will enable Current Word + 1 to be disambiguated, in turn allowing the correct resolution of *la.*

The essentially pragmatic nature of the SYSTRAN programs should be emphasised. They might make an academic computational linguist shudder, but they work. They aren't trees, they have no taxonomic structure, they don't rely on Artificial Intelligence, they simply run up and down the sentence like a kitten on the keys, asking themselves the childishly simple question 'What might we find here in an everyday sentence in the real world?'

When it has finished its homograph resolution, HOMOR stores in the homograph area the alternatives which it has rejected, in case SYSTRAN wants to come back to them during one of the later stages and change its mind.

At this stage HOMOR transfers certain syntactic information found in the morphology tables into some of the bytes reserved for synthesis.

HOMOR also resolves any NFW which is capitalised, making it a proper noun, masculine singular, with the semantic code 'geographical location', and also resolves as a year any number of four digits starting with 19.

A major change was introduced about two years ago, with the creation of two new types of dictionary expression, the HLS, or Homograph Limited Semantic expression, and PLS, or Parsing Limited Semantics expression.

Broadly speaking, whereas LSs and CLSs are concerned with obtaining the correct meaning on the target side, HLS and PLS expressions are designed to handle difficulties on the analysis side. As such, they deal with particularly awkward or specific cases, often in fact the hundredth case where the source-language strays from the rules it follows in the other ninety-nine.

For example, when HOMOR is testing a French word that

might be a noun, one of the tests it can apply is whether the word to its left is a verb. If it is, the word under test cannot be a noun. *Je prends son* cannot possibly mean 'I take sound,' it has to be 'I take his . . .' A noun in French *cannot* be preceded directly by a verb. And then comes the hundredth case: *Je prends note.* Rather than coding this one oddity into the homograph routines themselves, however (some of them by now being very complex), it is preferable to write an HLS expression on *note.* These expressions override homograph resolution programs, and in effect say that just this once the rules don't apply. Observe, however, that no meaning has to be given in this expression to *note* (although it can be). All that the expressions has to do is to get the part of speech right for subsequent analysis stages.

In order to be able to override the homograph resolution, HLS expressions are called immediately *before* the program HOMOR. Parsing expressions, on the other hand, can be called at any one of five points of entry; immediately after HOMOR and before the first structural pass through the sentence; before any of the following structural passes; or after the last one but before the transfer stage.

A case where a PLS might be used is with a construction such as *éviter que l'argent soit dépensé,* which has to be rendered not as 'avoid that the money be spent', but 'prevent the money being spent'. Such a radical restructuring could be handled by analysis in the conventional sense, but would be unnecessary clutter for such a specific case. The new development made it possible instead to handle it by writing a dictionary expression, in which one single line of programming can check whether *éviter* governs *que,* delete the *que* from the analysis, and change the subordinate verb to a gerund. Once again, it should be emphasised that no meaning need be given, and indeed that the meaning (not the syntax) of the words in this expression are still open to modification if a CLS expression called subsequently finds an appropriate match. (Which will in fact be the case – further down the line, while leaving the grammatical structure untouched, a CLS will change *argent* governed by *dépenser* from its Stem meaning of 'silver' to a less poetic but more correct 'money'.)

Homograph resolution is followed by a series of structural passes, progressively establishing the syntactic structure of the sentence.

PASS–0, scanning from left to right, undertakes the definition of clause boundaries, which has to be done at this early stage in order to establish the syntactic relationship within the clause and to determine the limits of the subject/predicate search. On the other hand, it cannot be done any earlier, before HOMOR, because the words delimiting the clauses, such as *que,* are themselves homographic.

It places a marker at the last word of each clause and pointers on either side of embedded clauses to be skipped when parsing the embedding one, and indicates the clause type in one of the bytes of each word of it.

PASS–0 also branches into different subroutines according to whether the current word (CW) is a relative pronoun, comma, dash, parenthesis or conjunctive adverb, and seeks the rightmost limit of the clause which this CW is introducing.

There is also a special subroutine for subordinate clauses, called when PASS–0 finds a word which is flagged as being a clause opener. Here we see again the all-pervading importance of the dictionary, because the type of clause which a conjunction can introduce has to be coded explicitly when the conjunction is entered in the Stem, and if this information is missing in the dictionary, the flagging byte will be empty and PASS–0 will incorrectly resolve the clause boundaries, with disastrous consequences for the rest of Analysis!

PASS–1 scans from right to left, establishing basic syntactic relationships between modifying words.

Relationships are set up between a noun and its adjective, by writing the sequence number of the noun into a specified byte of the adjective and the sequence number of the adjective into the corresponding byte of the noun, between a preposition and its object, by linking two other bytes, two nouns by linking two more, a relative pronoun and its antecedent by linking two more, and so on.

PASS–1 remembers the part of speech encountered by setting switches on and saving the location of the word. The switch is turned off when no more relationships can be set with a word.

On the other hand, all switches are turned off whenever a relative pronoun, subordinate clause, comma, or coordinate conjunction is encountered, to avoid the risk of setting syntactic relationships across different clauses. If the comma or the conjunctions are not in fact clause separators but enumeration links, PASS–2 will recognise this and correct the relationship.

PASS–1 also sets translated a number of features such as auxiliary verbs, articles, adverbs of negation, adverbs of comparison and so on, and instead puts the information they carry into the main word itself, marking the verb as negative, the adjective as comparative, the noun as modified by a particular type of article, and so on.

All this information will subsequently be interpreted by the

SYN(thesis) program, which will store the information in the bytes 40 to 50 of the target language equivalent, for use by the various subroutines which follow it.

In addition, PASS–1 detects capitalised words which are being used as proper nouns, and keeps them in their source-language form. Mrs King may remember early SYSTRAN translations of the minutes of meetings she had attended, when she came out as *'Madame Le Roi'*. The head of translation at the Council of Ministers, Mr Duck, of course used to be *'M. Canard',* and the head of the interpreting service for Commission and Parliament, Renée Van Hoof, used to be rendered as *'Madame Sabot de Camionnette'*. There are those who feel that SYSTRAN lost something when it stopped making such charming howlers.

After PASS–1 comes, logically enough, PASS–2, also from right to left. This one deals with enumeration, looking for commas or coordinate conjunctions which are not acting as clause separators (those which do separate clauses, you will remember, were identified a moment ago by PASS–0). When one is found, PASS–2 looks at the words to the left and right of it and branches to different subroutines, depending on the parts of speech it has found.

When words are detected as being in enumeration, on the basis of these part-of-speech codes, syntactic codes, or semantic codes, the pass puts this information into two interlinked bytes of the words concerned. 'Road and bridge construction', for example, will thus be correctly resolved as meaning 'construction de routes et de ponts', and not 'route et construction de ponts', as the correct inference will be drawn from the fact that 'road' and 'bridge' are both coded 'concrete' whereas 'construction' is coded 'abstract'.

PASS–2 also extends the government relationship established by PASS–1, so that each word of the enumeration is pointing to the same syntactic modifier. Whereas PASS–1 could establish only a relationship between 'fried' and 'eggs' in the phrase 'fried eggs and bacon', at this later stage an adjective/noun relationship is also set up between 'fried' and 'bacon', because by now it has been detected that 'eggs' and 'bacon', both semantic-coded 'food product' are in enumeration with one another. To the standard objection: 'What would happen if the menu had "scrambled eggs and bacon",' we reply that 'scrambled eggs' would already have been detected as an LS, or noun phrase, several stages ago.

Next comes PASS–3, which undertakes the subject/predicate search. This one runs from left to right, scanning each clause separately, skipping embedded clauses to come back to them later.

First, it searches for predicates, looking for words which have bit

10 on in byte 12 (this information having been provided from PASS–1) and writes the sequence number of this predicate into the appropriate byte of each word of the clause.

It used to be regarded as one of the weaknesses of SYSTRAN that it would try too hard to find a verb, and interpret as a sentence what was in fact only a string of nouns. Having recently seen the limpidly clear sentence *Tout est fini* rendered as 'Any finished East', I think one may safely say that this weakness has been over-cured!

Provided a predicate has been found, PASS–3 then searches for the subject in the clause, which it defines as a noun or a pronoun not functioning as a complement of any type (again, information on complements was provided from PASS–1). The first subject of the clause is indicated in the appropriate byte of each word of the clause.

Normally, of course, PASS–3 looks leftwards for the subject, since that is where subjects are to be found, but it also has to cover the cases in which the subjects may be on the other side, namely questions and subordinate clauses starting with the verb.

Last of the structural passes is PASS–4, which has three main goals: first, it carries out an operation which SYSTRAN calls generalising relationships, but which would be more commonly called searching for the deep subject and deep object and marking these in the appropriate bytes, for subsequent use by CLS expressions; secondly, it decides on gender and/or number of ambiguous words in the source-language, on the basis of the other words now detectable as agreeing with them.

Lastly it tackles the thorny problem of the government of prepositions, giving first priority to any word coded as actually requiring the preposition, second priority to a verb which might be governing the preposition, and third priority to the noun closest to the preposition. Prepositions are divided into three different groups, depending on their syntactic behaviour and each sub-group has a slightly different routine.

Special problems are posed by *y, en,* and *dont* (problems quite apart from their being homographs in themselves), namely that they hide within themselves another preposition, *à* and *de* respectively. Several bites are needed at these particular cherries: here in Pass 4, later in the PREP program and also in individual lexical routines which we will come to later.

Where the words on either side of a preposition are in the same LS, the government is resolved as being with the word to the left, even if the normal resolution, in the absence of an LS would be the opposite. Thus SYSTRAN will correctly translate *Il parle à la femme de ménage* as 'He speaks to the charlady', and not as it would in the absence of the LS *femme de ménage* as 'He speaks to the woman about household.'

A further refinement is that if two prepositions are identical, they will *not* be resolved as governed by the same word, unless the words they themselves govern have already been resolved as being in enumeration.

Thus in: *Il donne à sa petite amie de jeunesse l'ancienne lampe à gaz,* the words *amie* and *gaz* are not in enumeration (since SYS-TRAN fortunately does not know that a former girlfriend may also be an old flame!), and therefore the sentence will be correctly resolved. On the other hand, in: *Il donne la lampe à sa femme et à sa soeur, femme* and *soeur* are in enumeration, and therefore the second *à* is resolved as depending on *donne.* It is also, in the interests of elegant English, deleted at the Synthesis stage: 'He gives the lamp to his wife and his sister.'

After the passes comes CLS LOOKUP. It is at this point that the influence of the dictionary coder becomes most apparent. Several stages ago, we had LS lookup, lookup of noun phrases consisting solely of two or more nouns, or of nouns with adjectives. But in each of these cases, the constituent words have to appear in the text in exactly the sequence in which they have been coded in the dictionary. *Commission Européenne, poignée de porte,* and so on. Cases where the words are not contiguous are simply not recognised, so that in a sentence such as *L'assiette est tombée pendant le vol* the system will not make the mistake of thinking that this is one of those cases where *assiette* means 'altitude' – here it is indeed a dish.

On the other hand, however, the sentence *Le pilote a noté l'assiette et la direction de vol* will not be correctly resolved, as 'direction' comes between *assiette* and *vol.* This is a case where a simple string of contiguous words will not do, and where the coder should have written a more complex expression, rather than a simple LS, to say that if *assiette* and *vol* are in an adnominal relationship, then *assiette* should have its special aviation meaning.

Such discontinuous phrases are handled by what SYSTRAN calls CLS or Conditional Limited Semantics expressions. Broadly speaking, these are expressions to which conditions are applied – 'if this . . . then that' – the conditions being expressed by the bits and bytes of the words concerned.

It is in CLS expressions, too, that phrases involving verbs are handled. *Accuser,* for example, with the surface object or deep object *reception,* will be given the meaning 'acknowledge' instead of 'accuse'. The theory being that a sentence such as *J'accuse la Réception d'avoir perdu le paquet qui y a été déposé,* will never occur. Which is fine until you have worked for a large bureaucracy for a time . . . On the other hand, *accuser* with the object *hausse* or *baisse* will probably be turned right round to give the translation 'to be up', or 'to be down', respectively.

Incidentally, such expressions also work across relative pronouns: *'La hausse que les prix ont accusée en 1983 . . .'* will be correctly matched.

CLS expressions are also used to give a different meaning depending on semantic coding (which may surprise those academics who have always maintained that SYSTRAN doesn't have any semantics.) Thus *inscrire* where the object is coded 'financial' means 'to enter on a budget sheet', and not to 'inscribe'; the adjective *brut* modifying something coded 'financial' means 'gross' and not 'rough'; *marcher* where its subject is coded 'device' means 'function' and not 'walk', and the noun *cadre* in enumeration with something coded either 'human' or 'profession' means 'manager' and not 'frame'!

Other minor things that can be done with CLS expressions include giving a different meaning if the word starts with a capital letter: *Commission* with a capital *C* is always to be translated 'Commission' and not 'sub-committee'; a different meaning depending on the position of the word in the sentence: *ainsi* as the first word means 'thus' and not 'in this manner', and so on.

Both on this tidying-up level and with the more significant expressions, it is the CLS expressions which turn the output from what is evidently a translation into something approaching the appearance of a target-language original.

The next subprogram to be called is PREP, which handles the translation of prepositions according to any specific or non-standard meaning given at the coding stage. 'About' when governed by 'talk', for example, is not *autour de,* but *au sujet de,* but since only one meaning can be given for each entry in the Stem (in this case *autour de*) any alternatives such as this are fixed by preposition coding on the word 'talk'.

Next come lexical routines. A number of words (40 in French–English, 115 in English–French and 120 in English–Italian) behave in such a complicated way that they have to have their own little subroutines all to themselves. Examples are: 'Any', which might be translated in French as *un peu, quelques, pas de, en, tout, n'importe quel,* and no doubt several other possibilities as well; or 'in', as in the *au Canada/en France* problem.

Some routines are permanently loaded in central memory, because they will be called frequently, having been written for categories of words rather than individual words. Examples are: the FAIRERTN, which is called for all infinitives preceded by *FAIRE* to handle the difference between *'faire pleurer l'enfant'*, *'faire marcher la machine'*, and *'faire aux gens voir la différence'*, or the NATIONRTN, called for any word of nationality, to ascertain whether it is a human being or a language.

Other routines, however, are loaded only when a word requiring one of them is encountered. Examples of such routines are: the ILRTN, called only when *il* or *elle* is encountered, or the IGNORRTN, called only when some form of the verb *ignorer* occurs.

The routines themselves work by scanning the syntactic context of the word under investigation, asking questions about what is found. For example, the routine to deal with the construction *il y a, il y aura,* etc., is called when the system encounters a third person form of *avoir* in any tense. The routine then asks

'Is the subject *il* or *-t-il*' (for questions).

If no, exit.

If yes, is the verb modified by an adverb?

If no, exit.

If yes, is the adverb *'y'*?

If no, exit.

If yes, go to the verb's object. Is the object coded 'Time Period'?

If yes, add the word 'ago', and delete all meaning for the words *il, y* and *a.*

If no, translate as 'there is/there are/there won't be . . .'

We are now on the home straight, the synthesis of the target language. The program which carries this out is called, astonishingly enough, SYN, and runs from left to right.

Words which have been set translated by any previous program do not go through SYN. The rest go through various subroutines, depending on their part of speech, and the SYN program will put synthesis information into the bytes which as far back as HOMOR were reserved for this purpose.

If an earlier program (such as LEXRTN or a PLS) has already filled those bytes, the word goes out of the subroutine immediately.

The program VERBR decides on the voice, number, person and tense of TL predicates, based on subject, context and dictionary codes, and handles reflexive forms, while NOUNR handles articles and also checks whether the noun-to-noun relationship set by PASS–1 is still valid, as it may have been cancelled by subsequent analysis stages.

If the relationship is kept, the French preposition is set translated, and one of the subroutines of the REARR program, coming shortly, will cause the nouns to be reversed – *lampe de table* becoming 'table lamp', for example.

If the relationship is *not* kept, having been cancelled by a coding specifying a preposition or a genitive instead of rearrangement, or by SYN itself on the basis of semantic tests (such as a 'container' with a 'food product'), then the words are not rearranged, and *verre de lait,* for example, is translated correctly as 'glass of milk' and not as 'milk glass'.

When translating into English, a check is also done by the same subroutine for a possible possessive case, so that the apostrophe can be added correctly.

INFR adds the infinitive particle when necessary and makes a gerund out of an infinitive whenever necessary, according to special meaning codes.

RELPRONR changes the Stem dictionary meaning, 'which', into 'who/whom' when the antecedent is human. Provided that PASS–1 has correctly identified the antecedent – in real text, often not easy to do!

The program RTNS consists essentially of tables of endings, and uses the information loaded by SYN into bytes 40 to 50, in order to synthesise the target-language text, adding the endings and conjugating verbs according to the synthesis codes which our industrious coders put on the target side of the Stem all that time ago.

REARR, running from R to L, takes care of the word order in the target language, unless it has already been attended to in LEXRTNS.

The REARR program has a number of subroutines, which are called in sequence: adnominal expressions of which we spoke a moment ago; adjectives, placing them after the noun in French, for example, unless coded otherwise; apposition rearrangement, to make *bombardiers B12* come out as 'B12 bombers', but *le roi Léopold* remain in the order 'king Leopold'; subject/predicate (for impersonal patterns translated as passive patterns); direct/indirect object pronouns; rearrangement of certain objects to become subject: *il est établi un régime* – 'a regime is established'; adverbs; and lastly questions, making use of the information stored so long ago at the beginning.

There are two more subroutines which take place here. While not part of the overall process of rearranging a sentence, they nevertheless have to fit in here, after everything else has been completed.

On the other hand, any antecedents in the current sentence of an *il/elle* in the next sentence are checked for animateness or inanimateness and this information is stored in a couple of bytes of word zero of the *following* sentence, to help with the resolution of the pronouns. If the antecedents in Current Sentence of pronouns in Current Sentence + 1 are themselves pronouns, then *their* antecedents, in Current Sentence − 1 are examined and this information stored in Current Sentence + 1.

A similar procedure is followed for predicate complements. The information that the complement was a noun or an adjective respectively is also stored, as this may be needed to resolve a pronoun in the next sentence. As in:

> *Son père était boulanger. Le fils l'est aussi* – 'The son is also one' but

> *Son père était horrible. Le fils l'est aussi* – 'The son is too.'

Earlier on we attacked the myth that SYSTRAN has no semantics. I hope to have shown at this point, the very last point of the whole translation procedure, that another of the reproaches made of SYSTRAN is not true either, namely that it cannot look beyond the bounds of its current sentence. Admittedly, it cannot yet look very far, and cannot look at many things, but it is learning all the time.

THE MT SYSTEM SUSY

Parsing and machine translation of natural language have a long tradition at the University of the Saar. The first parser was built in the 1960s for the analysis of modern written German, followed by a prototype for Russian–German translation in 1970–1. At the same time, several teams in different institutes of the university started to work on machine-readable dictionaries and mathematical linguistics. In 1972, this research was combined within a new institution, the Sonderforschungsbereich Elektronische Sprachforschung, subsidised mainly by Deutsche Forschungsgemeinschaft. The main goal was, and still is, research in the field of machine translation. Several teams are working together, each one having its specific interest but contributing to the common task.

The Russian–German prototype was the starting point for the development of a more general, multilingual translation system, called SUSY (Saarbrücker Übersetzungssystem). By generalisation of the algorithms and addition of language-specific components to the first prototype it was possible to implement a system which now is able to deal with German, Russian, French, English and Esperanto. SUSY is, like many other systems, as e.g. ARIANE–78 (GETA, Grenoble) (cf. chapters 10 and 16) and EUROTRA (European Community) (cf. chapter 19), a transfer-based system, i.e. translation is achieved through the three main processes: analysis, transfer and synthesis.

The result of SUSY's analysis of a text is a sequence of so-called 'basis structures', each representing a formal description of a sentence in the form of a tree structure. If for one and the same sentence different valid basis structures are delivered, the sentence is ambiguous and may have different translations. The basis structure is mainly a dependency tree in which the leaves represent original text words. There is no direct representation of consti-

209

tuency. If a node C is an immediate constituent of some non-terminal N, the node contains in its labelling the relation to C, say *rel* C. The relation *rel* may express a (surface or deep) syntactic function, e.g. NOM C (of N) indicates that C is the subject of N, RETR C (of N) indicates the governor (the 'nucleus') of N. If C has no syntactic relation to N, a constituent relation between C and N is established, e.g. PRP C (of N) indicates that C is a prepositional constituent (of the prepositional noun group N). This kind of organisation allows an easy access to dominated nodes whose relation to the dominating node is known. On the other hand, the identification of a node in a certain position within a complex construction is quite difficult, e.g. the access to the first (leftmost) constituent of some given noun group. However, this kind of structure is very suitable for testing valency frames or semantic disambiguation (especially in languages with relatively free order of constituents, like German and Russian), because in these problem areas constituency plays only a minor role.

It has to be stressed that basis structures are language-specific, especially with regard to the leaves which carry information on the lexical units. The only exception are prepositions which after semantic analysis are either artificial words – if they are semantically contentful – or can be regarded as deleted – if they serve as a syntactic casemarker. This shows that the task of TRANSFER is at least the translation of the lexical units of the source language. In SUSY the replacement of lexical units is achieved through a bilingual dictionary. After lexical transfer the above-mentioned relations are replaced by target language relations by using a table for standard mapping for the given language pair and – in exceptional cases – information stemming from the entries of the transfer dictionary. This mapping effects only the relations which indicate syntactic roles and does not change the geometrical shape of the basis structure.

The resulting basis structures are now handed over to SYNTHESIS, which has to generate out of them strings of words according to the grammar of the target language. Because the basis structure represents constituency only with difficulties, the generation is not achieved by consecutive transformation – reordering and flattening – of the trees, but by a totally new production of target language structures. After a preparatory phase, in which the target lexical units receive the necessary morpho-syntactic information from the dictionary, the basis structure is used only as a source of information for the process. The grammar of the target language building contains the construction plans, and the building materials are taken

from the basis structure and the target dictionary. The main characteristics of SUSY can be summarised in the following way:

(1) The translation process SUSY can be divided into three
consecutive subprocesses: ANALYSIS, TRANSFER,
SYNTHESIS.

(2) Each subprocess hands its results over to the following
subprocess. If some process fails, i.e. if it does not deliver a
result for a certain text unit, the translation process is blocked
and will also fail for this piece of text.

(3) SUSY expects as input a character string and will output a
character string as well. The initial character string can be a
whole text which in the course of processing will be broken
down into smaller units. (There exist binding conventions for
the seizure of texts.)

The analysis process

The task of SUSY's analysis process is the mapping of German,
Russian, etc. sentences on to their possible structures. The quality
of analysis is essentially determined by the completeness of the
source language dictionaries: The necessary entries appearing as
word forms in the text should be present and their description
should be complete. Of course, missing or wrong entries will not
necessarily cause failures. Moreover, the quality of analysis cruci-
ally depends on the elaboration of the analysis grammars for the
different languages. Because our main interest in the past was
focused towards Russian and German, the dictionaries and the
grammar systems of these languages are well developed, and much
better results can be expected from their analysis than for other
languages.

ANALYSIS is not simply an unstructured set or sequence of
rules, but is subdivided into eight subprocesses (called 'operators'
in SUSY terminology), each of which has to fulfill a specific and
well-defined task. These operators work in a strictly sequential
mode, which partially was imposed upon the implementors of
SUSY by restrictions of the computer on which SUSY was
developed (TR 440).

The analysis process is purely stratificational: Each operator
needs the results of its predecessor, and if one operator fails to
analyse a given sentence, the following operators will not treat it.
However, this does not mean that ANALYSIS totally fails for
that sentence. In the worst case the result of morphological analysis
will be delivered, but for transfer this counts as failure. The results
that analysis returns have a quality label which is assigned during

processing. If for a sentence the quality of its analysis structure is too low, it will not be processed by transfer and no translation will be produced. The single operators of ANALYSIS proceed in the same way. For a given sentence S the $(i+1)$-th operator selects those interpretations of S which, after having been produced by the i preceding operators, are marked as best possible. If a certain degree of quality is not reached, the actual operator and its successors will not operate on structures proposed.

In order to simplify orientation in the rest of this section, we will give a short presentation of the subprocesses of ANALYSIS. The eight operators are:

LESEN	Text input, identification of words
WOBUSU	Morphological analysis
DIHOM	Homograph disambiguation
SEGMENT	Segmentation of sentences into subclauses
NOMA	Analysis of noun groups
VERA	Analysis of verb groups
KOMA	Combination of verb and noun groups
SEDAM	Semantic disambiguation

LESEN, the first subprocess of SUSY, reads the initial character string. It divides the text into sentences and isolates the individual words. Because the user has to mark periods which indicate sentence boundaries, the division of the text into sentences is a trivial task. Single words are defined as being separated by blanks or by punctuation marks. They are assigned a sentence number and a word number, a normalised string (only capital letters) information on their typographic aspects, etc. Special signs for subject field, text type (e.g. 'title'), proper names, precoded word-classes ‹optional›, etc. are taken into account. The result of LESEN is a structured and purified 'text', consisting of a sequence of sentences, which may be clauses, noun groups or even single words. Each sentence is composed of a sequence of items which describe individual words. There is no restriction on the length of a text or the number of sentences, but a sentence may contain only up to 100 words. The maximum length of a word is 36 characters. In fact, these two restrictions cause many difficultes in practical application. The process LESEN is a purely technical device without any linguistic content. It represents the link between the outer world and the linguistic part of SUSY.

WOBUSU, the second subprocess, takes LESEN's results – the string of 'words' with typographic information, etc. – and deals with all morphological phenomena and some kinds of fixed phrases. Its task is the identification of possible lexical units together with their morpho-syntactic information. This process can be regarded as broken down into two subprocesses, IDENTIFY SINGLE WORD and IDENTIFY FIXED PHRASE. The first process deals with single, isolated words, i.e. the normalised strings identified by LESEN. Each string is submitted to morphological analysis and produces at least one interpretation, which in SUSY terminology is called 'variant'. This means that a 'variant' describes the properties of an interpretation of a string. Such properties are e.g. the word-class, gender, number, person, tense and many other specific features. Amongst these features, the property FSY plays an important role in IDENTIFY FIXED PHRASE, because sequences of strings are considered as possible fixed phrases only if at least one interpretation of each string is labelled with FSY = YES. Such sequences are dealt with in the process IDENTIFY FIXED PHRASE, which looks them up in the dictionary.

The main characteristics of WOBUSU are:

(1) WOBUSU consists of two sequential subprocesses:
     IDENTIFY SINGLE WORD
     IDENTIFY FIXED PHRASE

(2) Input to WOBUSU are the single sentences, considered as sequences of words, i.e. strings with some additional information.

(3) The second subprocess works on the results of the first, i.e. in stratificational mode.

(4) The result of WOBUSU can be represented in the form of a chart for each sentence. For each string (in the case of a single word) or for a sequence of strings (in the case of a fixed phrase) exists at least one variant, one possible interpretation. The number of variants associated to a string is called its degree of ambiguity, DA. If $DA = 1$ for some string S, we call S unambiguous, otherwise ambiguous.

It is evident that the definition of ambiguity depends on the properties and their types. For example, case, number and gender of German adjectives are considered as one complex property in which all possible combinations can be represented. Therefore *guten* is unambiguous, although it has many interpretations with respect to case, number and gender. On the other hand, if a string has more than one interpretation with respect to word class, it is always ambiguous. If a string is ambiguous in the above mentioned sense, it is called a homograph. The data structure which contains the results of WOBUSU is a chart (see chapter 8) of a special shape. After the termination of IDENTIFY SINGLE WORD we have the following situations: If a sentence consists of n strings, the chart contains $n+1$ nodes. The descriptions of the i-th word are attached to the arcs between the i-th and $(i+1)$-th node. IDENTIFY FIXED PHRASE does not change this kind of structure. If, e.g., the strings i and $i+1$ can be considered as a fixed phrase, the arcs between the nodes i and $i+2$ are deleted and new arcs (not paths!) are inserted between these nodes, according to the ambiguity of the fixed phrase. This means that overlapping ambiguities cannot be represented. If, e.g., *peu à peu* is in the dictionary, WOBUSU will interpret *peu à peu* unambiguously as a fixed phrase. The sentence *Il donne peu à peu de gens* would not be analysable.

WOBUSU's subprocess, IDENTIFY SINGLE WORD, deals with single normalised strings, i.e. a text word like *Speichern* would now have the form SPEICHERN, an indication that it has a capital letter, a word number, etc. The identification of single words can again be broken down into smaller tasks. First, we try to find all possible divisions of the word into two pieces: a possible stem and a possible inflectional ending (including the empty ending). The possible stems are looked up in the dictionary, compared with the endings, etc. If they fit together, the resulting information can be computed. This process is called INFLECTION. If INFLECTION fails, we can assume that the given word is a compound or a derivation or a combination thereof. In this case, we enter the process COMPOUND which computes all possible divisions of the string into prefixes, suffixes, stems and inflectional endings according to a grammar of compound words. If COMPOUND succeeds, the information associated with the complex word is identical to the information linked to its rightmost constituent, except the form of the lexical unit. If INFLECTION and COMPOUND fail to identify the string submitted to them, we call this word unknown. Because the syntactic analysis expects a certain minimum of information for every word, we have to submit unknown words to a safety net process trying to extract some information, e.g. possible word class, from the typographic form and the final graphemes. This process (UNKNOWN) is built in such a way that it never fails: it will always produce at least one variant per string. The main characteristics of IDENTIFY SINGLE WORD are:

(1) It consists of three sequential subprocesses:

    INFLECTION
    COMPOUND
    UNKNOWN

(2) Input to IDENTIFY SINGLE WORD is one item,

containing the normalised string, etc.

(3) The subprocesses work in the preferential mode:

    If INFLECTION succeeds, the process stops; otherwise it invokes COMPOUND.

    If COMPOUND succeeds, the invoking stops; otherwise UNKNOWN is invoked.

(4) Output is a non-empty set of items, the so-called variants. The original string is attached to the resulting variants, which are used to construct the chart.

INFLECTION can also be subdivided into simpler tasks. The first task is the segmentation of a given string into possible stems and endings. The corresponding process DIVISION produces a set of pairs of the form PST + FLEX by looking up the list of inflectional endings, such that all FLEXs are possible inflectional endings. In the case of our example SPEICHERN we would receive

    SPEICHERN   + 0

    SPEICHER   + N

    SPEICHE   + RN

    SPEICH   + ERN

The following step is the identification of the possible stems (PST) by looking up the morpho-syntactic dictionary. Those PSTs which cannot be found are disregarded. In the positive cases, we have to consider that some stem may belong to different entries. In the case of SPEICHER we will find two entries, one belonging to the verb *speichern,* and the other for the noun *Speicher.* We will also find the noun stem for *Speiche.* The result of the dictionary look-up will be a set of pairs

    (entry of *speichern* for SPEICHER, N)

    (entry of *Speicher* for SPEICHER, N)

    (entry of *Speiche* for SPEICHE, RN)

The third step checks for each such pair whether the ending is in agreement with the paradigmatic scheme of the entry. If not, the pair is deleted from the set (in our example SPEICHE, RN is deleted). Otherwise the morphological information is computed and added to the syntactic information of the entry. This material forms a 'variant'. In our case, the verb SPEICHER plus N produces two results: a reading as infinitive, and another as finite verb. As in German infinitives can always be considered as nouns, a third variant is produced. As the example shows, for a given pair of dictionary information and inflectional ending more than one interpretation can be derived. The pair (SPEICHER, N) finally is interpreted as a form of the noun SPEICHER (dative plural). Up to now, typographic information and the position in the sentence were not taken into account. Moreover, we must know whether the text was written according to standard German orthography with capital letters for nouns. All this information is used in the last step in order to discard some invalid intermediate interpretations. This may eventually cause a total deletion of all results. The process which checks these typographic restrictions is called TYPOTEST. The process INFLECTION has the following characteristics:

(1) It consists of four subtasks:

      DIVISION
      DICTIONARY LOOK-UP
      DEFLECTION
      TYPOTEST

(2) Input to INFLECTION is a normalised string together with its typographic and positional information.

(3) The subtasks work in the stratificational mode: each one deals with the results of its predecessor.

(4) The task TYPOTEST is different from all processes we have described up to now, because it does not produce new material, but checks its validity. Therefore TYPOTEST is a rather complex filtering device.

We will not go into further details in the description of the four subtasks, because further refinement would lead immediately to a description of a large amount of rules. We may add that DEFLEC-TION could perhaps be split up into four subtasks: treatment of wordforms (stored as such in the dictionary), of nouns, verb and adjective stems.

WOBUSU's subprocess, IDENTIFY FIXED PHRASE, does not work on single strings, but on sequences of variants. The sentence, i.e. now the chart which represents the actual state of analysis, is traversed from left to right in order to find the first variant with FSY = YES. If this item is followed by a sequence of items of length n, all having FSY = YES, we have found a candidate for a fixed phrase. Initially, the variable n is set to 3. By concatenation of the strings of the single elements (with one blank between each pair) a new string FSW is built and then looked up in the dictionary. If it is not there, we reduce n by 1 and try the shorter FSW, until $n = 0$. If FSW is found, a fixed phrase is detected, and the dictionary information replaces all previously computed variants in the domain of the chart covered by FSW. The algorithm proceeds then with the right neighbour. If we do not find a fixed phrase starting with the focused item, we continue to traverse the chart to the right. The process stops when the sentence end is reached. The main characteristics of IDENTIFY FIXED PHRASE are:

(1) The traversal from left to right is important.

(2) The principle of longest match is applied.

(3) Fixed phrases which consist of more than four elements cannot be identified.

(4) New material *replaces* old material immediately. Interleaved fixed phrases or the ambiguity between a fixed phrase and its regular interpretation therefore cannot be represented.

DIHOM, the homograph analysis, is given for each sentence a data structure in the form of a chart in which all information for the subsequent syntactic analysis is present. The decision to start syntactic analysis by a homograph disambiguation module was suggested by the promising experiences with the syntactic parser for German which had been implemented and improved by the former Syntaxgruppe until 1969. The architecture of DIHOM, however, is very different from this approach, because DIHOM makes an extensive use of different kinds of tables which contain linguistic knowledge and experience on compatibility and distribution of linguistic items, whereas the homograph analysis of 1969 was a complex system of interrelating programs. The data structure these procedures operated on was a simple sequence with one block of information for each word. The main categories were the homograph classes plus the word classes, such that only unambiguous words had a word class and all others a homograph class. The goal of homograph analysis consisted in the replacement of homograph classes by word classes. This representational approach has some evident deficiencies: The assignment of homograph classes to strings can be done in a reasonable way only if a word form dictionary is used. Each word form has to be carefully inspected by an expert in the field of homography, and the structuring of the information contained in dictionary entries is rather a difficult task. Finally, the elaboration of the homograph programs and the establishment of the flow of control requires a very specialised expert on homograph reduction procedures. All these difficulties were the reason for elaborating a new concept for homograph disambiguation, based on traditional linguistic description and avoiding the dependency on a homograph specialist. In its actual state, the homograph disambiguation module of SUSY consists of five consecutive tasks.

The first subprocess of DIHOM (INHIBIT) is based on the knowledge that in an actual text only certain pairs of words, characterised by their syntactic features, are possible. According to linguistic considerations, the most important features have been isolated, such that each word can be characterised by an 'inhibition class' representing the bundle of these features. This leads to the definition of (a maximum of) 150 classes, the definition of which depends on the SUSY word classes and additional characteristics (e.g. 'relative', 'interrogative', 'modal', etc.). On this basis, we can state that a word of class A can never be followed immediately by a word of class B, e.g. 'article + finite verb', 'noun + relative pronoun' for German are impossible pairs within a correct sentence. This leads to the elaboration of a binary matrix with 150 rows and columns, which describes which pairs of 'inhibition classes' may appear in a syntactically correct text. The subprocess INHIBIT is associated with an algorithm, which applies the information of the 'inhibition matrix' to the data structure in the following way: If, e.g., some word A has three readings, i.e. three different classes u, v, w, and is followed by B with the unambiguous interpretation p, and if the matrix element (u, p) is zero, then the interpretation u of A can be deleted. This procedure is carried out for all pairs of words taking into account all combinations of classes. Experience showed that the process INHIBIT reduces the ambiguity n of a word to an average ambiguity of $n^{0.6}$. This means that the average ambiguity $n = 2$ per word is reduced to approximately 1.5 by this procedure.

Specific homographs, for which a general scheme for resolution could not be established, are treated by means of special programs called SPECIAL CASES. Hereby, a general principle is obeyed, according to which homograph routines can either delete or confirm a reading within the bundle of variants they operate on. The framework of SPECIAL CASES, in which these procedures are embedded, makes sure that the data structure is traversed from left to right. Within the single routines the whole sentence surrounding the homograph can be inspected. Examples for special cases in German homograph analysis are, e.g., the word *als,* circumpositions (discontinuous prepositions), postpositions, nominalised modal verbs, detached verbal prefixes, relative pronouns.

The third subprocess of DIHOM, WEIGHTING, is again a module, in which algorithm and linguistic data are separated. The idea behind WEIGHTING is the fact that a reader of a sentence, having read and identified a word, can guess with high probability which kind of word will follow. If, e.g., an article has been read, we can be quite sure that the next word will be a noun or – less probable – an adjective. For the realisation of this idea, a definition of 25 weighting classes was worked out, such that each variant of a word carries such a class. The next step consisted in finding appropriate (integer) values for characterising pairs of weighting classes with respect to their acceptability. A high score for a pair (A, B) means that if a word is interpreted as A then its successor is expected to be in class B. A low score for (A, B) means that A is probably not followed by B. If (A, B) is high and (A, C) is low and if the successsor of a word of class A can belong to the classes B and C, the interpretation B is preferred. The set of values for pairs of weighting classes forms a matrix of 25 rows and columns, the values ranging between 0 and 200. The initial matrix was established on the basis of some statistic work, but afterwards refined by intensive testing. The process WEIGHTING is associated with an algorithm which applies the knowledge accumulated in the weighting matrix to the actual data structure, the whole sentence in chart representation. If we imagine that the chart is a condensed form for the set of all sequences of word variants, we can say that the weighting process calculates a score s(p) for each such path p, and isolates the paths with maximal scores. If p is considered as a sequence of weighting classes

$$p = (p1, p2, \ldots, pn)$$

and the weighting of a pair (x, y) denoted by w(x, y), the score s(p) is defined by

$$s(p) = w(p1, p2) \times w(p2, p3) \times w(p3, p4) \ldots$$

(resp. by the sum of the corresponding logarithmic values). The output of WEIGHTING consists conceptually of maximally 10 paths with highest scores, but the paths are still preserved in the chart structure. Each variant has now information which allows identification of which path it belongs to. We want to remark that the weighting process is somewhat more difficult than sketched above, because there exists a possibility of characterising word forms as more or less obsolete. If such a form appears in a homograph bundle, the weighting process has to reduce the score of the path in which it occurs or even to reject it. Experience showed that there are still some possibilities left to detect inadequate paths. As INHIBITION and WEIGHTING work only on pairs of words, it can happen that a sequence A + B + C is accepted because A + B and B + C are possible pairs, but nonetheless the triple A + B + C is impossible. WEIGHTING is therefore followed by a process NTUPLE which discards those paths which contain a subpath specified in a list of frequent erroneous n-tuples.

A last filtering device deals with necessary conditions on discontinuous sequences. This module, called PLAUSIBILITY, checks very simple rules, as e.g. 'if you find a relative pronoun, then somewhere to its right a finite verb must be present'. This means, however, that even paths with highest score may be destroyed. In fact, there is often no other possibility to distinguish at this stage of analysis between a finite verb and an infinitive, because in German subclauses the finite verb is shifted to the end of the clause and then has the same distribution as the infinitive.

The architecture of DIHOM can be sketched as follows:

(1) DIHOM consists of five sequential subtasks:

> INHIBITION
> SPECIAL CASES
> WEIGHTING
> NTUPLE
> PLAUSIBILITY

(2) All subtasks work on the whole sentence. Each subprocess uses the material which the preceding subprocess returns, i.e. a stratificational mode. In contrast to formerly described stratifications, where the main process is blocked and fails if one of its subprocesses does not produce new material, the subprocesses of DIHOM continue to work even if some predecessor did not make changes in the data structures.

This feature allows the implementer or user to parametrise DIHOM in such a way that certain subprocesses are switched off, which is very helpful for debugging purposes.

(3) The output of DIHOM consists of up to ten paths per sentence, ordered according to their scores. The user can set a parameter for the number of paths he wants DIHOM to produce. The default value is two. The output still has the form of a chart.

Finally, we want to mention that DIHOM is controlled by some global parameters, the most important of which are LANGUAGE and TEXTTYPE. According to these parameters DIHOM totally fails, i.e. no path is returned for a given sentence. The quality of the disambiguation depends on the language treated. Best results are achieved for German and Russian – the right path is in ninety-five per cent of the cases amongst the first two proposals, because the linguistic work was concentrated on these languages.

SEGMENT has the task of identifying subclauses within a sentence and establishing a structure which describes the relations between these clauses. This analysis is carried out in several consecutive stages. The initial data structure which SEGMENT works on still has the form of a chart, but this time the chart represents individual readings of sentences, as identified by DIHOM. The chart structure is still necessary, because paths which differ only with respect to simple word ambiguities are collapsed into one structure. For example, the ambiguity of the word *gedacht* (participle of *denken* and *gedenken*) is considered as simple. The definition of simplicity takes into account that ambiguities of a certain kind do not affect the flow of analysis and the resulting structures within SEGMENT.

The first subprocess of SEGMENT, DELIMITATION, makes a partitioning of the sentence which results in a sequence of subsegments. The limits between subsegments are indicated by punctuation marks and coordinating conjunctions.

In the second step, IDENTIFY SEGMENT, we have to check whether the limit between two subsegments is really a limit between clauses or simply a coordinator (between noun groups, etc.). The remaining limits are classified as certain or ambiguous, and all segments receive a set of labels computed on the basis of the information attached to the words they consist of. The resulting sequence of segments is still preliminary and will be refined in the next stage of analysis.

This submodule (SEGMENT CLASSIFICATION) deals especially with the position of verbs within a segment and assigns the definite labels to the segments. Before termination, the segments are checked for validity. This may cause that the sentence variant treated is marked as wrong. The resulting structure of SEGMENT CLASSIFICATION is again a chart (because the segmentation itself and the labelling may be ambiguous), although it is not represented in this form. The software system extracts the individual paths out of the given segment structure and submits them one after the other to the parsing module SEG-ANALYSIS.

SEG-ANALYSIS works on sequences of classified segments using the segmentation grammar, a set of rules which describe under which conditions which relations between segments can be established and which segments constitute complex segments. The rules are reduction rules, which on the right hand side always have a non-empty symbol. Many rules are context sensitive, but only in order to avoid too much backtracking. Each rule is connected with a programmed procedure, which checks additional requirements for its application and calculates the labels of the resulting element in the case of applicability. Before termination of SEG-ANALYSIS, the content of the push down store, in which pieces of structure and applied rule are stored, is translated into the corresponding linguistic interpretation, which has the form of a tree.

Finally, the resulting structure is checked for validity: the top node should have a label for main clause. If this goal is not reached, the analysis mode is switched to RESCUE, and the sequence of analysis modules is applied again to the initial data structure. The RESCUE-information causes these modules to operate in a different way: many restrictions are weakened or not at all checked, such that the system is forced to produce some output. The switching to the RESCUE mode can occur not only in the last submodule of SEGMENT, but at any stage. The output of SEGMENT receives a quality label for the distinction between regular and RESCUE results. For a given sentence, several results (also with different quality labels) can be returned. But whenever a regular result has been obtained for some interpretation of a sentence, the switching to RESCUE during the analysis of other interpretations of the same sentence will be blocked.

The module SEGMENT is used only for the analysis of German and Russian. For English and French, quite different processes (PHRASEG and SEGFRA) are used instead. The quality of SEGMENT is very satisfactory, and in the past five years only minor corrections were necessary. If we try to summarise the main characteristics of SEGMENT, it is useful to combine the three modules IDENTIFY SEGMENT, SEGMENT CLASSI-FICATION and SEG-ANALYSIS into one process, which we will call ANALYSE SEGMENTS. The flow of control within SEGMENT can best be described in a procedural way and using global control variables for parameterising the subprocesses. The control procedure which we sketch below gives only a simplified picture of SEGMENT. It presents, however, the general RESCUE-concept which is equally used in most other parts of the SUSY system. Each sentence, identified by its sentence number, can be characterised by a property called STATUS. As the results of the main subprocesses are stored altogether in an analysis result pool and all individual results have their own quality label, it is always possible to select at a given moment – even at the end of the whole translation process – the best result for some given sentence.

The user has a limited possibility to control SEGMENT by communicating how many DIHOM results should be treated and how many interpretations SEGMENT should produce at most. Moreover, a trace facility is offered for all important components of the system. These additional features do not appear in the following overview:

```
loop
    input next data structure S
    if S = empty then stop fi
    N : = sentence number of S ;
    if QUALITY of S = DIHOM then RESCUE : = false ;
        invoke DELIMITATION ; SUCCESS : = false
        loop until SUCCESS :
            invoke ANALYSE SEGMENTS ;
            if SUCCESS then
                if RESCUE
                    then add SEG-RESCUE to STATUS of N
                    else add SEG-REGULAR to STATUS of N
                fi
                output analysis result
            else (i.e. failure case)
                if not RESCUE and MAXIMAL STATUS of
                        N = DIHOM
                    then RESCUE : = true
                    else SUCCESS : = true
                fi
            fi
        pool
    fi
pool
```

Summarising, we want to point out the following features:

(1) The RESCUE mechanism, the user supplied restrictions on the number of input and output structures per sentence, and the demand for efficiency seem to imply a procedural description of SEGMENT.

(2) Input to SEGMENT are the results of DIHOM having the form of a chart. In most cases, they are degenerated to simple sequences.

(3) For each sentence, the output of SEGMENT is a sequence of structured entities with possibly different quality labels, and a STATUS which characterises the analysis quality of all results per sentence as a whole. Each entity consists of a DIHOM result and – if SEGMENT was successful – a parallel interpretation on the level of subclauses, which conceptually form a tree whose nodes are segments (or subclauses). The terminal segments can be considered as linked with charts which represent the subclauses on the level of words.

(4) If we neglect the procedural aspect of the control structure

of SEGMENT, we can state that SEGMENT consists of two sequential tasks:

DELIMITATION
ANALYSE SEGMENTS

which work stratificationally: the second module needs the results of the first.

ANALYSE SEGMENTS in turn can be considered to consist of two tasks:

REGULAR SEGMENT ANALYSIS
RESCUE SEGMENT ANALYSIS

which are also sequential, but work preferentially: the second operation is called only if the first fails. The architectures of these two processes are identical, only the rules they apply are different: for each regular rule r exists a rescue rule r' which differs from r only with respect to the conditions under which it is applicable. The rescue rules are simply less restrictive. In the following, we will sketch SEGMENT ANALYSIS without taking the RESCUE feature into account. We will remember that this process has three tasks:

IDENTIFY SEGMENT
SEGMENT CLASSIFICATION
SEG-ANALYSIS

These tasks are carried out stratificationally, i.e. each one needs the output of its predecessor as input for further processing.

The input to DELIMITATION is an interpretation of the sentence on the level of words, represented as a chart which contains only harmless ambiguities. In many cases, the chart is degenerated to a simple sequence of word interpretations. Within DELIMITATION, the single word variants receive an additional classification on the basis of the information derived from the dictionary. This new classification is a set of nine boolean properties, e.g. 'word is finite verb', 'word can introduce a subordinate clause, a relative clause', 'word can be a limit between segments', 'word is a unique limit', etc. After the general classification, some special cases are treated dealing with possible limits, i.e. coordinating conjunctions and punctuation marks and aiming at their disambiguation as non-limits or unique limits. Hereby, the preceding or following word of a possible limit is taken into account.

The process IDENTIFY SEGMENT builds segments between words which have been classified as limits by DELIMITATION. At the same time, a disambiguation of possible limits is attempted, e.g. for expressions in parenthesis. A segment is built up by starting with the first word after a limit and adding one word after the other to the segment (left to right traversal). At each step, a consistency check takes place such that obviously wrong segmentation is avoided. After termination of this first segmentation phase, the limits which are not yet unique are looked at again and eventually disambiguated on the basis of the information collected in the preceding and the following segment.

The segment classification of IDENTIFY SEGMENT is only a rough and preliminary characterisation of segments. A refined characterisation is achieved in the first subprocess of SEGMENT CLASSIFICATION, which delivers a set of features represented as a bit string and called INVENTORY. The final content of INVENTORY gives information on the nature of the segment (e.g. main clause, relative clause) and possible relations to other segments (e.g. that it can govern an object clause). In the case of German analysis, SEGMENT CLASSIFICATION uses several rules dealing with finite verbs, in order to distinguish between dependent and main clauses. Finally, each segment is checked for consistency on the basis of its INVENTORY. If a contradiction is discovered, SEGMENT CLASSIFICATION outputs a failure message which possibly activates the RESCUE mechanism.

SEG-ANALYSIS is a rather complex process which uses context free and context sensitive rules written in a special formalism and connected with additional subroutines. The rules are interpreted and applied by a special algorithm based on a general concept for context free parsers. Conceptually, SEG-ANALYSIS can be broken down into three consecutive processes, working in the stratificational mode:

    START SEG-ANALYSIS
    DO SEG-ANALYSIS
    TERMINATE SEG-ANALYSIS

The third subprocess has to check whether the desired result was achieved, if at all a result was produced. A wrong result leads to a failure message and possibly to a RESCUE operation.

DO SEG-ANALYSIS is the main parsing operation within SEG-ANALYSIS and consists of seven subtasks:

    RULES 1
    SEG-COORDINATION
    BOUND-SEGMENTS
    MAIN-CLAUSES
    DISCONTINUITY
    SUBCLAUSES
    DETACHED VERBAL PREFIXES

These subtasks are *not* carried out in a sequence, because the output of any subprocess can serve as input to any other module. Moreover, their application has to be iterated until the production of new structures terminates. The subtask RULES 1 can be described by a set of relatively simple rules, which operate mainly on single segments by inspecting their labels. A subset of these rules forces the parser to analyse from right to left. The remaining six subtasks are structured, i.e. they can be broken down into smaller tasks, all being invoked in sequence. SEG-COORDINATION deals with coordinated segments. In the positive case, a new segment which covers the constituting (two) segments and the coordinating conjunction is produced. BOUND-SEGMENTS treats pairs of segments in order to establish a dependency relation between them. Relative clauses can be attached to the preceding segment, adverbial clauses to the preceding or the following segment. The so-called SBJ-OBJ-segments – starting with a subordinating element (*dass, ob,* interrogative) – can be attached to a preceding or a following segment, if this is marked as possible governor of such a clause. The necessary information can be found in the above mentioned INVENTORY. SUBCLAUSES recognises whether a segment can be a subclause, and MAIN-CLAUSES does the same for main clauses. DISCON-TINUITY operates on a sequence of three segments with punctuation marks (commas, dashes, parenthesis) between and tries to identify the segment in the middle as an interruption of the first and the last segment. In the positive case, these two segments are combined and form a new, continuous clause. DETACHED VERBAL PREFIXES is a special process for the treatment of discontinuous German verbforms. The prefix, which is placed somewhere to the right of the finite verb, but within the same segment – and therefore the DISCONTINUITY process is crucial – is attached to the normalised string marked as finite verb and looked up in the dictionary. If it is found there, the lexical unit of the verb is changed and the prefix is marked as dead. If at the end of SEGMENT the consistency check TERMINATE SEG-ANALYSIS discovers a surviving verbal prefix, it produces a failure message.

NOMA is the process for the analysis of noun groups. The first task within NOMA is to input the best possible analysis results for each sentence obtained up to now. The analysis quality must be at least SEGMENT-RESCUE. If a sentence failed to be analysed by SEGMENT, it will not be treated by NOMA. If there are results marked as SEGMENT-REGULAR, one of those will be dealt with. But NOMA does not operate on a sentence as a whole, but on terminal segments, which are connected to a chart-structure describing the interpretation of a piece of text on the level of words. These pieces have to be identified by NOMA's input device using the results of SEGMENT. If a segment contains verbs (finite verb, infinitive, participle), these are used to break the segment down into analysis units which contain only nominal words. Exactly these units are the data NOMA operates on. The control structure of NOMA is much more difficult to describe than SEGMENT because of several reasons. One is the fact that NOMA uses two RESCUE-parameters, one for the invocation of a safety net in the case of failure within the analysis of a single segment, and another for the case where the analysis of the sentence as a whole is contradictory. Another important reason is the fact that NOMA attempts a new segmentation under certain circumstances, and then has to start processing again from the beginning. A third reason is SUSY's data structure which is not powerful enough to deal with ambiguities on the level of the description of noun groups. Important ambiguities, e.g. genitive attribute versus dative complement, must be explicitly identified and lead to the production of separate analysis results for the whole sentence. If we abstract from all these complications, we can say that NOMA consists of the following subprocesses:

NOMA-INPUT
NOMA-ANALYSIS
NOMA-OUTPUT

NOMA-INPUT produces the analysis units which NOMA-ANALYSIS operates on and NOMA-OUTPUT stores the results and marks the analysis quality for each sentence. The control structure of NOMA-ANALYSIS is quite complex again because of the reasons mentioned above. Apart from these difficulties, we can distinguish two components: a structure building component and a filtering or consistency checking device. Therefore NOMA-ANALYSIS consists of

NOMA-STRUCTURES
NOMA-CONSISTENCY

As it is not possible to construct only well-formed structures during the building phase, we have to check before termination where the analysis goal is reached, e.g. we must not find a noun group consisting of an article only. If an ill-formed structure is detected, a failure message will be produced, and this may led to a RESCUE operation.

The task of NOMA-STRUCTURES is the production of

noun group structures. We can identify approximately twenty subprocesses which in turn are highly structured. In general, the input structure is traversed from left to right. As soon as a new noun group is built up, we try to attach it to its nearest neighbour (N) to the left. In certain cases, this implies a traversal of the left neighbours, starting with N and climbing up to the top node T, searching a new neighbour of T, etc. Normally, this procedure leads to the adjunction of a new tree at the right corner of the tree with top T. The traversal, however, is quite difficult, because the nodes of the trees are not explicitly ordered, so that the original order of the constituents has to be computed, if necessary. The subprocesses of NOMA-STRUCTURES deal with the following phenomena:

> *Recognition of simple groups,* e.g. preposition + article + noun, article + adjectival group + noun, etc. The procedure is based on two tables containing the description of simple groups for the different languages. The longest match principle is used, restrictions on case, number and gender agreement are observed.
>
> Treatment of *postpositions.*
> Treatment of groups of the form *'Hand in Hand'.*
> Treatment of special kinds of *appositions.*
> Treatment of special phenomena with *numerals.*
> General conditions on *attributes and coordinations.*
> Special conditions on *coordinations.*
> Special conditions on *attributes.*
> Treatment of nouns having information on possible *oppositions.*
> Construction of *appositional structures.*
> Conditions on the coordination of *prepositional noun groups.*
> Complex *adjectival groups* (left recursion in German).
> *Genitive attributes* on the left of a noun group (for German, English).
> Special cases of coordinated *adjectival groups* (adjectives in singular, noun plural).
> Construction of noun groups with *complex adjectival groups.*
> Coordination of *embedded structures* (left recursion).
> Recognition of *English compounds.*

It should be mentioned that SUSY knows only two types of non-terminals: noun groups and predicates. Adverbial groups are a subclass of noun groups, and adjectival groups a subclass of predicates (clauses are predicates as well, but these do not yet appear in NOMA). The recognition of attributes (except appositions) depends on the valency frame of the governor of the dominating group, which is derived from dictionary information. For the different languages, a default case is added to the valency frame, if the governor is a noun (genitive for German and Russian, 'of' for English, *de* for French). If the intersection of the valency frame of the presumed governor and the possible cases of the presumed attribute is empty, the attribute grammar will not be applied. If a noun group is attached to an adverbial group, the rules for passive transformation will be invoked, such that a *von*-phrase may be considered as a deep subject. It might be interesting to notice that our syntactic cases are not the traditional ones, but constitute a list of approximately thirty values (this differs from language to language), which is a combination of the inflectional cases with values which indicate the prepositions used as case markers. For German and Russian the prepositions which imply different inflectional cases are considered as different case markers.

A check for linguistic consistency, NOMA-CONSIST-ENCY, has to be added, since we cannot be sure that at the end of NOMA-STRUCTURES all trees produced are valid linguistic descriptions. There are approximately ten rules (or ten sets of conditions) which the results have to obey. The result must be a sequence of trees in the area of the input text unit (segment or piece of segment). This condition has some consequences: if the sequence is interrupted by a word (which could not become a noun group or a constituent of a noun group), we have either

(1) a coordinating conjunction or a punctuation mark. In this case, a new segmentation is attempted and analysis starts again.

(2) another word class. In this case, analysis failed, a failure message is produced, and RESCUE is activated.

In a German main clause, the sequence of trees left from the finite verb must consist of exactly one tree (there are some very specific exceptions, however). If this condition is not met in the actual data structure, we try to attach possible attributes (i.e. prepositional or adverbial groups) as non-valency-bound constituents to preceding groups. If this fails, analysis of the input data failed. There is no restriction on the number of noun groups, i.e. on the length of the sequence of trees. However, there is a limit on the number of 'free' groups which are not prepositional or adverbial, because these will be valency bound constituents of the clause. If this restriction is not met, analysis failed. All noun groups must be 'complete', i.e. groups consisting of a single article or a preposition and an article, etc. are ill-formed. This may happen especially in German analysis when noun groups with complex adjectival groups have to be dealt with, e.g. as in

'die *von der Kommission im vergangenen Jahr* beschlossenen Massnahmen'

During analysis, the article 'die' forms a provisional noun group, the italic groups should be attached to the participle 'beschlossenen' which forms an adjectival group within the noun group for 'be-schlossenen Massnahmen' and only then the noun group which represents 'die' is combined with the complex group to its right. If one of these analysis operations fails, the article will still be present as a spurious result. However, if the system is in the RESCUE mode, such inconsistencies will be accepted. Another kind of completeness has to be tested for coordinated groups. If a discontinuous coordinator is found (e.g. *weder . . . noch,* neither . . . nor), both constituents must be present. Moreover, coordinated groups must agree with respect to their cases. After the structure building phase, the case information is present and can now be used to reduce case–number–gender ambiguity of the single groups. This check may lead to a detection of a wrong coordination. In this case, the coordination is destroyed and a new segmentation attempted. Finally, rules for the reconstruction of deleted elements (prepositions, nouns) within a coordination are applied.

VERA, the following process, has to treat the verbs, as NOMA dealt with nominal elements only. The task of VERA is to identify the single verbal groups within each segment, assign structures to them, connect them with verbal structures in depending or governing segments and reconstruct deleted elements in segment coordinations. At termination of VERA, we will have several descriptions for clauses in parallel: the level of words, of noun groups and of verbal groups. The first operation in VERA is the input of the NOMA results. For each sentence only those results which have highest quality (at least NOMA-RESCUE) are read, the others are discarded. Although VERA does not deal with noun groups, we have to use the NOMA results, because these are bound to the segmentation structure which might have been changed by NOMA. The input data structures offered to VERA for processing are complete sequences of coordinated segments. They are treated by seven subprocesses linked together by a rather complex control scheme, which has to take the RESCUE mechanism into account. If we abstract from RESCUE and computational tricks for accelerating processing, we can distinguish six subprocesses in VERA, apart from the input module, which are arranged sequentially:

PARTICIPIAL REL-CLAUSES
SEGMENTS WITHOUT VERBS
VERBAL GROUPS
VERA-COORDINATIONS
VERA-CONSISTENCY
VERA-STRUCTURES

The first subprocess deals with participial relative clauses, checking the agreement of the morphological information of the participles in case of coordination and builds an artificial relative noun group. This ensures that later on all relative clauses have the same structure and can be handled by the same procedures. For Russian, we need a special process for inserting an artificial copula for 'to be' in sentences which miss a verb. This is done in the process SEG-MENTS WITHOUT VERBS. In the third step, VERBAL GROUPS, we restrict the analysis area to single segments, iden-tify the verbal elements per segment and construct an intermediate structure for each. Within this process we access language specific lists containing the different kinds of admissible verbal groups together with information on their structure. In VERA-CO-ORDINATIONS, we treat sequences of segments again, i.e. sequences of verbal groups, in order to detect incomplete groups and reconstruct the deleted elements. VERA-CONSISTENCY checks single segments for the completeness of the verbal groups. If inconsistencies are detected, a failure message is produced and the RESCUE mechanism activated. VERA-STRUCTURES produces the final trees by deriving them from the intermediate description of the verbal groups. This means that auxiliary verbs (for perfect, passive) disappear and information on tense, number, gender and person is established. All non-terminal nodes in these trees are labelled as predicates and dominate exactly one terminal element, the governor, which is labelled with the valency frame information on necessary subject agreement, etc. In the case of a relative clause, we have a pointer to the relative noun group. We will not got into further details of these six subprocesses, but a final remark may be desirable: the most difficult process is VERA-COORDINATIONS because of the need to identify missing elements, which are deleted in the surface structure, but must be detected for establishing a deeper structure, on the basis of which the appropriate tenses and dependencies have to be computed. Moreover, the reconstruction is necessary for the attachment of noun groups to the appropriate governor. As the reconstruction strategy aims at completing structures as far as possible, it may happen that too many elements are rebuilt. But in these cases also a human reader would have difficulties.

KOMA (complement analysis) expects for each sentence at least a result with quality VERA-RESCUE, i.e. a result which describes the individual segments in terms of a predicate structure and a noun group structure. Furthermore, the relations between the predicate nodes have now labels which indicate the syntactic relationship. Apart from the input and output modules, we can distinguish seven subprocesses in KOMA, which operate sequentially:

COLLECT FREE GROUPS
COMPLETION
REDUCTION
ATTRIBUTIVE CLAUSES
REFLEXIVES
INFINITIVE CLAUSES
SENTENTIAL ADVERBS

The process COLLECT FREE GROUPS operates on single elements and has two different, but nearly related tasks, whose sequence is irrelevant. The first task consists in the identification of noun groups which are still 'free', i.e. not governed by the nucleus of another noun group. For each segment, these groups are collected in the so-called A-array. As VERA establishes relations only to infinitives which are considered as constituents of a complex verbal group, we have to complete this analysis by treating (finite and infinite) subclauses, the dominance relation between which has been established by SEGMENT. Clauses which are presumed as subjects are tested for agreement with the governing verb, those which certainly depend on a noun will be treated in a later process. As a result of this operation, the non-attributive subclauses are attached to their governors together with information on their syntactic function.

The process COMPLETION operates on sequences of the above-mentioned A-arrays, which represent coordinated clauses. For each A-array, noun groups from the left and right neighbours are borrowed and marked as provisionally reconstructed elements. This is only a preparation of the following process.

The process REDUCTION operates on single segments again. Its task consists in attaching the groups of the A-array to the verbal governor of the segment. This is achieved by the use of four submodules. The first submodule inspects only those groups of the A-array which are not prepositional or adverbial. Their cases are compared with the set of valencies, i.e. for a given group G the intersection I of its possible cases and the valencies is computed. If

I is empty, we distinguish two cases:

(1) If G is a reconstruction, it is deleted from the A-array, and analysis continues with another G.

(2) If G is real, the process stops with a RESCUE message.

If I contains exactly one element, we have detected a non-ambiguous syntactic function of G. This result is preserved, the valency frame reduced, and the intersecting operation starts again with the beginning. It should be mentioned that subject agreement is tested, if the nominative appears in I. If this procedure does not exhaust the A-array, a preferential case assignment takes place, e.g. the left-most group, which can be a subject, is definitely accepted as such, and then we can start the intersection operation again, etc. The second submodule of REDUCTION operates on whole segments as well and uses the results of its predecessor for attaching the noun groups with their syntactic functions to the governor. At the same time, number and gender ambiguity is reduced. The third submodule is a counterpart of the two preceding ones for prepositional and adverbial noun groups. Those groups which fit into the valency frame are considered as valency bound objects, the others as free circumstantials. A final operation deals with the mode of the verb by inspecting its subject.

The process ATTRIBUTIVE CLAUSES operates on sequences of coordinated clauses which are not attached to a governor. The dominating segment is identified and its noun groups inspected from right to left. If the sequence is marked as a relative clause – this information stems from SEGMENT – we have to test number/gender agreement between the actual noun group and the relative noun group of the subclause. If they agree, the subclause is considered as a relative clause of the actual noun group G. In this case, the relative group is replaced by G, such that relative pronouns disappear from the structure. For interrogative clauses or those which are introduced by a subordinating conjunction (e.g. *dass, ob, that, que*) we have to find a noun group whose governor has appropriate information on the kind of subclauses it can dominate. If a positive result is obtained, the subclause is attached to the noun group. The conjunctions are deleted, because the clause is sufficiently subspecified.

In the REFLEXIVES process (possessive and normal) reflexive pronouns or noun groups are treated. For each reflexive, a pointer to the corresponding subject of the clause of which it is a constituent is set. This rule is not applicable if the clause has no subject. As the reason may be that the clause is infinite and the subjects of infinite clauses will be reconstructed in the following process, INFINITE CLAUSES, the module REFLEXIVES will be invoked again in this later analysis stage. It should be mentioned that REFLEXIVES checks the agreement between the reflexive and the subject (with respect to number, gender, and person). Moreover, the governor of the clause is checked for whether it allows a reflexive object and which case relation exists between them. These tests are used to reduce a possibly existing person/number ambiguity. If REFLEXIVES detects an inconsistency, it produces a failure message for the RESCUE mechanism.

The process INFINITE CLAUSES treats infinite clauses (i.e. clauses which have an infinitive as governor) and aims at reconstructing their subject. We have to distinguish two cases:

(1) The infinite clause is dominated by a predicate (i.e. by some kind of a verb, not by a noun). The governor of the predicate contains a label which indicates the case relation to a complement on the same level of the tree (a sister node), which is the subject of the depending infinitive clause. If such a complement exists, it becomes the subject of the infinitive. If the voice of the infinitive is passive, the subject is turned to an accusative object. If the infinitive clause is coordinated, the reconstruction has to be extended to all members of the co-ordination.

(2) The infinite clause is dominated by a noun, i.e. it is a constituent of a noun group. Normally, we do not try to reconstruct the subject, except in a case like this:

> . . . *die Gruppe, deren Aufgabe es its, . . . zu prüfen*

The clause with *zu prüfen* depends on *Aufgabe,* which has a possessive relative pronoun pointing to *Gruppe.* In this case, the noun group with the governor *Gruppe* becomes the subject to *prüfen.*

The process INFINITE CLAUSES has to traverse the predicate trees from top to bottom, because infinite clauses can be constituents of infinite clauses. Consequently, if a reconstruction was successful, the reconstruction in a depending infinitive clause can be attempted.

In SENTENTIAL ADVERBS, sentential adverbs like *gerne* which in NOMA were treated as normal adverbs forming adverbial noun groups are now changed by transforming the noun group into a predicate which dominates the clause which it was a constituent of. The new relation is considered as accusative. The analysis result of e.g. *ich arbeite gerne* will be as shown in figure 14.1, which is approximately the structure of the corresponding 'I like to work'. At termination of KOMA the analysis results are stored on a file

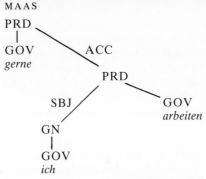

Figure 14.1.

(just as in the preceding processes like NOMA, VERA, etc.) and a quality marker for each sentence is preserved.

SEDAM is the semantic disambiguation phase of SUSY's analysis. Like the preceding modules it inputs for each sentence only the results with highest quality. The minimum quality which is required by SEDAM is NOMA-RESCUE, i.e. we need at least noun groups. SEDAM consists of two components: the program SEDUS and a dictionary which is chosen according to the language treated, e.g. DEUSEM for German or FRASEM for French. The dictionary contains the rules or sets of rules, and SEDUS the mechanisms for control, application and interpretation of rules. As the rules require syntactic structures to operate on, the minimum analysis quality must be NOMA-RESCUE.

The control structure of SEDUS is quite simple. When processing starts, the sets of noun groups and predicates are known, and the prepositional groups receive an additional structure as predicates, the preposition being the governor of the rest of the noun group. The relation between preposition and a noun group corresponds to the rection (case requirement) of the preposition, which is normally accusative or dative. The governor of the original prepositional group is considered as the subject of the preposition. SEDUS operates first on the set of noun groups, and, if this is exhausted, on the set of predicates. In many cases, the result of rule application is only a modification of labels within a noun group or a predicate, but it is also possible that new pieces of structure are inserted. The new pieces will be inserted into one of the two sets mentioned and will also be processed in a following iteration. This means that the sets of noun groups and predicates are in fact queues.

The semantic dictionaries contain two types of entries: feature entries which allow the assignment of semantic features to lexical units (only nouns!) and rule blocks, i.e. sequences of rules, operating on syntactic structures. All these rules are accessible only via a lexical unit, represented as a character string. Examples for such strings are RULE (VRB), RULE (SUB), COMPLETE (ADJ), COMPLETE (VRB), etc.

Feature entries exist only for nouns and some pronouns (like 'I', 'he'). The associated information consists of two separated sets of features:

(1) a kind of Chomskyan classification with values like 'human', 'abstract', 'concrete', 'animate', 'collective', etc., which we simply call 'semantic features'. Experience shows that this classification is of restricted use, because it is rather rough. The German semantic dictionary contains approximately 80000 entries which have this kind of classification (automatically derived from a dictionary produced by the former project on lemmatisation).

(2) a language specific set of features, called 'P-classes'. These P-classes were defined originally for the identification of the meaning of prepositions in circumstantial noun groups, but they are very useful also for other purposes, especially for the description of conditions on valency-bound constituents or for the analysis of compound words.

As mentioned before, the P-classes are language-specific, but it might be possible to describe most of them by sets of universal values. Up to now, a mapping between P-classes of different languages has not been attempted.

To each (primarily morphologically defined) lexical unit we can assign a rule block, i.e. a sequence of rules R1, R2, . . ., Rn. If a block is to be applied to a piece of structure, the rules are tested subsequently, and only the first one for which the conditions are met will actually be applied. This means that rule blocks have much in common with previously described analysis processes in which the constituting subprocesses are invoked sequentially in the preferential mode. From this point of view we can interpret a rule block as an analysis process associated with a lexical unit. In this sense, the task of SEDUS is only the control over the SEDAM-processes. The aim of a SEDAM-process is primarily the disambiguation of syntactic structures, especially of lexical units, and in certain cases a structural transformation. There exists also a possibility to produce a failure message which will cause SEDUS to stop further processing of an actual interpretation of a sentence. The intermediate result will be deleted in this case. For the formulation of rules, we dispose of a special language which is clearly adapted to SUSY's basis structure. It allows navigation in trees vertically (up and down) and horizontally (left and right), starting at one of the implicitly focused nodes, namely the node where the actual lexical unit is attached (which is always a governor) and its father node. At the same time, conditions on the properties of nodes can be checked. If a semantic property is to be tested, the system automatically accesses the dictionary for providing the information, if it is not yet present. The rule formalism allows us to describe what should happen, if the conditions are met: mostly this results in a change of the originally focused lexical unit, but also new pieces of structure can be produced. The formalism offers the possibility to define abbreviations for operations – conditions and/or assignments – which are frequently used. These abbreviations can be used like subroutines in a programming language or like primitive, predefined operations in the SEDAM formalism. This facility is especially useful for the description of structural synonymy or fixed phrases where a limited number of very frequent patterns can be identified, which differ from each other only with respect to the lexical units.

Although the formalism is far more procedural than descriptive, the use of abbreviations can simulate a purely descriptive notation. For example, we could write a rule for *Übersetzung* consisting of the statement ADJSUBXYZ (AUTOMATISCH, MASCHINE), the evaluation of which could be interpreted in several ways – depending on an actual definition of ADJSUBXYZ. It could mean, e.g.: if you find a noun group with governor *Übersetzung* with an adjectival group consisting of the governor *automatisch* interpret it as *Übersetzung mit Hilfe einer Maschine* or replace the adjectival group by an instrumental noun group with governor *Maschine*, etc.

As mentioned above, SEDAM is designed for the resolution of semantic ambiguities. A special problem class are the prepositions which either are considered as case markers of valency-bound constituents or as representatives for circumstantial relations. For the latter case, we dispose of a SEDAM process for each preposition whose interpretation can only be identified by inspecting its syntactical context. In a first approximation, this meaning can be deleted by using the P-classes of the noun governed; in more difficult cases, more complex conditions are established. As a result of the disambiguation the preposition receives a new, artificial lexical unit, which can be considered as an element of an interlingua, because it will not be translated. Only in the semantic generation phase will it be replaced by a target language word. For modal constructions, we dispose of a SEDAM process for each modal verb, which on the basis of the structure and tense and voice information assigns new lexical units and perhaps changes the structure. Whilst continuous and uninflectable fixed phrases were treated in WOBUSU, inflectable and discontinuous idioms are dealt with in SEDAM. As idioms normally have a well-formed syntactic structure, their recognition is not necessary in earlier analysis stages. However, it seems preferable to recognise idioms as early as possible, because this can lead to a considerable reduction of ambiguity and consequently to an improvement of the analysis quality. The recognition of a fixed phrase in SEDAM always leads to structural changes, because several words will be combined to form a new lexical unit. An interesting possibility is the use of the synonymy relation, which can be established between structures. For example, nominalised structures like *die Lösung von Differentialgleichungen* can be related to their verbal counterparts with the appropriate syntactic function labels, e.g. *Differentialgleichungen lösen* with accusative relation to the noun group for *Differentialgleichungen*. Similarly the internal structure of a compound noun can be represented via the synonymy relation. If necessary, the synonymous structures are used in transfer or generation. With the termination of SEDAM, the end of SUSY's analysis is reached. For each sentence we have now a set of results, each of which is marked by a quality label and an indication whether it was processed by SEDAM. Moreover, we can access status information which shows the maximal quality of the analysis for each sentence.

Transfer

The task of TRANSFER is the translation of the SL basis structure into a corresponding TL basis structure. The main transfer component is the bilingual dictionary, by means of which the SL lexemes are replaced by TL expressions. In exceptional cases, the geometry of the basis structure is changed, but normally it remains untouched. This means especially that the order of the nodes in the analysis trees is not changed. This would even be very difficult, because the structures are very much dependency-oriented and order can be identified only through rather difficult procedures. Looking at TRANSFER in detail, we can distinguish the following main operations:

TRANSFER-INPUT
TRANSLATE LEXICAL UNITS
TRANSFER NEGATIONS

TRANSFER NOUN GROUPS
TRANSFER-SAFETY-NET
TRANSFER PREDICATES
TRANSFER-CONSISTENCY
TRANSFER ADJ/ADV
TRANSFER-OUTPUT

These subprocesses are invoked in the sequential mode, and each uses the results of its predecessor(s). TRANSFER has no RESCUE-mechanism.

TRANSFER-INPUT is the device which offers structures to be treated to the following subprocesses. For each sentence, the maximal analysis quality is known, and on this basis the best results are selected from the database. At least a result of DIHOM quality is required.

TRANSLATE LEXICAL UNITS is linked with the bilingual dictionary and operates only on single terminal nodes. The string associated to a terminal is looked up in the dictionary and replaced by a target language string. There are no conditions on the surrounding syntactic context, because it is assumed that the semantic disambiguation has delivered the right interpretation of ambiguous words. Only the condition on the special field of the text is evaluated, if it is present at all. Just like in the SEDAM dictionary, we can imagine that each source language lexeme has an associated rule block in the transfer dictionary, the rules being applied preferentially. The only difference is that a transfer dictionary entry can have only a simple condition on the special field. If we have to translate on the basis of a sentence description which was not processed by SEDAM, the translations of the words will be quite arbitrary (if the special field information does not help). As a result of this process we receive a basis structure whose terminal nodes are labelled with the new lexical units and some information associated to them as well as with the 'old' information, except the source lexeme. It should be mentioned that artificial lexemes (especially the artificial prepositions) are not touched by this process. Pronouns and articles are translated, if they have entries in the dictionary, but these translations are overwritten in generation (if everything goes well – otherwise they are part of the safety net). If a noun, adjective or verb is missing in the dictionary, the system tries to translate it by using two subsequent (preferential) processes. The first one identifies the structure of which it is a governor and checks whether a synonymy relation is present. If it is, the synonymous structure is evaluated and integrated into the structure to which it is attached. This is the only place where important structural changes happen. The second process checks whether the lexical unit contains hyphens or affix/composition markers, and if this is the case, the individual strings are translated.

TRANSFER NEGATIONS is a process activated if the source language has discontinuous negations (like French and Russian, e.g. *aucun . . . ne*) and the target language does not.

TRANSFER NOUN GROUPS deals with the syntactic functions and the number of noun groups and with articles. It is broken down into three subsequent tasks, which operate on single noun groups.

The task of TRANSFER GN-CASES consists in translating the syntactic functions within noun groups. The governor of a noun group may be associated with a set of pairs (s, t), where s is a source and t a target case. If the governor has the relation s to a dominated noun group or predicate, this relation is changed to t. If a relation s' is not found as first element in the set of pairs of cases, s' is looked up in a default table and replaced by the corresponding target case in the actual noun group. Of course, this procedure is applied only to valency bound relations. It should be mentioned that most of the case translations can be carried out by using a default table. Only exceptions are coded in the transfer dictionary.

The number of a noun group is considered as a universal, which does not change from one language to another. There are, however, a few exceptions connected with pluralia or singularia tantum, i.e. nouns which in one language always are plurals, even when they actually have a singular meaning or vice-versa. Such nouns have special information in their dictionary entries, which is evaluated by TRANSLATE GN-NUMBER. (If the target word is a singulare or plurale tantum, generation will solve the problem – so this case need not be treated in TRANSFER).

The process TRANSFER ARTICLES is invoked only if the source language is Russian (or Esperanto), because in this case, we have to decide which article a noun group should have. If a determiner is present or if the group is an apposition, nothing will happen. In the remaining cases, a definite article (for Russian) or an indefinite article (for Esperanto as source language) is added. But this is only a default decision, which may be revised in a following transfer process or even in generation.

A TRANSFER-SAFETY-NET makes a consistency check, especially concerning attributive subclauses and coordinations. In the case of an obvious analysis error, we try to remedy the structure such that no catastrophe will happen in generation.

TRANSFER PREDICATES is a subgrammar which operates on predicates only; i.e. on clauses and adjectival groups. It has two components. TRANSFER REFLEXIVES is used when a verb of the source language has a necessary reflexive pronoun, but its equivalent in the source language is not reflexive. This difference is stated in the entry of the transfer dictionary. The information causes a deletion of the pronoun. Also the reverse case is foreseen, but it does not cause the production of a noun group for the reflexive pronoun. Only a label for obligatory reflexive is added to the governor of the predicate, which will be used in generation. The TRANSLATE PRD-CASES procedure is exactly the same as in TRANSLATE GN-CASES. Only a different list of default case translations is used. Only when verbs behave exceptionally do they need additional information, e.g. for *gefallen* – 'like'; ((SBJ, ACC), (DAT, SBJ)), which means 'if you translate *gefallen* by 'like', change the subject relation to accusative and the dative relation to subject'.

TRANSFER-CONSISTENCY is analogous to TRANS-FER-SAFETY-NET. Crucial errors are detected and remedied as well as possible.

TRANSFER ADJ/ADV is used when translating from French to German (e.g.); groups of predicative adjectives – being considered as predicates – are transformed into adverbial noun groups, e.g. 'elle est *heureuse*' – 'sie ist *glücklich' (We consider glücklich* as an adverb in this case!).

TRANSFER OUTPUT produces the transfer results on a file together with a label for each sentence showing that it is a transfer result.

Synthesis

The task of SYNTHESIS consists in constructing target language sentences on the basis of their description delivered by TRANSFER. The generation phase is determined only by the target language and has no knowledge from which source language the transferred structures are derived. Superficially, SYNTHESIS can be regarded as a reversed ANALYSIS. It is split up into three sequentially operating subsystems, namely

SEMSYN
SYNSYN
MORSYN

i.e. a series of semantic, syntactic and morphological components.

SEMSYN is the semantic generation component. It is realised through the software component SEDUS (which we mentioned in connection with the semantic analysis) and the semantic dictionary of the target language. The task of SEMSYN is to a large extent the reverse of SEDAM, namely the production of idioms and the translation of artificial words into target lexical units. As TRANSFER treats idioms exactly in the same way as simple words, namely as character strings without structure, there will be no difference between simple and complex expressions after TRANSFER. What is a simple word in one language may turn out to be complex in another, and vice versa. Because fixed phrases may be discontinuous and inflected at different places in a sentence, we have to write generation rules for them, which produce their syntactic structure. These generation rules which are accumulated in the semantic dictionary are applied by SEMSYN – just in the same way as rules are applied by SEDAM. The prepositions with artificial lexical units are also treated as in SEDAM. The nouns they govern receive their appropriate target language P-classes, and on this basis, the target lexical unit is computed. The semantic dictionary must contain as well rules for the production of modal verb constructions, i.e. for the transformation of the deep syntactic description into a more surface-oriented structure.

SYNSYN uses the output of SEMSYN or of TRANSFER to produce a sequence of terminal nodes, labelled with the character strings of the stems of lexical units and all morphological information necessary for the production of the final strings. SYNSYN is carried out in several consecutive steps:

SYN-DICTIONARY
SYN-COMPOUNDS
SYN-NUMBER/GENDER
SYN-PRONOUN-REFERENTS
AGING-PROCESS
LINEARIZATION

In the SYN-DICTIONARY phase, the syntactic dictionary of the target language is looked up, in order to assign to the lexical units the corresponding stems and all information needed in the following processes. The majority of the additional information is morphological. If a word is unknown, i.e. not found in the dictionary, a safety net procedure tries to provide some tentative labels. For example, unknown German nouns are considered as feminines, having a plural in -*en*, for English nouns, the plural ending will be '-s', as well as for French nouns, etc.

It should be mentioned that the SUSY generation dictionaries do not have a direct connection to the corresponding morpho-syntactic analysis dictionaries. Each lexical unit has only one entry in the generation dictionary, whereas it may have several in the analysis dictionary, according to the number of the stems to be considered. Of course, also the content of the entries is quite different.

The SYN-COMPOUNDS component treats compounds and derivations by using three sets of rules:

(1) The first set of rules transforms certain syntactic structures into compounds, e.g. the structure of *système de traduction*, in which TRANSFER replaces only the lexical units (which would then correspond to *System von Übersetzung*) is transformed into a single node carrying the stem *Übersetzungssystem*. Under which conditions such compounds are produced, depends on the structure with the associated labels and the dictionary information.

(2) The next set of rules deals with the transformation of prepositional attributes into attributive adjectives, i.e. a noun group can be changed into an adjectival group under certain conditions. Example:

Zitronen aus Israel = israelische Zitronen.

(3) The third set of rules deals with derivations. It may happen that a lexical unit has to be realized as an adjective, but actually is marked as a noun. This happens in our example above: We have an adjectival group with the noun governor *Israel*. We consider the following derivations:

ADJ–SUB: *heiss–Hitze,* 'hot'–'heat'

VRB–SUB: e.g. *berechnen–Berechnung,* 'compute'– 'computation'

SUB–ADJ: e.g. *Israel–Israelisch*

The SUB–ADJ derivation can also produce left elements of compounds, e.g. *Übersetzung–Übersetzungs-.*

Also derivations from verb to adjective are possible, and this is done in two steps: first VRB–SUB and then SUB–ADJ. So, the VRB–ADJ derivation, applied to *übersetzen* would deliver *Übersetzungs-.* It should be noticed that this is different from the production of a participle, which is also a VRB–ADJ derivation.

For appropriate functioning of the derivation mechanism, verbs, noun, and adjectives need a derivation code in the dictionary which either tells how to build the derived form or indicates irregularity (e.g. for *heiss*). In such a case, an artificial lexical unit is built (e.g. HEISS*AS) and looked up in the dictionary, which returns us the right form (e.g. HITZE). More complex derivations, e.g. the production of nomina agentis, are not foreseen.

SYN–NUMBER/GENDER assigns number and gender to noun groups. We distinguish three cases:

(1) If the governor is a noun, the values are copied on the noun group, observing possible number restrictions of the noun.

(2) In pronominal noun groups, number and gender are taken from the referent.

(3) Adjectival noun groups receive their number and gender from the subject or from another constituent (in the *attribut de l'objet* case).

A rather complex process, SYN–PRONOUN–REFERENTS, deals with pronouns which did not receive a pointer to their referents in analysis. The process uses number and gender information of the *source* language, identifies a set REFN of possible referents, and assigns a weighting to them on the basis of the sequence of the syntactic relations which connect the referent with the pronoun. There exists as well a set called REFA, in which noun groups of the preceding sentences are stored together with their weightings. Also the members of REFA are compared with the actual pronoun, and those which fit are added to REFN. Finally, the element of REFN with the highest score is considered as referent of the actual pronoun.

In the AGING PROCESS members of the above-mentioned set REFA receive a new weighting which is lower than the weighting they had before. Elements whose score is less than a given limit, will be deleted from REFA.

In the next step, the actual sentence structure is traversed, the noun groups are identified and weighted on the basis of their syntactic relations (subjects have a higher score than objects), and added to REFA. REFA is now prepared for the next sentence.

The process of LINEARISATION is in fact the core of SYN-SYN, because up to now we have applied only minor modifications to the actual data base. The present process will use the actual state of the description of a sentence for the production of a linear structure, by starting with the root node and expanding the levels of the tree subsequently.

The linearisation is not carried out by operating within the given tree, but a totally new structure is produced in a separate place. The new structure is linked to the old tree structure by pointers. The tree is not changed, but only inspected, if necessary.

The new structure is held in a push down store, called UPTG, whose elements are highly structured. Each such element consists of the constituents of a noun group or a predicate, which are ordered according to the grammatical restrictions of the target language. Where the target language does not imply a fixed order, we use the order of the constituents in the SL text.

The linearisation process operates always on the last element of the push down store, which is a sequence of constituents. At some stage of processing, the constituents are expanded and new elements are added to the store. If a constituent is terminal, it is attached to the sequence of terminals and disappears from the push down store. When the push down store is empty, the sequence of terminals is complete and can be passed to MORSYN.

The linearisation process itself consists again of several sub-processes, which are applied in sequence. The last one attaches the terminals to the sequence of terminals, by inspecting the last push down store element from left to right. If it finds a non-terminal, this will become the nucleus of a new push down store element, and then the sequence of linearisation processes will start again from the beginning, in order to expand the new constituent.

MORSYN receives the results of SYNSYN, i.e. a sequence of terminals for each sentence, and deals with morphological generation. The terminals are in fact structured entities, which contain precise information on the kind of operations to be applied, in order to produce the word-form desired. Verbs, nouns and adjectives have sufficient information for the generation of inflected forms, because the assignments were carried out by the dictionary look-up in SYNSYN. For articles, prepositions and pronouns, the appropriate character strings are produced by consulting the generation dictionary again. The procedures are the following:

(1) Articles: MORSYN constructs an artificial word using the information on definiteness, case, number and gender and looks it up. As a result, the word form is returned.

(2) Prepositions: The prepositions which indicate circumstantial relations received their target language string either in TRANSFER or in SEMSYN. What is left are the prepositions which indicate valency boundedness. For such a preposition, MORSYN builds an artificial word containing the case value and looks it up in the dictionary.

(3) Pronouns: For all kinds of pronouns, artificial words are built which characterise the pronoun with respect to its function (e.g. relative, interrogative, possessive, genuine, etc.) and morphology (inflectional case, number, gender, person). This word is looked up in the dictionary and the corresponding word form is returned.

These examples of using the dictionary – and there are still more instances which we did not mention – show a general strategy adopted especially in the implementation of the generation phase of SUSY. There is a distinction between normal, regular phenomena and exceptions: the regular cases are implemented as programs, the exceptions are put in the dictionary. Some exceptional cases appear also in the last submodule of MORSYN, e.g. the combination of French *de le* to *du,* the modification of articles (English 'a'–'an', French *le–l*), etc. The necessary rules are also dictionary entries. Finally, the typographic form of the individual strings is fixed, e.g. capital letters at the beginning of sentences, capital letters for German nouns, and then the resulting text is printed out in two columns (or displayed on the screen of a terminal, or stored on a file). The left column contains the original text, divided into sentences, the right column the translation results.

MACHINE TRANSLATION AT
THE TAUM GROUP

In 1965, the Université de Montréal began to develop expertise in natural language processing, with the creation of the CETADOL centre (Centre d'études pour le traitement automatique des données linguistiques), under the sponsorship of the Canadian National Research Council. The new centre brought together linguists and computer scientists, under the direction of Prof. Guy Rondeau. Machine translation (MT) gradually emerged as an important focal point in CETADOL's research.

Between 1968 and 1971, under the direction of Prof. Alain Colmerauer, the group made MT research its main objective, renaming itself TAUM (Traduction Automatique, Université de Montréal). Using Colmerauer's newly developed Q–SYSTEMS software ([532] Colmerauer 1971) a first complete MT prototype (TAUM–71) was assembled, already featuring a typical second-generation design. Under the leadership of Prof. Richard Kittredge, the group continued, until 1977, the development of research prototypes, exploring possible ways to overcome the complexity barrier that faced systems of a realistic scale. It then appeared that one promising approach was to have MT systems specifically tailored for restricted types of texts.

This idea, which developed into the sublanguage approach would soon be given a real test. Since 1973, the group had been funded by the Canadian Secretary of State, in the hope that near-term applications would result. In 1975, TAUM was given a contract to develop a system capable of translating public weather forecasts. The resulting system, TAUM–METEO, was delivered the following year, and has been in daily operation since May 1977.

This success led the group to accept a second, and more ambitious contract: the next three years would be spent, under the direction of M. Marcel Paré, on the development of a prototype system for translating aircraft maintenance manuals.

Because many unanticipated difficulties were encountered in the first two years, the prospect was just getting off the ground when the deadline arrived for an independent evaluation of the current prototype, TAUM–AVIATION ([186] Gervais 1980). The evaluators concluded that the scope of the system was still too narrow to envisage cost-effective production in the near term. Shortly after, the sponsor decided to halt the project and look for a broader funding base. Meanwhile, the TAUM group had to be disbanded. (For more details on the demise of the TAUM group, see [312] Macklovitch 1984).

In the next section of this chapter we examine the general model which the TAUM Group was led to adopt. The section following that describes the implemention of this model in two specific systems: TAUM–METEO and TAUM–AVIATION. Finally, in the last section, we express some views on the current state of the art and on the future of MT.

TAUM's machine translation model

By 1971, TAUM had laid down the outlines of an MT model which would form the basis for the development of several large-scale systems in the following years. Defined in close interaction with other research groups, such as the GETA of the Université de Grenoble, the model embodied the fundamental principles of the second-generation approach to MT. It incorporates the three basic design features of second-generation systems: indirect approach, transfer scheme and separation of algorithms and linguistic data.

In early MT systems, translation rules are applied more or less directly to the input string: no attempt is made to characterise this input in terms which are independent from the target language. By contrast, in TAUM's systems, an analysis process assigns a complete structural description to the sentences of the input text before any translation rule is applied. This is justified by the fact that, in general, translation rules have to be framed in terms of structural correspondences rather than in terms of string-to-string correspondences.

Conceptually, the simplest way to realise the indirect approach would be to map the input text into a language-independent semantic interlingua, from which a text could be directly synthesised into any target language. Unfortunately, this old idea of a universal semantic language (cf. Leibnitz's Characteristica Universalis) suffers from deep philosophical and technical problems that, to this

day, remain insuperable. Even if 'neo-nativist' linguistics has renewed hopes for a broad universal base to natural languages, universal semantics is still an uncharted territory. This is not to say that
investigations in this territory should not be pursued seriously. The
work of Wierzbicka ([623, 624] 1972, 1980, etc.) is, in this respect,
among the most fascinating and important. In fact, even putting
aside the question of universality, we still know very little about
how to adequately represent the meaning of realistic subsets of a
single natural language, and about how to establish the mapping
between such a representation and the texts (e.g. discourse structures). It is therefore no surprise that the few attempts to develop
experimental MT systems using language-independent semantic
representations (e.g. [633] Carbonell et al. 1981; [644] Lytinen
and Schank 1982) have met with very limited success. To our
knowledge, no such system can currently analyse more than a
handful of examples; moreover, the deeper the analysis, the more
difficult it becomes to synthesise an adequate text in the target
language, because discourse structures are among the least understood phenomena of natural languages. There is no question that
research on deeper semantic representations, universal or not, is
crucial to the future of MT. But in the meantime, MT systems of a
realistic scale have to use relatively shallow, language-dependent
structural descriptions of natural language texts. A transfer component is therefore required to effect a mapping between the
structural descriptions (SDs) of the source language and those of
the target language. The complexity of this component is an inverse
function of: (a) the depth at which the SL and the TL SD's are
stated (that is, the degree to which they manage to abstract away
from superficial language-specific constraints); and (b) the degree
of relatedness of SL and TL. In most systems, the SDs retain SL
and TL lexical units, and the transfer component includes a bilingual dictionary.

In early MT systems, each specific rule was directly programmed
for the computer, with the result that algorithms and linguistic facts
were inextricably intertwined in huge and non-modular programs.
In second-generation approaches, such as TAUM's, specific rules
are written using special purpose metalanguages. The linguistic
data is described as sets of rules; implicit in these rules is a fixed set
of algorithms which are described separately, as rule interpretation
programs. Grammars written in this way are much more perspicuous for the linguist/translator, and much easier to modify. One of
the best examples of these metalanguages is TAUM's Q–SYS-
TEMS ([532] Colmerauer 1971). This formalism enables the lin-

guist to write grammars as a set of transformation rules which map strings of trees on to other strings of trees. The rules can incorporate variables, and thus state correspondences between classes of strings of trees. For example, the rule (1):

$$(1) \quad V(I^*) NP(J^*) = = VP((I^*), NP(J^*)).$$

(where starred letters represent variables) maps a string consisting of any verb I^* followed by any noun phrase J^* on to a string consisting of a verb phrase made up of the verb I^* and the noun phrase J^*. The data structure on which these rules operate is a chart (a particular type of graph, with arcs labelled with trees; cf. also chapter 8). The left-hand side of each rule is matched against the paths of the chart, variables thereby becoming instantiated. When a match is found, a new arc labelled with the string of trees described in the right-hand side of the matched rule is added to the chart. All rules are applied in parallel to all paths of the chart, including newly added paths. We will briefly examine some examples of rule application below. This formalism is of an extreme simplicity; yet, it is powerful enough to express the rules of a complete MT system, dictionaries included. Over the years, TAUM developed a number of other metalanguages. The most recent examples are SISIF, REZO and LEXTRA; these tools were intentionally designed so as to trade off some of the generality of Q-SYSTEMS for efficiency in more specialised tasks. SISIF ([327] Morin 1978) is a very simple finite-state automaton, suitable for such tasks as automatic pre- or post-processing of the texts. REZO ([559] Stewart 1978) is an adaptation of Wood's ([626] 1970) transition networks used for syntactic-semantic analysis. LEXTRA ([535] Gérin-Lajoie 1981) is a transformational system especially designed to perform lexical transfer.

We have stated above that TAUM adopted an approach in which transfer rules relate two language-dependent structural descriptions. Generally speaking, these descriptions are semantically annotated deep structures of SL and TL sentences. Of course, it is well-known that to capture the full meaning of a text, it is necessary to identify various types of textual links across sentence boundaries: anaphora, foci, etc. TAUM did conduct a few experiments on textual processing (e.g. [20] Hofmann 1971). It soon became evident, however, that too little was known about textual phenomena to make text grammars practical in large-scale NLP systems. At that point, TAUM hypothesised that, in some restricted types of text, sentential representations could fare reasonably well. The deep structures used in TAUM's systems were occasionally quite different from those used in transformational grammars. The criteria on the basis of which they were chosen included that they: (a) be effectively computable on the basis of our current knowledge of language; (b) provide as far as possible an unambiguous characterisation of natural language sentences; and (c) facilitate the formulation of transfer rules. In this last respect, it is to be noted that in most of TAUM's systems, the set of tree structures admissible as intermediate representations is the same for the source and the target language, lexical items aside; both languages share the same set of context-free 'base rules' but have different lexicons. TAUM's intermediate language (IL) is thus essentially similar to Vauquois' ([477] 1975) 'pivot language'. TAUM's IL is language-dependent: it preserves SL and TL lexical items: no lexical decomposition is performed (but a partial meaning specification is given by means of semantic features). The representations are syntactic, since they encode information concerning the deep syntactic constituents of the sentences (with some traces of surface syntax). Finally, the IL also encodes a good deal of semantic information: (a) predicate/argument patterns (in a sense similar to [568] Bresnan 1982) are highlighted; (b) tree constituents are marked with semantic features which have been projected from lexical items through a fairly elaborate semantic calculus, as described later. Of course, this IL is by no means a complete semantic representation. But, at least at the time TAUM's systems were designed, this level was about as deep as the state of the art permitted large-scale systems to go.

The analysis component

In TAUM's model, SL analysis is further subdivided into three sub-components: pre-processing, morphological analysis and syntactic–semantic analysis.

Since the data structure on which the main components of the linguistic processing will operate is a chart, the input string must first be converted into a (degenerate) chart. This is accomplished by taking each successive word form as the label of an arc linking successive chart nodes. At the same time, since the processing unit is the sentence (or a lower-rank element such as a single item in a parts list), the input text is segmented into successive units. Pre-processing is sequentially ordered before the rest of the processing and realised by means of a simple finite-state automaton such as SISIF.

In TAUM's systems, this process converts an input chart into a chart whose arcs are labelled with lexical items, together with their various morpho-syntactic, syntactic and semantic properties. The linguistic data needed for morphological analysis is a set of morphological rules and a source language dictionary. There are four sub-types of morphological rules:

(1) inflectional morphology deals with syntactically conditioned alternations in word forms (e.g. number, gender, tense, etc.);

(2) derivational morphology consists of rules for forming new words through affixation over existing words (e.g. special–specially);

(3) compositional morphology accounts for words formed by concatenation of other words (e.g. engine-driven);

(4) category assignment rules identify classes of words which are not listed in the dictionary, such as numbers in digit notation.

As an example of morphology rules, consider the input chart in (3) and the Q-SYSTEMS rules of (4):

(3) $\xrightarrow{\text{BUY}}$ • $\xrightarrow{\text{SUPPLIES}}$ • $\xrightarrow{\text{PONC(.)}}$ •

(4)   R1: $A^* == WORD(\$\$A^*)$.
      R2: $WORD(U^*, S) == STEM(U^*) + S$.
      R3: $WORD(U^*, I, E, S) == STEM(U^*, Y) + S$.
      R4: $WORD(U^*) == STEM(U^*)$.

The variable A^* matches any label (here any word); $\$\$$ is an operator which relates any label and the corresponding list of characters; and the variable U^* matches any list of daughter nodes in a tree (here, lists of characters). Rule R1 'explodes' the words into trees representing lists of characters. Rule R2 states that a word ending in -S may be made up of a stem plus an -S suffix (e.g. *ties*). Rule R3 states that -IES may signal the addition of the suffix -S to a stem ending in -Y (e.g. *supplies*). Finally, rule R4 signals that any input word may also be a basic lexical form (e.g. *series*). These rules transform (3) into (3′), in which we have annotated each new path with the number of the rule which created it.

(3′)

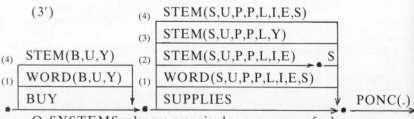

Q-SYSTEMS rules are organised as a sequence of subgrammars;

at the end of each subgrammar, all paths which have been used up
by the rules (that is, on which a match with the left-hand side of
some rule has been found) are pruned. Thus, (3′) becomes (3″):

(3″)

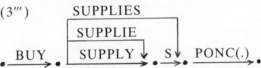

In the next subgrammar, rule R5 and consequent pruning will
produce (3‴), ready to be submitted to dictionary lookup:

(5)   R5: STEM($$A) == A*.

(3‴)

In the next subgrammar, rule R5 and consequent pruning will
produce (3‴), ready to be submitted to dictionary lookup:

SL dictionary rules will then be applied to the chart; for example,
the rules of (6a) would, after pruning of used paths and of spurious
stems, produce (6b):

(6a)   BUY == V(BUY).
        SUPPLY == V(SUPPLY).
        SUPPLY == N(SUPPLY).

(6b)

Generally speaking, SL dictionary rules add much more informa-
tion to the chart than a simple syntactic category. Lexical informa-
tion can be classified along at least two different axes: syntactic
versus semantic and inherent versus contextual:

(a) inherent syntactic information:
syntactic category (N, V, ADJ, ADV, ART, etc.)
e.g. 'small' is an ADJ
syntactic subclasses (mass or count for nouns, preadjectival
for adverbs, etc.)
e.g. 'oil' is a mass noun
(b) contextual syntactic information: strict subcategorisation
frames
e.g. the verb 'check' takes:
a first argument which is an NP
a second argument which is either an NP or a clause ('check
that S', 'if S', etc.)
an optional third argument which is a PP ('check x for y')
an optional particle ('check out', 'check up')

(c) inherent semantic information: semantic subclasses e.g. 'oil' denotes a fluid, 'check' an action, etc.

(d) contextual semantic information: selectional restrictions e.g. an NP that acts as second argument for 'install' must denote a physical object.

All this information is encoded in the chart, and will be used in subsequent processing.

We group syntactic and semantic analysis together, because in TAUM's processing model, both are fully integrated, and semantics is subordinated to syntax. More precisely, the analysis process is syntax-driven, but receives some semantic guidance: syntactic paths which lead to semantic anomalies are pruned as soon as possible. For example, no VP structure will be built if the verb and its object are semantically incompatible. No distinct semantic data structure is built, beyond annotating the syntactic trees with semantic features. Syntactic descriptions are usually of a fairly standard type. More or less traditional syntactic categories are used in conjunction with a complex set of subclasses. Most standard syntactic transformations (or rather their inverses) are performed so as to produce a canonical ordering: predicate + arguments + 'circumstancials'. This is required both for the checking of semantic constraints and for the production of an output tree which belongs to the IL expected by the transfer component. While the syntactic tree is being built, a semantic calculus takes place. The semantic features attached to lexical items are projected on to higher nodes, by means of a set of rules working in compositional fashion. These rules specify structural constraints on the 'feature percolation' process. Here are some examples of rules which might be used for the semantics of noun phrases:

(a) An NP node receives all the semantic features of its head noun.

(b) If noun phrase NP_1 has a head noun which: (1) belongs to such classes as partitive (e.g. portion, piece) or measure unit (e.g. pound, gallon), and (2) has an 'of NP_2' complement, then it (NP_1) receives the semantic features of NP_2. This rule accounts for the fact that *a pound of cake* has most of the selectional properties of *cake*.

(c) Certain semantic features of attributive adjectives are raised to the mother NP node. This mechanism accounts for facts such as the following: in technical manuals, nouns denoting 'defects' (e.g. leak, damage) exhibit certain selectional patterns, and the same patterns also occur when a non-'defect' noun is modified by a 'defect' adjective (e.g. *defective pump*).

(d) In conjoined NPs, the higher NP node receives the *intersection* of the semantic features of the conjuncts.

These examples give an idea of the sort of semantic regularities which can be described in systems such as TAUM's. TAUM's analysis grammars have been implemented by means of various parsers. In recent years, two contrasting approaches have been tried: (a) Q-SYSTEMS, in which the basic algorithm is bottom-up, all paths, parallel parsing (although some other parsing algorithms can be simulated); and (b) REZO, in which the basic algorithm is top-down serial parsing (expectation-driven rather than data-driven); (cf. chapter 7). The relative perspicuity of these two approaches is largely a matter of taste and opinion (large-scale grammars always tend to be opaque). Their comparative efficiency, however, is highly dependent on the linguistic data: expectation-driven parsers fare better where we can have strong expectations at the outset (e.g. clear statistical tendencies), or where there is systematic ambiguity in the lexical data (e.g. between nouns and verbs in technical English). Data-driven parsers, on the other hand, are preferable in cases where expectations are weak (e.g. conjunctions, left substructures shared by several constituent types, left recursion, etc.). There is no space here to examine the details of the formalisation of the rules.

The transfer component

The transfer component maps an SL intermediate structure on to a TL intermediate structure. The most obvious task is to translate lexical items; but there is also a need to transfer certain structures.

As we have seen, the intermediate language is based on the fact that SL and TL can be assigned deep structures by means of a single set of base rules. In spite of this approach, lexical transfer cannot simply substitute lexical items, leaving the tree structure unaffected. This is because, generally speaking, corresponding lexical items of both languages can differ as to their subcategorisation frames. It follows that lexical transfer has to incorporate powerful transformational mechanisms. It should be noted that comparable lexically-driven transformations would also be needed in a hypothetical interlingua model, both for analysis and synthesis; these rules would be comparable to the lexicalisation transformations of generative semantics, which relate lexical structures to configurations of semantic primitives. If transfer is to produce acceptable TL structures, various types of lexical transformations ([247] Labelle 1981) will have to be provided for, as illustrated by the following examples:

(8) Syncategorematic prepositions
check x against y → *comparer x à y*
x depends on y → x *dépend de y*
(9) Permuted arguments
supply x with y → *fournir y à x*
bleed ⟨tank⟩ of ⟨air⟩ → *purger ⟨l'air⟩ du ⟨réservoir⟩*
(10) Incorporation of arguments or adverbials
reinstall x → *remettre x en place*
cantilever x → *monter x en porte-à-faux*
bond x electrostatically → *métalliser x*
swim across x → *traverser x à la nage*
(11) Incorporation of head of argument
deenergise x → *couper l'alimentation de x*
service x → *faire l'entretien de x*

More complex situations arise with certain classes of verbs. Consider, for example, the verbs of checking (check, inspect, etc.). When one of these verbs occurs with a *for* argument denoting a 'defect', a negated existential has to be added in the translation, as in example (12):

(12) Check for leaks → *S'assurer qu'il n'y a pas de fuites*

It is important that transfer rules provide for an adequate constituent structure in the target language. For example, translating *deenergise* as an unanalysed string *couper l'alimentation de* creates no problem in (13), but produces an incorrect result in the corresponding passive:

(13) Deenergise test stand.
→ *Couper l'alimentation du banc d'essai.*
(14) Test stand must be deenergised.
→ **Le banc d'essai doit être coupé l'alimentation de.*

It is legitimate to ask how lexical rules can be constrained in a principled way. It does seem to be the case that most rules relate predicate/argument patterns. It has been proposed in transformational grammar that lexical insertion rules have a strictly local domain; one could hypothesise that lexical transfer rules can only relate frames that are admissible as subcategorisation frames in each language. This of course would not mean that a larger context has no role to play in lexical transfer: the elements described in these frames frequently have to be disambiguated or recovered (cf. anaphora and ellipsis) on the basis of textual and pragmatic rules.

Lexical transfer as described here turns out to be awkward to implement in Q-SYSTEMS. The main problem is that all pattern-matching in Q-SYSTEMS is anchored to the root nodes of the trees. Since lexical items occur at any arbitrary depth in the trees, one is forced to recursively decompose the tree into a string of degenerate trees before applying rules involving lexical items. But this is most unsatisfactory, since the tree is built in the first place as a means of expressing all structural relations which might be relevant in stating transfer rules. For this reason, TAUM was led to develop a special metalanguage called LEXTRA, which makes it easier to state the type of tree transformations required by lexical transfer. LEXTRA enables the translator to associate with a given lexical item L_s any number of contextual patterns: all these patterns are rooted in L_s. A matched pattern triggers a corresponding set of tree transformations which will insert a TL equivalent into its proper environment. In accordance with the essentially local nature of these operations, all three variables in LEXTRA are bounded.

LEXTRA incorporates another feature worth mentioning. As pointed out above, transfer effects a mapping between structures defined by a single context-free grammar. This constraint is formally enforced by LEXTRA: it takes as *data* an *explicit* description of the admissible tree structures, and guarantees that any tree it receives or creates is indeed an admissible tree.

This notion should be related to computational formulations of transformational grammar such as that of Petrick ([612] 1973). In Petrick's system, an explicit base component filters out spurious structures produced by inverse transformations. By contrast, typical parsers written with most current formalisms (including Q-SYSTEMS and transition networks) the base is merely a by-product of the operation of tree-building and tree transformation rules. If you try to figure out the exact description of the base component in Wood's LUNAR system or in TAUM-METEO, using nothing but the grammar, you will see that it is by no means easy. With some luck, you might even discover cases where tricky interactions between a whole set of transformations produce a most unlikely tree in the output – a tree not intended by the authors of the grammar.

The construction of a large-scale transfer dictionary is a very demanding and error-prone task, which of necessity involves a number of people, primarily translation specialists. In the longer term, translators should be provided with tools that make it possible to dispense with explicit tree manipulations. But in the meantime, it is certainly helpful to constrain lexical transformation tools in such a way that they only accept correct intermediate language. Since LEXTRA is a compiled metalanguage, it was possible to enforce some of these constraints on lexical rules at compile time, which is by far the most interesting way of doing it ([535] Gérin-Lajoie 1981).

Since both languages share the same base component, the structures produced by lexical transfer are very close to being acceptable deep structures for the target language. However, in TAUM's systems, some adjustments still have to be made. For example, SL tenses have to be replaced with TL tenses, since they have not been analysed into language-independent semantic tenses. Thus, we need to state contrastive rules, which, in this case, are far from obvious, and require a subcategorisation of verbs and adverbs. Of course it would be even more difficult to calculate language-independent semantic tenses. Both languages can also differ as to the use they make of optional ordering transformations (passive, extraposition, etc.). Ideally, these phenomena would be handled in language-specific text grammars, using such notions as focus, theme, etc. Since little is currently known about these phenomena, TAUM's strategy is to rely on the frequent parallelism of SL and TL with respect to their use of ordering transformations. To this end, the intermediate language retains some indications of the surface constituency in SL. However, when we do observe some contrast in the use of a particular device, the relevant facts can be encoded in transfer rules. For example, a number of rules can be given to translate English passives with other constructions in French.

The synthesis component

In TAUM's systems, synthesis comprises two sub-components: syntactic synthesis and morphological synthesis. Syntactic synthesis performs a function which, broadly speaking, is the inverse of that performed by the syntactic-semantic analysis component. There is one important difference, however. The analysis component maps an input chart (produced by morphological analysis) into an intermediate representation. The complexity of this task is largely due to the necessity of solving numerous lexical and structural ambiguities before arriving at the correct intermediate structure. Since synthesis starts from an unambiguous structural description of the target language, its task will be much easier. This does not mean that synthesis is a simple problem. While we know relatively well how to produce an acceptable sentence out of an unambiguous structural description, we know much less about how to generate a coherent discourse. Discovering these textual rules is obviously the key to better automatic synthesis of texts. However, this problem cannot be separated from the analysis problem: as long as the structural descriptions provided by the analysis component do not contain more global discourse information, the synthesis component cannot resort to text grammars to produce a more coherent text. Thus, in TAUM's systems, synthesis is done on a sentential basis; some degree of textual coherence is nevertheless obtained by taking advantage of the fact that languages such as French and English use to a large extent the same means of expressing that coherence. Syntactic synthesis attempts to preserve the properties of the SL surface structure, to the extent that these properties are compatible with target language syntax.

Morphological synthesis takes as input a string of lexical items annotated with: (1) their syntactic categories; (2) the morpho-syntactic categories that are to be actualised (gender, number, tense, etc.); and (3) the morphological class of the stem (e.g. Bescherelle classes for verbs). This input data enables the morphological synthesiser to produce the final form of each word.

Applying TAUM's model to sublanguages

Over the years, TAUM has realised various implementations of the model that we have just described:

> TAUM-71 ([8] Colmerauer et al. 1971); TAUM-73 ([234] Kittredge et al. 1973); TAUM-76 ([235] Kittredge et al. 1976); TAUM-METEO ([134] Chevalier et al. 1978); TAUM-AVIATION ([214] Isabelle et al. 1978).

All of these systems were developed for English to French translation, because of the peculiarities of the Canadian environment. However, the modular structure of the systems makes them easily extensible to other language pairs. Currently, only TAUM-METEO is operational. This section will be devoted to a detailed examination of the last two systems: TAUM-METEO and TAUM-AVIATION.

At least since Bar-Hillel ([3] 1960) and ALPAC ([1] 1966), it has become clear that fully automatic high-quality translation of unrestricted text (FAHQTUT) can only be a long-term goal. Much research will be necessary in order to overcome some of the limitations of current systems. However, it has also become clear that, short of FAHQTUT, properly engineered suboptimal solutions can sometimes produce valuable results.

There are basically three places where compromises can be made: (a) the level of quality (cf. the use of the Georgetown system for intelligence purposes); (b) the level of automation (cf. machine-aided human translation and human-aided machine translation): and (c) the scope of the system. Of course, all types of blends among these strategies are possible. Here, we will only examine the third approach.

Limited-scope systems can be tailored either for an arbitrarily prescribed subset of the SL (as in the TITUS (cf. chapter 12) and SMART systems), or for a natural *sublanguage* ([604] Kittredge and Lehrberger 1982), as in TAUM's systems.

The notion of sublanguage is not easy to define theoretically. However, it usually turns out to be easy to identify and characterise particular sublanguages. In many cases, texts to be translated are written within a very narrow subject matter: technical manuals, medical reports, scientific articles, financial market reports, etc. In such cases, the texts will exhibit only a part of the whole language's vocabulary and syntax; even more important, the restrictions on the domain of discourse will be mirrored by parallel restrictions on the possible semantic patterns that can be found in the texts.

When a language is considered in its entirety, without any restriction on the domain of discourse, its formal description becomes most elusive. From the point of view of syntax, 'all grammars leak' (Jespersen); as for semantics, very few safe generalisations can be made: the most unlikely sentence will turn out to be acceptable, given the appropriate context. However, when we restrict our attention to particular sublanguages, linguistic descriptions become much more tractable. The inventory of admissible syntactic patterns is sometimes drastically reduced (although it frequently includes patterns that are prohibited in the standard language). Clear-cut semantic constraints are more easily stated. Statistical analyses, which are of very little use in describing a whole language, become a valuable tool ([415] Slocum 1984).

From the standpoint of MT, sublanguages provide a powerful handle on the translation problem. For simple sublanguages, reasonably good parsers can be developed with relative ease. Moreover, transfer and synthesis are made easier by the fact noted in Kittredge ([232] 1982), that parallel sublanguages tend to exhibit parallel structures across related languages.

Criticisms invoking the lack of generality of this approach can hardly be taken seriously at a time when what has to be shown is the possibility of high-quality MT in any real-life domain whatsoever. This is precisely what TAUM set out to demonstrate, with its TAUM-METEO system.

The TAUM-METEO system
In 1974, under some pressure from its sponsor, TAUM was looking for possible applications of its current technology in limited

domains. At that time, only part of the weather forecasts issued by Environment Canada were being translated, and there was already a problem: translators were generally dissatisfied with a task that was both boring and subjected to time pressure. When the Canadian government decided to make bilingual forecasts available all over the country, TAUM was presented with an ideal opportunity to show that MT could become a practical reality.

Feasibility studies, design, development and on-site implementation of an operational version of the system took less than two years (approximately 8 person/years).

The weather reports issued by Environment Canada include two different types of texts: general synopses and regional forecasts. General synopses describe the movement of air masses across the continent and their influence on local weather conditions. Regional forecasts deal only with predictions about weather conditions in relatively small areas. Since synopses are written in a more complex sublanguage and represent a relatively small volume of text, it was decided not to translate them automatically.

Regional forecasts are written in a highly telegraphic style (in both SL and TL). Sentences usually lack a tensed verb. A typical example is given in (12):

(12) FPCN11 CYHZ 100900
FORECASTS ISSUED BY THE MARITIMES WEATHER OFFICE AT 5 AM AST SUNDAY MARCH 10TH 1974 FOR TODAY AND MONDAY HALIFAX AND VICINITY
SOUTH SHORE
VALLEY.
CLOUDY. SNOW OCCASIONALLY MIXED WITH RAIN BEGINNING LATE THIS MORNING AND ENDING LATE THIS AFTERNOON. CLEARING THIS EVENING. MOSTLY SUNNY BUT COLD ON MONDAY. WINDS INCREASING TO STRONG NORTHWESTERLY THIS EVENING. HIGHS TODAY LOW TO MID THIRTIES. LOWS TONIGHT IN THE TEENS.

Each sentence (or fragment) can be interpreted as an instance of one of a small number of semantic patterns: list of place names, atmospheric condition, minima and maxima, outlook for the following day, etc. Frequently, these semantic types have no straightforward correspondence with syntactic structures. For example, the pattern 'atmospheric condition' consists of the expression of a weather condition optionally modified by a locative or temporal specification; but the 'condition' itself cuts across syntactic categories:

(13) (a) MAINLY SUNNY TODAY.
(b) A FEW SHOWERS THIS EVENING.

TAUM-METEO is based on the same techniques that were used in the TAUM-71 and TAUM-73 experimental systems. The overall translation model has however been somewhat simplified, so as to trade some of its generality for efficiency in a very restricted domain.

The most important departure from previous systems is that there is no transfer component in the system. Most of the processing usually done at transfer is integrated into the analysis component, and the rest into the synthesis component. This does not mean that TAUM-METEO uses an interlingua model: the analysis component is TL-dependent, and the synthesis component is to some extent SL-dependent. Nor does it mean that the system is based on a direct approach comparable to first-generation systems. TAUM-METEO does produce a full structural description for input sentences, and uses it in translating.

The system deals with translation problems just as if they were SL-internal ambiguity problems. In multilingual systems, it is important to separate carefully cases of real SL ambiguity from cases of multiple TL equivalents. Real ambiguity is a fixed property of certain linguistic forms, and for this reason, is more properly described in a language-specific analysis component. Linguists have devised various tests to determine whether or not a given expression is (language-internally) ambiguous ([629] Zwicky and Sadock 1975). 'Translational ambiguity', on the other hand, arises from discrepancies between the lexical structures of a given pair of languages, and is most easily described in a transfer component specific to that language pair. Of course, combining analysis and transfer in a multilingual system could only mean this: one would try to have the analysis component select word senses in such a way that there is a trivial (one to one) mapping between these word senses and the lexical terms of each TL. Early investigations of semantic fields have made it clear that the lexicons of different natural language frequently break down the conceptual space in different ways. Encoding as SL 'word senses' the specific breakdowns of several TLs would entail (in the worst case) creating $n*m$ word senses, where n is the number of TLs and m the average number of corresponding lexical items in these TLs.

However, these questions lose much of their importance in systems which, like TAUM-METEO, are meant to be applied to a single language pair. When a given SL lexical item has several possible equivalents in TL, it is described as ambiguous in SL; subcategorisation frames are then used to select applicable 'word senses', which happen to be labelled with TL lexical items. Thus, even if language-internal ambiguity tests would fail to distinguish two senses of *heavy* in *heavy rain and heavy fog,* separate entries are created in the SL dictionary: one for a 'sense' labelled *abondant* and selected by nouns denoting a 'precipitation' *(pluie abondante)* and the other for a 'sense' labelled *dense* and selected by nouns denoting a 'stationary condition' *(brouillard dense).*

As we saw, lexical transfer generally requires complex transformational operations to insert the TL equivalents into an appropriate environment. This might be an additional source of difficulty in the fusion of transfer and analysis, because the subcategorisation frames in TAUM's dictionary entries are primarily filtering devices for the structures built by independent syntactic transformation rules. Since there is no lexical decomposition, the dictionary entries do not provide for 'semantic mapping rules'. Thus, while the 'structural description' part of lexical transformation rules is easily encoded, the mechanisms available for their 'structural change' part are very weak.

In the case of TAUM-METEO, this difficulty did not turn out to be serious, because cases where complex rearrangements (such as those exemplified earlier) are required seldom occur; and when they do, they are handled by ad hoc means (e.g. 'rule features').

The analysis component includes a dictionary and a syntactic/semantic parser. No morphological analysis rules are used: they would be inefficient given the morphological simplicity of the sublanguage. The complete dictionary, including morphological variants, idioms, several hundred place names and multiple entries resulting from the no-transfer strategy, contains approximately 1500 items.

The parsing strategies used in TAUM-METEO are largely dictated by the peculiarities of the sublanguage. In most of TAUM's systems, verbs constitute the core element around which the parsing rules are organised and the output tree is built. The fact that weather forecasts lack tensed verbs has led to the adoption of a different strategy in TAUM-METEO. More or less conventional deep structures could be built if extensive use was made of rules for recovering deleted material; but then, most of this material would have to be deleted again at synthesis, since SL and TL use similar deletion rules.

For these reasons, the designers of TAUM-METEO decided

to adopt an approach which can be compared to Burton's ([569] 1976) 'semantic grammars'. In these grammars, the rules used are essentially of the syntactic type (e.g. phrase structure grammars). But they state co-occurrence patterns in terms of semantic classes (e.g. person, event) rather than in terms of the usual syntactic classes such as noun and verb phrase. In TAUM-METEO, the grammars and the output trees are based on a mixture of syntactic and semantic categories. Broadly speaking, the low-level categories are syntactic (but with semantic subcategorisation) and the higher level categories are semantic. The topmost node in the tree is labelled with the name of one of the five basic semantic patterns of the sublanguage mentioned earlier. A typical intermediate structure is given in (14).

(14)

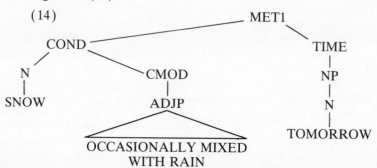

(In actual trees, the nodes are annotated with features; moreover the lexical items are already those of TL at this stage.)

In tree (14), MET1 is the label for the semantic pattern 'atmospheric condition', COND the label for a basic condition, CMOD the label for an accompanying condition. These categories result from the projection of semantic features (which often cut across syntactic categories) from lower nodes. For example, a COND is defined as a noun phrase or an adjective phrase whose head bears the feature 'weather condition'; in the dictionary this feature is assigned to such words as rain, snow, cloudiness, cloudy, sunny, etc.

The analysis component also includes some rules which could be considered to belong to structural transfer; for example, the am/pm time notation system is converted into the twenty-four notation, which is preferred in TL. Similarly, temperature ranges such as 'low twenties' are given a numerical interpretation, such as *21 à 23*.

Syntactic synthesis applies a few movement rules (e.g. placement of French adjectives). But, generally speaking, the constituents of the input tree are already close to surface structure constituents. Agreement rules are applied so as to distribute the appropriate morpho-syntactic markings.

The second step, morphological synthesis, is also significantly simpler than usual, because of the absence of tensed verbs.

TAUM-METEO has been in daily use at the Canadian Meteorological Center in Dorval (a suburb of Montreal) since May 1977. It is embedded in a larger production system which includes the following components:

(a) a monitor program that runs on a CDC CYBER 720 (which is used as a front-end for the CYBER 176), and performs the following tasks:

extract from a communication network the relevant data and pre-process this data for the MT system,

call the MT system,

send the material rejected by the translation system to the human editors,

reinsert the complete translation in the communications network;

(b) the MT system proper which runs on a CDC CYBER 176;

(c) an interactive edition program for the human translators which also runs on the CYBER 720;

(d) maintenance facilities.

The current workload of the translation system reaches eight million words per year. The rate of success is better than 80 per cent, and these 80 per cent do not need any human post-editing. As far as we know, this constitutes the only case of FAHQT reported so far. Sample translations are included in Appendix I. A large percentage of system failures are caused by communications noise, and spelling mistakes. When the system is unable to produce a complete parse for a given sentence, it does not attempt to make a best guess by means of 'failsafe strategies', which would render the quality of the output unpredictable. Rather, the SL sentence is turned over to human translators.

The translators appear happy to be relieved of what they used to consider boring work ([128] Chandioux and Guéraud 1981); their intervention in the translation of regional forecasts is limited to the most difficult cases, and they are able to concentrate on more interesting texts, such as general synopses.

The TAUM-AVIATION project started shortly after TAUM-METEO began to run experimentally in Dorval. The Canadian Secretary of State Department had just been asked to prepare for the translation of the maintenance manuals of the CP-140 patrol aircraft, then estimated at a staggering 90 million words. It was decided to give TAUM a contract to develop a system which would help accomplish that ambitious task, three years later.

It soon became clear that the texts were much more complex than had been anticipated. Handling this complexity would necessitate the development of new software that could circumvent some of the limitations of Q-SYSTEMS. Massive injection of new personnel (especially translators) created important organisation and training problems. Thus, just setting up a proper working environment took a large bite out of an already short time frame.

Development work had just reached cruising speed when the deadline arrived for an independent evaluation of the results (March 1980). Nevertheless, by that time, an operational new prototype had been assembled, complete with a number of new metalanguages, maintenance facilities, large-scale grammars for analysing English and synthesising French, and sophisticated dictionary entries for the core vocabulary of technical manuals plus a part of the vocabulary of hydraulics.

A detailed description of the sublanguage of maintenance manuals is presented in Lehrberger ([262] 1982). We will only mention some of the most salient characteristics. The vocabulary of aircraft maintenance manuals is relatively large: probably well over 50000 words if we include all of the sub-fields (electronics, hydraulics, structure, etc.). However, compared to the language as a whole, this number is still very small. Most of this vocabulary is made up of precise technical terminology. Only a small portion of the 'general vocabulary', approximately 2000 words, can occur. These include most of the closed-class words (though no personal pronouns such as I or she), but exclude some very common semantic subclasses (e.g. affective vocabulary) in open syntactic subclasses.

From the point of view of syntax, it is important to distinguish between two sub-types of texts: descriptive sections which contain only declarative sentences, and maintenance procedures sections, which are mostly made up of imperative sentences. Interrogative and exclamatory sentences are totally absent. These facts are, of course, reflections of pragmatic constraints on possible messages. But this conclusion holds only in virtue of the fact that this sublanguage exhibits a relatively direct mapping between syntactic types and illocutionary forces. For example, in the general language, interrogatives can frequently convey commands; but (15) cannot be used in a technical manual:

(15) Would you observe all safety precautions?

Note, however, that some declaratives can have the force of a command:

(16) If error is eliminated, step (2) should be omitted.

Within these limitations, the inventory of syntactic patterns includes most stylistically 'unmarked' constructions (conjunctions, passives, clefts, raisings, extrapositions, etc.). Moreover, maintenance manuals exhibit a number of deletion patterns that are uncommon in the standard language, such as the following deletions of article, object or *be*:

(17) Remove hydraulic reservoir.

(18) Remove used filter and discard.

(19) Check reservoir full.

Maintenance manuals also contain an unusual abundance of complex nominal compounds such as:

(20) Hydraulic ground test stand pressure and return line filters.

Compared to TAUM-METEO, the TAUM-AVIATION system corresponds more closely to the basic model. A strict separation between analysis, transfer and synthesis is observed.

Pre-processing is carried out by a finite-state automaton implemented in SISIF. Morphological analysis is split into three different modules:

(a) Non-inflectional morphology rules are also implemented in SISIF; these are mainly category assignment rules (e.g. numerals and ordinals in digit notation). TAUM-AVIATION does not incorporate an extensive rule-based treatment of derivational morphology. Composition (e.g. nominal compounds) is handled at the syntactic parsing stage, because productive compounding often exhibits properties that are otherwise thought of as syntactic (e.g. conjunction is possible: *pressure and return lines, first- and second-stage*).

(b) Infectional morphology is handled in a separate ad hoc PASCAL program. Since it was possible to write an exhaustive and final algorithm, priority was given to efficiency considerations.

(c) Dictionary lookup is performed by a new dictionary tool, SYDICAN ([369] Poulin 1981). This tool retrieves lexical rules (in an arbitrarly large database) and applies them to the input chart. Two types of rules are distinguished. Equivalence rules relate synonyms, spelling variants, etc. Content assignment rules associate a string of base forms with a full lexical description; for any rule, a choice is made between two algorithms: 'longest match' and 'all paths'. A longest-match strategy is useful to handle strings which are always idiomatic, e.g. 'in order to', but it cannot be used for occasional idioms (e.g. 'a little'). SYDICAN is provided with dictionary management and maintenance facilities.

Syntactic–semantic analysis is implemented in REZO ([559] Stewart 1978), TAUM's version of Augmented Transition Networks ([626] Woods 1970). There are several differences between REZO and Woods' ATN's. In TAUM-AVIATION, morphological analysis is completed before syntactic analysis, rather than being hierarchically subordinated; therefore REZO has to operate on a chart structure in which lexical alternatives are represented. Another difference is that in REZO grammars, it can be specified that given states are deterministic rather than nondeterministic: this makes parsing more efficient in certain cases. Finally, REZO grammars are compiled, again for the sake of efficiency.

But as with Woods' ATN's, the parsing algorithm (top-down left-to-right serial parsing) makes it possible to order the search paths so as to reflect the statistical tendencies of the text; this feature was used extensively in TAUM-AVIATION's parser.

The grammar is very large. It can handle most of the constructions which appear with some frequency in maintenance manuals. Co-occurrence patterns are stated in terms of very extensive syntactic subcategorisation: 12 fairly standard syntactic categories are further subclassified using more than 75 features (excluding morphosyntactic features); to this, one must add the use of strict subcategorization frames comparable to those of transformational grammar. Extensive use is made of more or less standard transformations; these transformations are often triggered by 'rule features' so as to reduce the number of spurious rule applications.

The parser is syntax-driven, but semantic checking is effected as soon as possible. Semantic analysis consists mainly of projection rules for the semantic features that label the nodes of the tree structures and selectional restrictions operating on these semantic features. The set of semantic subclasses used in TAUM-AVIATION is fairly small: approximately 35 features.

The projection rules calculate the set of features of a given node on the basis of the features of its daughter nodes; in the case of deleted or pronominalised constituents, feature sets of deletion controllers, of certain pronoun antecedents and of selectional frames are copied, so that they can be used in the disambiguation of related constituents.

The semantic mechanisms are more decoupled from syntax than

they are in TAUM-METEO. A good example of this is the handling of polysemy. In TAUM-METEO, a word is given separate entries for each one of its senses. By contrast, in TAUM-AVIATION, syntactically indistinguishable senses appear in the same entry. The ambiguity of the word is then reflected in the fact that its set of semantic features contains mutually incompatible features ([262] Lehrberger 1982). Projection rules are devised so as to filter these feature sets: when there is a contextual expectation (e.g. because of selection or coreference) for some feature F_1, only those features that are compatible with F_1 are raised to higher constituents.

This approach is more complex than the multiple-entry strategy of TAUM-METEO, but it has important benefits. When multiple entries are created solely for semantic reasons, they invariably add spurious alternatives in the syntactic search tree. In large-scale systems, this extra burden can easily make the difference between fairly efficient parsing and dreadful combinatorial explosions.

Much of the strength of TAUM-AVIATION lies in a very sophisticated transfer component. Lexical transfer is implemented in LEXTRA. Lexical rules were written mostly by professional translators and terminologists specialising in technical manuals.

The rules for translating some classes of words, especially general verbs (e.g. *allow, provide*) turned out to be extremely complex. For a description of the work that was done in this respect, see ([135] Chevalier et al. 1981).

The structural transfer (together with syntactic synthesis) is written in Q-SYSTEMS. This component includes a number of high-level translation rules. For example, some SL declaratives which have the force of commands are converted into TL imperatives. Thus (21) is translated as (22):

(21) Quick-disconnect fitting should not be removed.

(22) *Ne pas enlever le raccord à démontage rapide.*

Note that the French imperative is here in fact an infinitive, as is the custom in French technical manuals (this is certainly a sublanguage-dependent translation rule). Once a sentence has been analysed correctly, it is not too difficult to state such higher-level translation rules, which greatly enhance the overall quality of the results.

Syntactic synthesis is implemented in Q-SYSTEMS. This introduces some heterogeneity in the software of TAUM-AVIATION, but the tight schedule of the project left no alternative. The grammar has approximately the same coverage as the analysis grammar.

Inflectional morphology, like the corresponding module in the analysis component, is an ad hoc PASCAL program. French inflection, which is rather complex, is described exhaustively, using TAUM's extended Bescherelle classification for verbs ([150] Desroches and Bourbeau 1980).

Finally, a post-processing component written in SISIF reformats the output text.

It is notably difficult to assess the performance of an MT system. For an independent evaluation of TAUM-AVIATION (as of March 1980), refer to Gervais ([186] 1980). In some sense, TAUM-AVIATION failed the cost/benefit test: according to the evaluators, it was not usable in the near term in a production environment. The problem was not so much with the quality of the output, as with the restricted scope of the system (hydraulics maintenance manuals), and the estimated cost of extensions.

In fact, it could be argued that previous delays in project schedule had made this evaluation premature. The system was still at a stage where straightforward development work resulted in rapid and dramatic improvements in the coverage of the system and the quality of its output. Unfortunately, there is no standard method for evaluating the potential of MT systems under development.

During the last year of the AVIATION project, some time was spent devising a sophisticated evaluation scheme by which one could get an idea not only of the current performance of the system on some given portion of text, but also of its expected performance after further development work. This was done in the following way.

First, the system was run on samples of text, and the machine output was examined in detail, so as to establish a record of all mistakes in the translation. At this stage, mistakes are described in terms of superficial deviations from an acceptable translation (e.g. incorrect agreement between noun and adjective).

The idea was to measure the performance of the system at various stages of development; the variations in the ratio of the absolute number of mistakes to the number of word tokens was a first clue to assess the rate of improvement of the system. However, as long as error analyses remain at the level of the symptoms rather than looking at the causes, the figures that are obtained can be misleading; for example, how many mistakes does one count in the cases where the system failed to produce any output for some sentence? Moreover no direct clue is given to system developers as to where efforts should be concentrated because there is no simple relationship between symptoms and their causes.

It was thus decided to analyse the results with enough depth to trace the causes of every superficial mistake. Each problem was carefully recorded with the help of a sophisticated classification grid. Types of errors were tagged either as 'known solution' (plus an indication of the required effort) or as 'unknown solution'. This scheme made it possible to determine that in the last few months of the project, the system had improved dramatically and that this rapid progression could have continued for several more months. Some of these results are reported in Lehrberger ([261] 1981).

The present and the future of MT

In this section, we will first examine how much knowledge is needed, in general, to produce high-quality translation. Then we will try to assess how well second-generation models have met these requirements. From there, we will submit some views on the prospects for research and development in MT.

Most people accept the idea that in general, MT requires access to a representation of the SL text that involves both linguistic and extralinguistic knowledge. On the linguistic side, detailed morphological and syntactic knowledge is necessary: in order to reproduce a given meaning in TL, one must first of all gain access to it, and this can only be done through a detailed examination of the rich and complex system which governs the forms of SL. But beyond the internal logic of linguistic form, one has to come to grips with the difficult problem of meaning. Large-scale syntactic parsers become plagued with combinatorial explosions unless they are semantically guided; moreover, by themselves, these parsers can only produce highly ambiguous representations which are not suitable for the correct statement of translation and synthesis rules. Among the phenomena that syntactic rules cannot characterise adequately, some of the most important are the following:

(a) lexical ambiguities: polysemy;

(b) structural ambiguities:

conjunction scope,

prepositional phrase attachment,

internal structure of compounds (especially

nominal compounds),

etc.;

(c) textual links:

anaphora,

ellipsis.

Each of these problems (and others as well) may cause mistranslations. It can be shown that solving all instances of problems of these types implies the availability of textual (as opposed to sentential) meaning representations, over which the following poperties (and probably others as well) can be defined:

(1) Logical consistency.
(2) Pragmatic acceptability (e.g. observance of Gricean maxims).
(3) Compatibility with common sense.
(4) Compatibility with domain-specific knowledge.
(5) Compatibility with situation knowledge.

Thus, resolving the ambiguities left by syntax may require access to almost any type of linguistic or extralinguistic knowledge. There is no space here to present a detailed argument in support of these claims, but we will give a few relevant examples from the domain of technical manuals:

(22) Remove fitting and cover pump.
(23) Check reference line length. This line extends from . . .
(24) If clean air filter is clogged, . . .
(25) The red wire terminal is red.
(26) Note all major component damage. If damage is serious
. . .
(27) Before filling, remove filler cap and ground fuel tank.
(28) Discard used filter and replace (it).

Example (22) shows an interplay between lexical ambiguity, conjunction scope and nominal compounding: it has a deviant interpretation where 'cover' is interpreted as a noun. The only way to block that spurious reading is to rule out 'cover pump' as an admissible nominal compound: but this requires sophisticated semantic knowledge (and probably domain-dependent knowledge too).

In technical manuals, *line* can refer either to an abstract object (French *ligne*) or to any of a number of concrete objects (.e.g. French *canalisation*). In (23) the decision on the second occurrence of this word requires that an inter-sentential link be established with the first occurrence, which itself has to be disambiguated by compound noun semantics.

In (24), determining the scope of the adjective *clean* requires a characterisation in terms of truth conditions: if *filter* is included in that scope, a logical inconsistency will result (a filter cannot be simultaneously clean and clogged). In (25), taking *red* to modify *terminal* causes a pragmatically unacceptable sentence: the sentence becomes 'analytic' and violates Grice's ([587] 1975) maxim of quantity by failing to be informative. In (26), a related but more complex process occurs across sentence boundaries. If the scope of 'major' is taken to include 'damage', a clash is caused with 'serious' in the following sentence.

Sentence (27) shows a complex interplay between conjunction scope, lexical ambiguity, situation knowledge and domain-specific knowledge. In order to decide that *ground* is a verb and that the conjuncts are sentences, not noun phrases, one would have to resort to the knowledge that no *ground fuel tank* is involved in the situation, and/or to the knowledge that grounding a tank before refuelling is an expected (mandatory) step in the refuelling procedure.

Finally, sentence (28) shows an example where a lexical disambiguation involves common sense. 'Replace' is ambiguous between French *remplacer* and *remettre en place*. This latter reading can be ruled out on the grounds that it would be a lack of common sense to throw something in the trash if you know that you are to use it again shortly.

A system capable of solving all translation problems would have access to all the types of knowledge that we have mentioned, and probably several others as well. Needless to say, to construct such representations, one would need an impressive array of devices: text grammars, semantic and pragmatic analysers, frames, scripts, etc.

Compared to older systems, second-generation systems have made important improvements possible in the quality of translations. Because these systems provide structural descriptions of the SL material, their translation rules can go beyond a mere word-for-word rendering; also, it is possible to guarantee that the target language text will exhibit a high degree of grammaticality (provided one is extremely careful with so-called 'failsafe' strategies).

The most serious limitations of these systems are obviously a consequence of their relatively weak semantic component. Few of them go much further than using selectional restrictions. This device can indeed solve several types of lexical and structural ambiguity, by encoding some limited forms of linguistic and extra-linguistic knowledge. But selectional restrictions operate only on low-rank semantic units: they enforce coherence between individual predicates and their arguments. The coherence constraints that are necessary to solve the various ambiguities exemplified in (22)–(28) have to be stated at a much higher level, across propositional and sentential boundaries.

In the absence of mechanisms for dealing with meaning at the textual level, even tasks which appear innocuous such as segmenting the text into (sentential) processing units are somewhat problematic. In second-generation systems, this segmentation is usually realised in a deterministic pre-processing component. But segmentation problems such as periods signalling an abbreviation rather than a sentence boundary clearly indicate that this process should in principle result from an interaction between text and sentence grammar.

Thus, it is obvious that high-quality translation is, for many types of texts, far beyond the capabilities of second-generation systems. But at the same time, it is an empirical fact (witness TAUM-METEO), that there are some types of real-life texts which this technology can translate with remarkable accuracy.

The market pull for usable MT systems is stronger than ever. However, it is clear that the post-ALPAC trauma has not disappeared yet. Before ALPAC, MT used to be a central problem in artificial intelligence research (e.g. it occupied a prominent position in the 1961 conference on the 'Mechanisation of Thought Processes' in Teddington). In the last twenty years, MT research has been rather marginal, even though market pressures have been much stronger for MT than for some more fashionable research topics in natural language processing and artificial intelligence.

The best medicine for the fallen angel of MT would probably be a few more uncontroversial successes like TAUM-METEO. Ten years after the initial design of this system, we now have enough knowledge to adequately solve somewhat more complex translation problems. A crucial question that has to be answered is how to measure sublanguage complexity, relative to a given MT technology; important work on this issue is reported in Kittredge ([233] 1983). We feel confident that an enhanced second-generation technology will produce a number of useful systems in the near future.

Of course, if MT is to have any future, more basic research is needed, and potential sponsors have to be persuaded that the enterprise is worthwhile. Research work on 'third-generation' systems should not ignore the achievements of second-generation technology, but rather use them as a base on which to build more powerful mechanisms for semantic, textual and pragmatic analysis.

Many recent results in other areas of computational linguistics and in artificial intelligence can profitably be used in the design of more powerful MT systems. However, MT research will specifically have to address a number of issues on the translation problem, of which the following are a sample:

(a) Are we likely to progress in the direction of interlingual semantics?

(b) If we are to continue using transfer models, what other types of rules should they involve?

(c) How can translation rules be constrained?

(d) Can we develop more natural ways of expressing translation rules? For example, can we infer complex tree transformations required for lexical transfer from dictionaries written in a quasi-conventional notation?

These and other research topics are suggested and discussed in Isabelle ([216] 1981), which, notwithstanding its title, appears to have been among the last in the series of more than 450 papers produced at the TAUM group.

Acknowledgements

Most of the ideas expressed here reflect the collective work of the members of the former TAUM Group. I would like to thank in particular Laurent Bourbeau, John Lehrberger and Elliott Macklovitch, who made detailed comments on drafts of this paper. Of course, all remaining mistakes are mine.

Appendix I

Unedited Translations Produced by TAUM-METEO
Input Text:

FPCN13 CWWG 170120
FORECAST FOR NORTHWESTERN ONTARIO
ISSUED BY ENVIRONMENT CANADA AT 9.30
PM EST THURSDAY FEBRUARY 16 1984 FOR
TONIGHT AND FRIDAY WITH AN OUTLOOK
FOR SATURDAY.
THE NEXT SCHEDULED FORECAST WILL BE
ISSUED AT 5.30 AM.
RAINY LAKE
RED LAKE
SIOUX LOOKOUT
PICKLE LAKE.
CLOUDY TONIGHT AND FRIDAY. PERIODS
OF LIGHT RAIN OR WET SNOW IN SOME
LOCALITIES. RISK OF LIGHT FREEZING
DRIZZLE. LOWS TONIGHT MINUS 1 TO MINUS
3. HIGHS FRIDAY ZERO TO PLUS 2. PROBA-
BILITY OF PRECIPITATION IS 70 PERCENT
TONIGHT AND 50 PERCENT FRIDAY.
OUTLOOK FOR SATURDAY . . . MOSTLY
CLOUDY AND SIMILAR TEMPERATURES.

Translation:

FPCN74 CWWG 170120

PREVISIONS POUR L'ONTARIO NORD OUEST EMISES PAR ENVIRONNEMENT CANADA A 21H30 HNE LE JEUDI 16 FEVRIER 1984 POUR CETTE NUIT ET VENDREDI AVEC APERCU POUR SAMEDI.

«Untranslated fragment»

LAC RAINY

RED LAKE

SIOUX LAKE

PICKLE LAKE

TEMPS NUAGEUX CETTE NUIT ET VENDREDI. PLUIE FAIBLE PAR MOMENTS OU NEIGE MOUILLEE DANS QUELQUES LOCALITES. RISQUE DE BRUINE FAIBLE VERGLACANTE. MINIMUM CETTE NUIT ENTRE MOINS 1 ET MOINS 3. MAXIMUM VENDREDI ENTRE 0 ET PLUS 2.

PROBABILITE DE PRECIPITATIONS DE 70 POURCENT CETTE NUIT ET 50 POURCENT VENDREDI.

APERCU POUR SAMEDI . . . TEMPS GENERALEMENT NUAGEUX ET TEMPERATURES SEMBLABLES.

Appendix II

Unedited Translations Produced by TAUM-AVIATION (May 1981)

Input Text:

35 Case drain fluid from the two ac hydraulic pumps in the No. 1 system is routed to the system No. 1 service center assembly through cartridge-type check valves. The combined flow from the service center assembly is directed to the hydraulic system oil cooler in the No. 2 fuel tank. Case drain fluid from the No. 2 ac hydraulic pump is routed through the system No. 2 service center assembly, and a cartridge-type check valve to the hydraulic system oil cooler in the No. 3 fuel tank. The cartridge-type check valves are installed directly into the service center assembly and become terminal fittings for line connection. The purpose of the check valves is to prevent fluid from back-flowing through an inactive ac hydraulic pump.

36 Separate ground test conditions, one for pressure and one

for suction, are provided for each main system. The connections are quick-disconnect, sealing-type couplings. The ground test connections permit connection of a ground hydraulic power source without loss of fluid or introduction of air into the aircraft hydraulic system.

Translation:

35 Le liquide de vidange du carter des deux pompes hydrauliques ca dans le circuit no 1 est acheminé au bloc collecteur du circuit no 1 en passant par les clapets anti-retour à cartouche. L'écoulement combiné du bloc collecteur est envoyé au refroidisseur d'huile du circuit hydraulique dans le réservoir de carburant no 2. Le liquide de vidange du carter de la pompe hydraulique ca no 2 est acheminé en passant par le bloc collecteur du circuit no 2 et en passant par un clapet anti-retour à cartouche au refroidisseur d'huile du circuit hydraulique dans le réservoir du carburant no 3. Les clapets anti-retour à cartouche sont posés directement dans le bloc collecteur et deviennent les raccords de la borne pour le raccord de canalisation. Le but des clapets anti-retour est d'empêcher le liquide de refouler vers une pompe hydraulique ca hors service.

36 «SEGMENT NON TRADUIT» Les raccords sont les raccords à démontage rapide, obturateurs. Les prises de parc permettent le raccord d'une source d'alimentation hydraulique au sol sans perte de liquide ou d'introduction d'air dans le circuit hydraulique de l'aéronef.

SIXTEEN : J-P. GUILBAUD

PRINCIPLES AND RESULTS OF A GERMAN TO FRENCH MT SYSTEM AT GRENOBLE UNIVERSITY (GETA)

As the ARIANE-78 system has already been presented many times, we will restrict ourselves to recalling its main overall structure.

(a) The translation of a text is carried out in the three main phases: analysis, transfer and generation.

(b) ARIANE-78 uses a single data-structure (a complex labelled tree structure) to represent linguistic phenomena throughout processing. Each node is provided with a decoration describing the properties of this node and its syntactic and semantic relations with other nodes.

(c) Separation of programs and linguistic data allows the linguist to work with familiar concepts such as dictionaries, grammars, linguistic features and so on. The linguist's work hence consists in organising linguistic data previously translated into the metalanguage of each component of the system.

(d) The linguistic approach is multilingual. In other words the analysis (generation) process is independent of target (source) language. Only the transfer phase is common to a given language pair.

(e) ARIANE-78 integrates a conversational monitor allowing for preparation of linguistic data, management of corpuses and execution of the translation process on texts. The monitor completely guides the linguist from the beginning to the end of each session at the terminal.

Finally the two diagrams (both well known!) in figure 16.1 illustrate the correspondence between the current linguistic principle and the organisation of the translation process.

The linguistic model presented here has been developed in

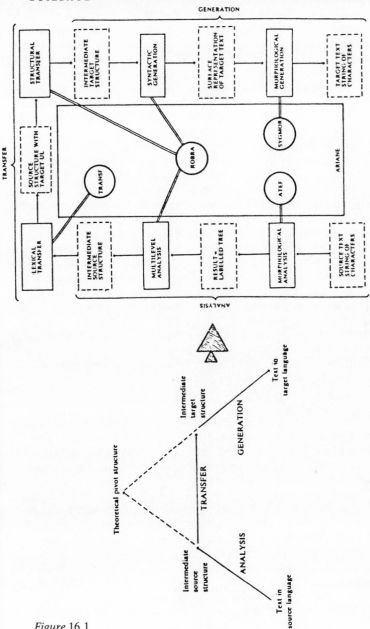

Figure 16.1

Grenoble and supported by the French War Office. It is still under completion. Some phases are more advanced than others. The dictionaries for analysis, transfer and generation contain respectively 1500, 400 and 7000 entries. Whereas the analysis module is in quite an advanced stage, transfer and syntactic generation modules are no more than mere outlines, while the module for French morphological generation is an exhaustive and finished module.

The German-French translation system

This model is tested on texts of a booklet edited by Siemens and titled: *Stromenergie fuer Heute und Morgen*. Some problems are not treated, such as compound words and separable verb particles. The complexity of the problem of compound words is purely of a linguistic nature. The separable verb particle, on the other hand, sets a purely technical problem connected with the organisation of phases in the current A R I A N E - 78 version: the opening of source dictionaries only occurs within the morphological process; and, as isolated verb particles are attached to the verb to which they belong during the structural analysis phase, it would, consequently, be useful at this stage to be able to consult a dictionary in order to find the new lexical unit and its new valencies (*angehen ≠ gehen*!). This defect is serious but not fatal since, more often than not, in texts the particle occurs attached to the verb (there are more *Nebensaetze* than *Hauptsaetze* and in most cases verbs make use of auxiliaries). Furthermore a new version of the system has been started upon which will eliminate this defect.

Morphological analysis

The morphological parser for German is at present sufficient to allow structural analysis to work correctly. It recognises all types of nominal, verbal and adjectival inflection, analyses the most productive derivational affixes, and separates compound forms if these do not already appear in a dictionary.

Every input-text is a sequence of character strings between blank spaces. The aim of morphological analysis is to identify words from such a sequence and assign to each word appropriate information (lexical reference, grammatical, syntactical and semantic properties). Units of treatment are generally strings of characters separated either side by two blank spaces. We call these 'forms' or 'occurrences'. Thus the text is a sequence of forms separated by blank spaces. Morphological analysis has to attribute the appropriate lexical information to each of these forms. This information is then used in the following stage of multi-level structural analysis.

ATEF, the algorithmic model used for morphological analysis, is a non-deterministic finite state transducer and thus imposes certain constraints on the linguistic model adopted. The lexical information calculated for each form is presented by VALUES of VARIABLES declared in the model. These variable values are contained in a register called a variable MASK. Each variable has a TYPE, a NAME and VALUES which are determined freely by the linguist. An EXCLUSIVE type variable (EXC) can at a given time have only one of the values of the declared set. A NON-EXCLUSIVE type variable (NEX) may have any subset of the values. The system, moreover, allows the distinction to be made between morphological variables (VARM) and syntactic variables (VARS). This purely formal distinction enables the linguist to class his variables into two groups according to the criteria which he chooses freely. Each variable mask associated with a form contains the calculated values of all the different declared variables plus an exclusive variable value UL (Lexical Unit) whose values are implicitly declared in the dictionaries. The UL variable may have, moreover, the values ULTXT (text), ULFRA (sentence), ULOCC (occurrence) and ULMCP (compound word), which permits the labelling of the tree resulting from morphological analysis (see figure 16.2). The principle of a finite state automaton is used to transmit information. The forms are segmented by searching in the dictionaries for the images of all or part of the character string of a form. These images, the morphs of the dictionaries, contain information and call one or more grammar rules. The rules are only applied if their condition of application is verified. These rules modify the register (mask) associated with a form of the text according to the information contained by the morph in the dictionary: the variable values of a mask associated with a form evolve as the segmentation process proceeds.

The lexical information is contained in the variable values found after morphological analysis in each mask associated with a form (or an expression assimilated with a form and treated specially). On the one hand there is information concerning the lexical references and the grammatical properties (syntactic and semantic) of the form, on the other the tactical aids necessary only during the process of analysis. We will give here an overview of the linguistic quantities handled:

(a) Terminal morphosyntactical classes are defined by two-level coding (general and a subclass) which simplifies their handling during the following phase of structural analysis (AS): the rules of AS may be written to operate at either level, depending on the case

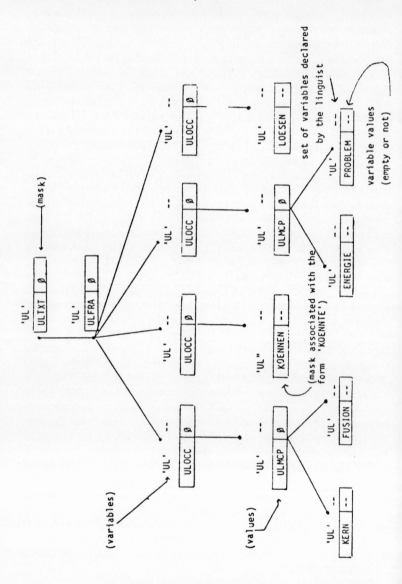

Figure 16.2. Example: Kernfusion koennte Energieproblem loesen.

involved. These morphosyntactic classes are terminal because they label the leaves or terminal nodes of trees both in AS and in morphological analysis (table 16.1).

TYPE	NAME	VALUE	REMARKS
NEX	KMS		Morphosyntactic class:
		VB	- verbal form;
		NM	- substantive or analogous class, with the exception of substantivized adjectives or participles;
		DR	- deictors or pronouns, with the exception of pronominal adjectives and adverbial pronouns of time and location;
		ADJ	- adjectives and participles, whether or not substantivized;
		ADIP	- invariable adjuncts (adverbs) and contracted pronouns (eg DAVON);
		COP	- subordinate and coordinate conjunctions as well as the contracted article (e.g. IM, AM);
		PC	- punctuation marks;
		NA	- non-alphabetic forms (various codes).

Table 16.1.

(b) Grammatical properties depend on the classes described above. They are calculated on the basis of the grammatical morphemes that govern word inflection and are indicated by such variables as those shown in table 16.2. The use of eight different variables allows us to express all the case values for number, gender and declension type (strong or weak). Thus, the adjective ending '-E' may give rise to:

$CSMB = (NOM)$ e.g. *der gute Mensch*
$+ CSFB = (NOM, ACC)$ e.g. *die gute Leistung*
$+ CSNB = (NOM, ACC)$ e.g. *das gute Ergebnis*
$+ CSFT = (NOM, ACC)$ e.g. *gute Milch*
$+ CPT = (NOM, ACC)$ e.g. *gute Freunde*

(c) Derivations: the analysis of inflectional endings allows various word forms that differ only in grammatical features to be related to a single lexical unit (UL). For example, the forms LERN, LERNE, LERNTE and GELERNT all have different grammatical features but belong to the same UL, LERNEN-V.

TYPE	NAME	VALUES	REMARKS
NEX	IC		Presence of an inflectional ending on a declinable adjective or deictor :
		1	- no
		2	- yes
NEX	NGM	NO	Morphological negation (e.g. UNLOESBAR)
NEX	GNR		Gender:
		M	- masculine
		F	- feminine
		N	- neuter
NEX	CSMT		Masculine singular, strong case;
	CSMB		Masculine singular, weak case;
	CSFT		Feminine singular, strong case;
	CSFB		Feminine singular, weak case;
	CSNT		Neuter singular, strong case;
	CSNB		Neuter singular, weak case;
	CPT		Plural, strong case;
	CPB		Plural, weak case;
		NOM	- nominative;
		ACC	- accusative;
		DAT	- dative;
		GEN	- genitive.

Table 16.2

These forms would be assigned the following masks of variables:

LERN: UL(LERNEN-V), KMS(VB),
            SUBV(HAB), MT(IMP), PSIP(2).

LERNE: UL(LERNEN-V), KMS(VB),
            SUBV(HAB), MT(IPR, SPR, IMP),
            PSG(1), PSIP(2,3).

LERNTE: UL(LERNEN-V), KMS(VB),
            SUBV(HAB), MT(IPA, SPA), PSG(1, 3).

GELERNT: UL(LERNEN-V), KMS(VB, ADJ),
            SUBADJ(RSTA), SUBV(HAB),
            MT(PPA).

Derivations allow the number of forms that are related to the same lexical unit to be increased even further. They relate to the same UL forms which may belong to different morphological classes. For example, the forms VERSORGTEST (KMS-VB) and VERSORGUNG (KMS-NM) are both derived from the same lexical unit VERSORGEN-V. Seeing that derivations are generally realised by means of affixes, and particularly suffixes, the analysis

TYPE	NAME	VALUE	REMARKS
NEX	DEG		Adjective degree:
		CP	- comparative (e.g. BESSER);
		SP	- superlative (e.g. BESTE).
NEX	PSG	1,2,3	Person singular;
	PPL		Person plural.
NEX	MT		Mood and tense :
		IPR	- present indicative (e.g. BIN);
		SPR	- present subjunctive (e.g. WISSE);
		IPA	- past indicative (e.g. WARST);
		SPA	- past subjunctive (e.g. WAERE);
		IMP	- imperative (e.g. GEH);
		INF	- infinitive (e.g. EINORDNEN);
		INFZU	- infinitive with ZU (e.g. EINZUORDNEN);
		PPR	- present participle (e.g. LERNEND);
		PPRZU	- present participle with ZU (e.g. EINZUORDNENDE);
		PPA	- past participle (e.g. GEGANGEN).

Table 16.3

of inflectional morphology can be coupled with a derivational analysis which interprets derivational morphemes. In this way, all the derived forms that can be analysed need not be indexed in the dictionary. In short, recognising derived forms both reduces the total number of ULs required (the system currently imposes a limit on the number of ULs, and saves indexing time by reducing the number of morphs that need to be indexed). Table 16.4 lists the derivations that are currently recognised for German.

(d) Syntactic properties of the lexical unit associated with the word form currently include those shown in table 16.5.

(e) Typographic information used so far concerns capital letters solely. The fact that a word appears entirely in capitals is solely of typographic importance and has no linguistic relevance. On the other hand, if a word begins with a capital letter, then it is either sentence initial or a noun.

Tactic variables simply allow some information to be calculated. They have temporary values.

Figure 16.3 provides an overview of the segmentation process. Each arc corresponds to the identification of a morph and links two circles which stand for classes of states that are particularly signifi-

```
+-----+--------+--------+----------------------------------------+
! TYPE! NAME   ! VALUES ! REMARKS                                !
+-----+--------+--------+----------------------------------------+
! EXC ! DRV    !        ! Derivations                            !
!     !        ! VN1    ! - substantivization of an              !
!     !        !        ! infinitive (e.g. das ESSEN);           !
!     !        ! VN2    ! - derivation from a verb to noun       !
!     !        !        ! with the suffix '-UNG' (e.g. die       !
!     !        !        ! VERSORGUNG) and other verbal           !
!     !        !        ! nominalizations (e.g. der              !
!     !        !        ! ANKAUF);                               !
!     !        ! VN3    ! - derivation from a verb to a          !
!     !        !        ! noun with the suffix '-BARKEIT'        !
!     !        !        ! (e.g. MACHBARKEIT);                    !
!     !        ! VN4    ! - derivation from a verb to a          !
!     !        !        ! noun with the suffix '-HEIT'           !
!     !        !        ! (e.g. BEDINGTHEIT);                    !
!     !        ! VA1    ! - derivation from a verb to an         !
!     !        !        ! adjective with the suffix '-BAR'       !
!     !        !        ! (e.g. MACHBAR);                        !
!     !        ! VA2    ! - present participle with an           !
!     !        !        ! adjectival ending (e.g.                !
!     !        !        ! ZUTREFFENDES);                         !
!     !        ! VA3    ! - past participle with an              !
!     !        !        ! adjectival ending (e.g.                !
!     !        !        ! GEFANGENER);                           !
!     !        ! AN1    ! - derivation from an adjective         !
!     !        !        ! to a noun with the suffix              !
!     !        !        ! '-HEIT' or '-KEIT' (e.g.               !
!     !        !        ! SCHOENHEIT, WICHTIGKEIT) or            !
!     !        !        ! without a suffix (e.g.                 !
!     !        !        ! SCHWAECHE);                            !
!     !        ! NN1    ! - derivation from a masculine          !
!     !        !        ! noun to a feminine noun (e.g.          !
!     !        !        ! LEHRERIN);                             !
!     !        ! DRA1   ! - derivation from a cardinal to        !
!     !        !        ! an ordinal (e.g. ZWEITE);              !
!     !        ! DRA2   ! - derivation from an ordinal to        !
!     !        !        ! an adverb (e.g. ZWEITENS).             !
+-----+--------+--------+----------------------------------------+
```

Table 16.4

cant. Each time a root form is identified and segmented, the inflectual or derivational ending is assigned a numerical prefix. This prefix, which stands for a paradigmatic sub-state, replaces several tactic variables (type of conjugation, declension, etc.) whose values define the form of a base and its lendings in terms of their paradigmatic properties. Thus, each prefix provides an additional condition on the interpretation of an inflectional ending. In short, inflectional endings are interpreted in terms of the inflectional paradigm to which the root preceding the ending belongs (thus, the

'-T' in GIBT and the '-T' in BEBT or WART receive different
interpretations).

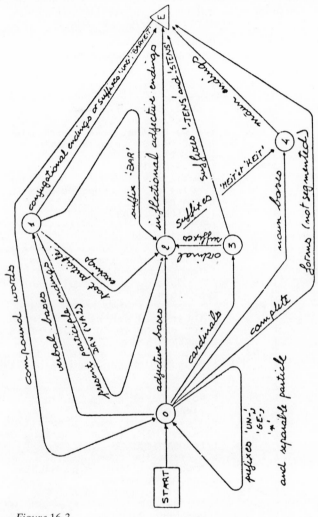

Figure 16.3

An example of the results of the morphological analysis is given
in figure 16.4.

In ARIANE-78.4, there is only one look-up in the monolingual analysis dictionaries and that is during the morphological analysis phase. This explains why the dictionaries of base forms – the only dictionaries of interest for the lexicographer involved in indexing – are dictionaries of MORPHS (for the segmentation of occurrences), and why these dictionaries include tactic, morphological and syntactic-semantic information. Dictionaries for machine

TYPE	NAME	VALUES	REMARKS
EXC	SUBV		Subclasses of verbs:
		RFL	- reflexive verbs;
		PAI	- verbs that take the double infinitive (e.g. KOENNEN, HOEREN);
		HAB	- verbs that only take the auxiliary HABEN in the active voice of the compound past tenses;
		RST	- other verbs.
NEX	SUBN		Subclass of nouns :
		PF	- nouns that do not require an article when they occur as thematic subject;
		IP	- nouns that do require an article;
		PSF	- personifiable nouns, i.e. those that can be substituted for JEMAND in 'JEMANDEM ETWAS GEBEN';
		LOC	- temporal or spatial locative nouns which can occur alone as circumstantials.
NEX	POS		Position of a hypotactic element:
		1	- before the noun, preposition (e.g. FUER, für die Rettung der Industrie);
		2	- after the noun, "post-position" (e.g. HALBER, Geschäfte halber);
		3	- first element in a "circum-position" (e.g. UM, um das Haus herum);
		4	- second element in a "circum-position" (e.g. HERUM, um das Haus herum).

Table 16.5

TYPE	NAME	VALUES	REMARKS
NEX	VAL1		Case governed by a verb (direct object in the nominative or accusative) or by an adjective (accusative);
	VAL4		Case governed by a verb (direct object in the dative or the genitive) or by an adjective (idem);
	VAL2A		Case governed by a verb, an adjective or a noun (indirect object with a preposition); also, case governed by a preposition:
		NOM	- nominative;
		ACC	- accusative;
		DAT	- dative;
		GEN	- genitive.
NEX	VAL2B		Governed preposition (also used for hypotactic lexemes in the dictionary); or second accusative:
		DOUB	- double accusative;
		ALS	- e.g. betrachten ALS;
		AN	- e.g. denken AN;
		AUF	- e.g. rechnen AUF;
		AUS	- e.g. bestehen AUS;
		FUER	- e.g. halten FUER;
		GEGEN	- e.g. kämpfen GEGEN;
		IN	- e.g. verwandel IN;
		MIT	- e.g. zufrieden MIT;
		NACH	- e.g. fragen NACH;
		UEBER	- e.g. froh UEBER;
		UM	- e.g. sich bemühen UM;
		UNTER	- e.g. leiden UNTER;
		VON	- e.g. müde VON;
		VOR	- e.g. Furcht VOR;
		WIE	- e.g. aussehen WIE;
		ZU	- e.g. wählen ZU.
NEX	VAL3		Type of governed subordinate clause or phrase :
		PHSUB	- tensed subordinate clause introduced by DASS;
		PHINF	- infinitive without ZU (e.g. ... muss gedeckt werden);
		PHINF1	- infinitive with ZU (e.g. Er braucht nicht ZU kommen);
		GPPA	- adjective phrase (e.g. die Meinungen sind kontrovers).

Table 16.5 *continued*

Figure 16.4.

```
                        -- TEXTE ORIGINE --

        Das Energieproblem ist zu einer grossen Herausforderung geworden.


        *** GROUPE D'ETUDE POUR LA TRADUCTION AUTOMATIQUE ***

                -- ANALYSE MORPHOLOGIQUE --

                                                        CODE LANGUE : ALY

RESULTAT DE L'EXECUTION, TEXTE : STREN  S1
```

```
                                          ULTXT
                                          .....1
                                          |---|
                                          ULFRA
                                          .....2
                                          |---|

    |---|   |---|   |---|   |---|   |---|   |---|   |---|   |---|   |---|   |---|
    ULOCC   ULOCC   ULOCC   ULOCC   ULOCC   ULOCC   ULOCC   ULOCC   ULOCC   ULOCC
    .....9  ....11  ....13  ....15  ....17  ....19  ....21
    |---|   |---|   |---|   |---|   |---|   |---|   |---|
    SEIN-V  ZU      EIN     GROSS-A HERAUSF WERDEN-
    ....10  ....12  ....14  ....16  O....18 V....20  .....22

    |---|   |---|
    ULOCC   ULOCC   ULOCC
    .....5          |---|
    |---|           ULMCP
    DER             .....6
    .....3          |---|
    |---|
                    |---|
                    *ENERGI *PROBLE
    .....4          E.....7 M.....8
    |---|
```

----SOMMET 4 --> unité lexicale = 'DER';
catégorie = déicteur; type de déicteur = relatif + représentant + normal; personne du singulier = 3;
cas fort neutre singulier = nominatif + accusatif; désinence = oui; typographie = majuscule; genre = neutre.
----SOMMET 7 --> unité lexicale = '*ENERGIE';
catégorie = nom; type de nom = parfait; personne du singulier = 3; typographie = majuscule; genre = féminin.
----SOMMET 8 --> unité lexicale = '*PROBLEM';
catégorie = nom; type de nom = imparfait; personne du singulier = 3;
cas faible neutre singulier = nominatif + accusatif + datif; typographie = majuscule; genre = neutre.
----SOMMET 20 --> unité lexicale = 'WERDEN-V';
catégorie = verbe + adjectif; type d'adjectif = normal; conjugaison = participe passé; désinence = non;
type de verbe = avec auxiliaire sein; complément direct = nominatif; cas du complément indirect = datif;

translation are very different from classical monolingual dictionaries are not initially easy for a linguist to understand. The lexicographer must keep in mind:

(a) that he is working in a dictionary that only lists base forms;

(b) that the indexing of a lexeme consists of assigning it a list of morphs (base forms), and assigning to each of these morphs an FTM code, an FTS code and a lexical reference (UL value). The various morphs that derive from a single lexeme are all assigned the same UL and FTS code. The FTM code is the name of an FTM format which describes the morphological properties of the base form. The FTS code is the name of an FTS format which contains the syntactico-semantic information on the lexeme. These formats are generally predefined in special files, and whenever a linguist introduces a new format, he must first declare it.

In summary, for each lexeme the dictionary contains:

(a) one or more base forms;

(b) an FTS code and a UL value, repeated if there are several base forms.

An example of a dictionary is given:

```
-------------------------------------------------------------+
+-------+         +---------+ +-------+ +----------+         !
!BAETE  !         !==WBAETE! !(HA3UZ! !,BITTEN-V! ).        !
!BAT    !         !==WFAND ! !(HA3UZ! !,BITTEN-V! ).        !
!BITT   !         !==WFIND ! !(HA3UZ! !,BITTEN-V! ).        !
!GEBETEN!         !==FCPPA ! !(HA3UZ! !,BITTEN-V! ).        !
+-------+         +---------+ +-------+ +----------+         !
     \               ↙        ╰──────╯ ╰──────────╯         !
  base forms         FTM       repeated UL and FTS code     !
  for 'bitten'       codes            for 'bitten'          !
                                                            !
   BLEIB           ==WSING      (SVRST    ,BLEIBEN-V ).     !
   BLIEB           ==WGLICH     (SVRST    ,BLEIBEN-V ).     !
   BRAUCH          ==VLEG       (PA2A1FSZ,BRAUCHEN-V).      !
   BREIT           ==AKLEIN     (VIDE     ,BREIT-A    ).    !
-------------------------------------------------------------+
```

The lexicographer has at his disposition an indexing manual as well as an automatic indexing assistance system, which makes his work much easier. Figures 16.5 and 16.6 are taken from the section of the indexing manual that concerns the *morphological indexing of German verbs* (how to determine the number and nature of the base morphs to be indexed and their corresponding FTM codes). For the verb GAEREN, for example, the lexicographer would index:

GAER along with the FTM code VLEG

For the verb AUSPRESSEN, on the other hand, he would index:

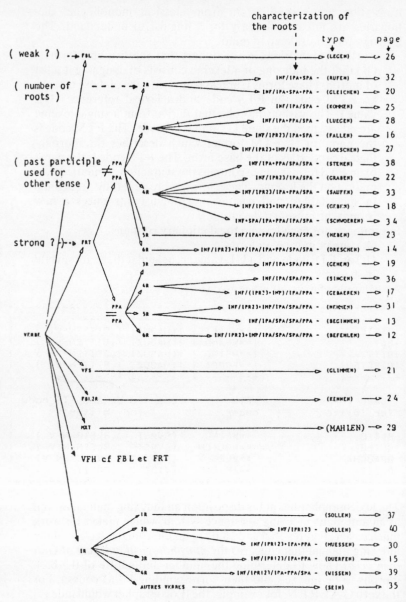

Figure 16.5.

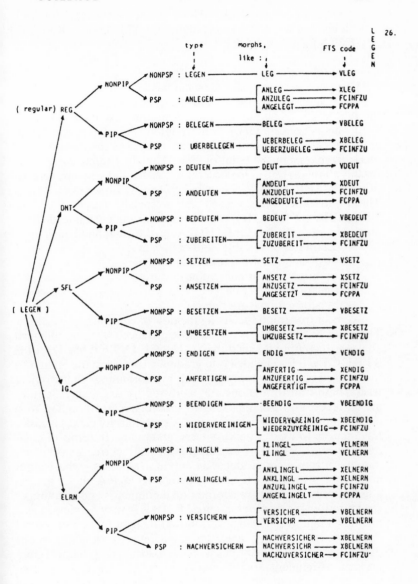

Figure 16.6.

AUSPRESS, AUSZUPRESS and AUSGEPRESST along with their corresponding FTM codes XSETZ, FCINFZU and FCPPA.

Structural analysis

The structural analysis of German has been developed by Professor G. Stahl ([426] 1980). This analysis operates essentially at the syntactic level and makes little or no use of semantic information. This is because German morphology and syntax are rich in grammatical information. Multi-level analysis is the most difficult phase from a linguistic point of view. The algorithmic component for this phase, as well as for structural transfer and syntactic generation, is ROBRA ([529] Chauché 1974; [88] Boitet et al. 1978). The three other phases each use a specific model (ATEF, TRANSF, SYGMOR), whose external data include dictionaries. The external data for ROBRA, on the other hand, include:

　a declaration of variables and of formats;

　a transformational system, comprised of:

　　condition and assignment procedures;

　　transformational rules;

　　transformational grammars;

　　a control graph specifying the flow of control.

The input to this phase is the flat tree produced by morphological analysis. Its output is called the SOURCE INTERMEDIATE STRUCTURE. The idea is to gradually transform the input into the desired output by applying local transformations to the object tree. These transformations are expressed by rules, and the rules are grouped into grammars, each grammar corresponding to a coherent linguistic process. ROBRA's modularity makes it easier for linguists to work with very large grammars. In terms of computer systems, ROBRA implements an abstract tree transduction model – a transducer produces an output in all cases – whose input and output structures are homogeneous. ROBRA allows the linguist to write modular transformational grammars and encourages structured linguistic programming. ROBRA may be viewed as a high-level algorithmic language, in that its data types and control structures are very high-level. The data comprise variables, masks of variables and labelled trees with the associated operators (comparison, assignment, transformation by rules). The control structure includes conditional assignments of variables, parallelism in rule application and non-determinism in the control graph. Rules may recursively call grammars or transformational sub-systems.

In addition to the morpho-syntactic classes already declared in

morphological analysis, the principal terms involved are either
syntagmatic classes (noun phrase, infinitival clause, etc.), syntactic
functions (subject, object, etc.) or logical relations (argument 0, 1
etc.) and certain semantic relations (goal, cause, etc.).

```
+-----------------------------------------------------------------+
| KSY --> Syntagmatic category (EXC).                             |
+---------+-------------------------------------------------------+
| VALUES  | REMARKS                                               |
+---------+-------------------------------------------------------+
|         |                                                       |
|         | 1) definitive values -->                              |
|         |                                                       |
| GN      | - noun phrase (das grosse Haus);                      |
| GP      | - prepositional phrase (für die Jugend);              |
| GA      | - inflected adjective phrase (sehr gutem);            |
| GA1     | - uninflected adjective phrase or adverb phrase       |
|         |   (sehr gut / nicht gern);                             |
| PHVB    | - sentence (main clause);                             |
| PHSUB   | - subordinate clause whose governor is                |
|         |   conjugated;                                          |
| PHREL   | - relative clause;                                    |
| PHINF   | - infinitival clause;                                 |
|         |                                                       |
|         | 2) intermediate tactic values -->                     |
|         |                                                       |
| GN1     | - incomplete noun phrase (grosse Haus);               |
| GPPA    | - past participle phrase (gebaut worden);             |
| GVB     | - verb group (may be transformed into a PHVB,         |
|         |   PHSUB, etc.);                                        |
| PHINF1  | - infinitive clause with ZU (will become              |
|         |   PHINF).                                              |
+---------+-------------------------------------------------------+
```

The non-exclusive variable FVB (verbal function) is used to indicate whether a sentence is in the passive (PASS), in the perfect
(PARF), in the future or the conditional (FUT); or whether its
governor is a modal verb (MDAL). The variable TPH (clause
type) is used for interrogative (INTER) or imperative (INVIT)
sentences. Important tactical information is provided by the non-
exclusive variable POSIB (hypothesis regarding the structure of
the sentence). In German, there are five possible forms for the
GVB (verbal group):

ALPHA→the standard form of the *Hauptsatz* (or main
clause),
e.g. *Paris liegt an der Seine*;
BETA→the subordinate form,
e.g. *die Ereignisse die Welt in Atem halten*;
GAMMA→the main clause form with the subject inverted,
e.g. *gestern verriet er ihn an seine Feinde*;

```
+---------------------------------------------------------------+
I FS --> Syntactic function (EXC)                               I
+---------+-----------------------------------------------------+
I VALUES  I REMARKS                                             I
+---------+-----------------------------------------------------+
I         I                                                     I
I         I 1) definitive values -->                            I
I         I                                                     I
I GOV     I - governor;                                         I
I EPIT    I - epithet (GROSSE Haus);                            I
I ATR     I - attribute of the subject (ist GROSS /wird EIN     I
I         I   ARZT);                                            I
I ATROB   I - attribute of the object (ihn EINEN ESEL           I
I         I   nennen);                                          I
I SUJ     I - subject;                                          I
I SUJF    I - formal subject ( ES kommen viele Leute);          I
I OBJ1    I - first (or direct) object (er hat SIE              I
I         I   getroffen);                                       I
I OBJ2    I - second (or indirect) object (er gab IHNEN ein     I
I         I   Buch / er hat ÜBER IHN gelacht);                  I
I CIRC    I - circumstantial ( AM ABEND kamen sie);             I
I DES     I - designation ( DAS Haus);                          I
I QUANT   I - quantifier ( VIER Protonen);                      I
I MF      I - modifier ( SEHR teuer / HÄUFIG kam er);           I
I COORD   I - coordinated phrase;                               I
I REG     I - introducing of a phrase or a clause ( DURCH       I
I         I   das Fenster);                                     I
I COMP    I - complement of a noun or an adjective (die         I
I         I   Windschutzscheibe DES AUTOS / günstig FÜR DIE     I
I         I   INDUSTRIE);                                       I
I AP      I - apposition (Hamburg, DER GRÖSSTE HAFEN IN         I
I         I   DEUTSCHLAND);                                     I
I AUX     I - auxiliary ( IST gegangen);                        I
I JUXT    I - juxtaposition (Frau FUCHS);                       I
I REF     I - reference to degree (grösser ALS DIESE            I
I         I   STADT);                                           I
I INCL    I - insertion (sagte - WIE ES SCHEINT - ).            I
I         I                                                     I
I         I 2) interim tactic values -->                        I
I         I                                                     I
I SUAT    I - subject or attribute;                             I
I SUOB    I                                                     I
I SUOBX   I                                                     I
I SUOBY   I - subject or first object or other values.          I
+---------+-----------------------------------------------------+
```

DELTA→the infinitive structure, a particular case of
BETA,
e.g. *die Welt in Atem halten*;
EPSILON→the participial form,
e.g. *nichts mehr von ihm gehört.*

```
+----------------------------------------------------------------+
! RL --> Logical Relation (EXC)                                  !
+---------+------------------------------------------------------+
! VALUES  ! REMARKS                                              !
+---------+------------------------------------------------------+
! ARGO    ! - argument O,                                        !
!         !   e.g.  Er kommt / die Elektrizität wird von         !
!         !         dem Kraftwerk erzeugt;                       !
! ARG1    ! - argument 1,                                        !
!         !   e.g.  Elektrizität wird erzeugt;                   !
! ARG2    ! - argument 2;                                        !
! ARGO1   ! - argument O or 1;                                   !
! ARG12   ! - argument 1 or 2;                                   !
! TRLO    ! - transfer to argument O,                            !
!         !   e.g. Er ist gross;                                 !
! TRL1    ! - transfer to argument 1,                            !
!         !   e.g. Er nennt ihn einen Esel;                      !
! ID      ! - same RL as father node;                            !
! SMARG   ! - semi-argument.                                     !
+---------+------------------------------------------------------+
```

At the outset, all five hypotheses are possible. Then, as the verbal group is gradually constructed, the number of possibilities is reduced.

The basic principle that governs our analysis of German is the following: we construct simple groups before constructing larger, more complex groups, and we give priority to those structures of which we are sure. The analysis component consists of 87 elementary grammars, each of which contains from one to nine rules (176 rules in all). The basic linguistic operations comprise:

(a) Only compound words made up of two lexical units are analysed, and of these AS handles only the following:

noun + noun ('HAUSTUER');
adjective + noun ('GROSSTADT');
verb + noun ('RENNPFERD');
noun + adjective ('HAUSHOCH');
adjective + adjective ('HELLGRUEN')

Combinations of preposition + noun, and particle + verb are not treated; nor are they segmented in AM. The first element of the compound is attached under the second, which is the governor of the group, and is assigned a syntactic function:

EPIT (GROSSstadt),
MF (HELLgrün),
or COMP (HAUStür).

(b) With the raising operation, the first simple groups are constructed. Each unambiguous word becomes the governor of a group:

Salz / Bilder → G N (*Salz / Bilder*);
Stuhl → G N1 (*Stuhl*);
grossen → G A (*grossem*);
gross / hier / heute → G A1 (*gross / hier / heute*);
aufzunehmen → PHINF (*aufzunehmen*);
ist / aufnehmen → GVB (*ist / aufnehmen*);
gebaut + worden → GPPA (*gebaut*);
zu + nehmen → PHINF1 (*nehmen*).

(c) In analysis of the formation of simple adjective phrases, the relevant strings of adjacent adjuncts become adjective phrases : this grammar handles such simple configurations as :

sehr hohes;
nicht sehr gern;
sowohl gut als auch billig;
guten, schonen und wertvollen; etc.

(d) Formation of simple noun phrases involves incorporating the adjectives that modify the head noun and the determiners that precede them. In the texts we translate, there are three types of adjective-noun sequences :

(i) The adjective-noun sequence constitutes a GN and does not need to be completed by a deictor,

e.g. *gutes Salz*;

(ii) The adjective-noun sequence constitutes a GN1 which needs to be completed by a deictor,

e.g. *gute Salz*;

(iii) The adjective does not modify the noun that follows it,

e.g. *grossem Kinder* (in '*mit grossem Kinder erschreckenden Lärm*').

For types (i) and (ii), we begin with a G A (which may be complex) and a GN or GN1, and build either a GN or a GN1. Case (iii) is recognised as not being relevant to noun phrase formation. Nouns that appear alone are handled as follows : In the AM dictionary, we distinguish between so-called *perfect nouns,* which do not require an article in the singular (such as WEIZEN, SALZ, etc.) and *imperfect nouns,* which do require an article in the singular (such as STUHL, BESEN, etc.). If the noun is imperfect and in the singular, it has weak case forms; otherwise, it has strong case forms. In AS, the three types of adjective-noun sequences are handled by checking for declension and for case compatibility among the elements (non-null intersection for case). If the elements are compatible, then we recalculate the new case and resulting declension as follows :

(1) strong + strong → STRONG

e.g. *grosse* Bilder (GN),
guter grosser Mann (GN1);
(2) strong + weak→STRONG
e.g. *grossen* Besen (ACC singular; GN);
(3) weak + strong→WEAK
e.g. *grossen* Untertanen (plural NOUN; GN), *guten* grossen
Mann (GN1);
(4) weak + weak→WEAK
e.g. *grossen* Besen (DAT singular; GN).

Once the adjective phrases have been incorporated, we can then search for deictors to the left, for example:

sieben + GN (*grosse Männer*)→GN (*sieben grosse Männer*);
sieben + GN1 (*grossen Männer*)→GN1 (*sieben grossen Männer*)
ein + GN1 (*Haus*)→GN (*ein Haus*)
das + GN1 (*Haus*)→GN (*das Haus*)

We then attempt to build the various types of prepositional phrases:

AUS *dem Haus*;
AUS *dem Haus* HERAUS,
des schlechten Wetters WEGEN;
IN *Text und Bild.*

(e) Noun and adjective complements: not every occurrence of two noun phrases may be considered to be a noun plus its complement, but there are some which can be analysed in this way with near certainty:

(*des Malers*) + (*Frau*)
(*der Kampf*) + (*der jungen Nationalstaaten*);
(*eine Meinung*) + (*über ein Verfahen*);
(*die Skala* (*der Forderungen*)) + *und* + (*Vorschlaege*).

On the other hand, we cannot group the nouns together in a syntactically ambiguous configuration like:

(*der Mann*) + (*der Besitzer*)(cf.: *dass der Mann der Besitzer ist*!!).

Complements to adjectives are less of a problem and are incorporated whenever they correspond to the adjective's valency specification or are part of an expression of the adjective's degree.

GA1 (*arm*) + GP (*an Vitaminen*)
→GA1 (*arm*, GP (*an Vitaminen*));
GP (*an Mineralsalzen*) + GN(GA(*reiches*), *Wasser*))
→GN(GA(GP(*an Mineralsalzen*), *reiches*), *Wasser*›.

(f) Formation of verbal groups: a sentence is composed of one or more propositions. Each proposition includes a simple or complex verbal taxeme, for example:

kommt
ist gekommen
wird gekommen sein
streiken wollen muessen wird, etc.

We call each of the elements that make up the verbal taxeme a GVB, or verbal group. At the end of analysis, not all of these GVBs will have the same status: some of these GVBs will have their status unchanged, whilst others will have become the governor of a PH or conjugation auxiliaries. We attempt to progressively extend each GVB that is a potential governor by incorporating the noun and adjective phrases that may be its complements to the left and to the right. Thus the sentence:

Das Energieproblem ist zu einer grossen Herausforderung geworden

will have two GVBs:

1. GVB (GN(*Energieproblem*), *sein*)
2. GVB (GP(*Herausfordern*), *werden*)

The complex sentence *Er will das Haus kaufen* will also have two GVBs. When two 'saturated' GVBs become adjacent, i.e. when all their complements have been located and nothing intervenes between them, they are combined into a simple or complex PH (PHVB, PHSUB, etc.).

simple PH:

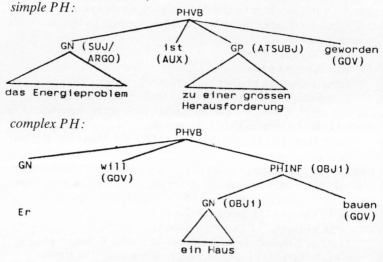

complex PH:

Notice that so-called modal auxiliaries are treated like main verbs (i.e. governors of propositions), unlike conjugation auxiliaries. Note too that not all GVBs are necessarily adjacent; here as elsewhere there are special cases, like the interweaving in such a sentence as: (*sicher will*) *ihm der Mann* (*ein Buch schenken*)

The PHSUBs and PHINFs, that result from the combination of GVBs are in turn combined with other PHs – either the matrix clause to which they are subordinated or another clause with which they are conjoined. At the end of analysis, every sentence in the text that is not a title is thus dominated (directly or indirectly) by PHVB.

(g) More accurate contrastive calculation of syntactic functions and logical relations: in extending GVBs out from the verb to the left or to the right, we do not always when incorporating a new element have the necessary information to decide whether it is a subject or attribute (. . . *der Besitzer ist* . . .), or a subject or first object (. . . *die Frau pflegt* . . .). (The problem is due to the use of rules that are often binary, and could perhaps be eliminated: it is a matter of strategy). A first approximation of the possible relations between a verb and its complements is provided by assigning bivalent temporary values. We then return to the resulting construction in order to study the various combinations of ambiguous values and deduce new values that are more accurate. For example:

(1) SUAT and SUOB following SUJ will be changed to ATR and OBJ1:

Der Mann *die Frau* pflegt.

(2) SUAT and SUOB before ATR or OBJ1 will be changed to SUJ:

Die Frau flegt den Mann, pflegt *die Frau* den Mann.

(3) SUOB following OBJ1 will be changed to SUJ:

ihn *die Frau* pflegt.

The grammars apply in sequence, though all grammars need not necessarily apply. Short cuts are provided, depending on the complexity of the sentence being analysed. Moreover, many rules apply recursively to the object tree, or recursively call other rules in the same or other grammars. The sequence in which the modules apply may be schematised as in figure 16.7, pages 302-3.

An example of the results of structural analysis is shown in figure 16.8, pages 304-5.

Figure 16.7

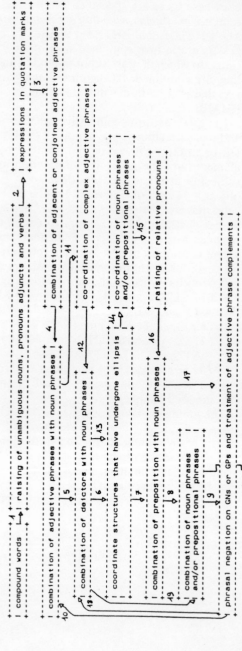

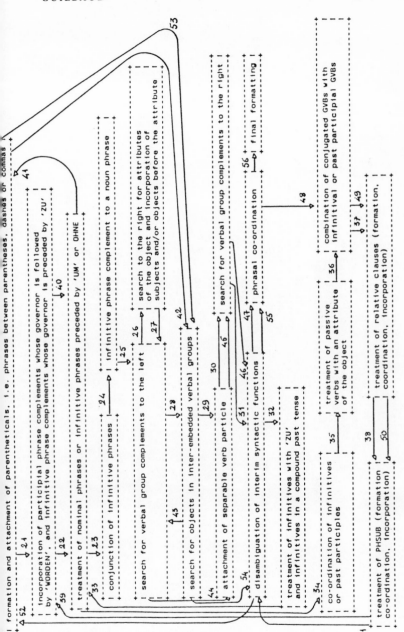

Figure 16.8

-- TEXTE ORIGINE --

Das Energieproblem ist zu einer grossen Herausforderung geworden.

*** GROUPE D'ETUDE POUR LA TRADUCTION AUTOMATIQUE ***

-- ANALYSE MORPHOLOGIQUE -- (cf. fig. 4)
****
-- ANALYSE STRUCTURALE --

CODE LANGUE : ALY

RESULTAT DE L'EXECUTION, TEXTE : STREN   S1

ULTXT
.....1

*PHVB
.....2

*GP
.....9

WERDEN-.
V....15 16

ZU   EIN   *GA   HERAUSF
....10 ...11 12 0....14

GROSS-A
.....13

SEIN-V
.....8

.
.....3

*GN
.....4

DER   *PROBLE
....5 M....6

*ENERGI
E.....7

```
-----SOMMET 1  --> unité lexicale = 'ULTXT'.
-----SOMMET 2  --> unité lexicale = '*PHVB';
type de verbe = avec auxiliaire sein; classe syntagmatique = phrase verbale; personne = 3; nombre = singulier;
catégorie = verbe;   personne du singulier = 3; conjugaison = indicatif présent; fonction verbale = parfait;
-----SOMMET 3  --> unité lexicale = '.'.
-----SOMMET 4  --> unité lexicale = '*GN';
classe syntagmatique = groupe nominal; fonction syntaxique = sujet; relation logique = argument 0; personne = 3;
nombre = singulier; catégorie = nom; type de nom = imparfait; personne du singulier = 3; genre = neutre;
typographie = majuscule.
-----SOMMET 5  --> unité lexicale = 'DER';
fonction syntaxique = désignateur; personne = 3; nombre = singulier; désinence = oui; catégorie = déicteur;
personne du singulier = 3; genre = neutre; typographie = majuscule; type de déicteur = normal.
-----SOMMET 6  --> unité lexicale = '*PROBLEM';
fonction syntaxique = gouverneur; personne = 3; nombre = singulier; catégorie = nom; type de nom = imparfait;
personne du singulier = 3; genre = neutre; typographie = majuscule.
-----SOMMET 7  --> unité lexicale = '*ENERGIE';
fonction syntaxique = complément de nom; personne = 3; nombre = singulier; catégorie = nom; type de nom = parfait;
personne du singulier = 3; genre = féminin; typographie = majuscule.
-----SOMMET 8  --> unité lexicale = 'SEIN-V';
type de verbe = avec auxiliaire sein; fonction syntaxique = auxiliaire; personne = 3; nombre = singulier; catégorie = verbe;
personne du singulier = 3; conjugaison = indicatif présent.
-----SOMMET 9  --> unité lexicale = '*GP';
type de verbe = avec auxiliaire "haben"; dérivation = verbe-nom; classe syntagmatique = groupe nominal prépositionnel;
fonction syntaxique = objet 2; relation logique = argument 2; personne = 3; nombre = singulier; catégorie = nom;
type de nom = imparfait; personne du singulier = 3; genre = féminin; typographie = majuscule.
-----SOMMET 10 --> unité lexicale = 'ZU';
fonction syntaxique = régisseur; catégorie = conjonction; position = avant ou élément 2 de circumposition;
type de conjonction = préposition; cas du complément indirect = datif.
-----SOMMET 11 --> unité lexicale = 'EIN';
fonction syntaxique = désignateur; personne = 3; nombre = singulier; désinence = oui; catégorie = déicteur;
personne du singulier = 3; genre = féminin; type de déicteur = normal.
-----SOMMET 12 --> unité lexicale = '*GA';
classe syntagmatique = groupe adjectival; fonction syntaxique = attribut du gouverneur; désinence = oui; catégorie = adjectif;
type d'adjectif = normal.
-----SOMMET 13 --> unité lexicale = 'GROSS-A';
fonction syntaxique = gouverneur; désinence = oui; catégorie = adjectif; type d'adjectif = normal.
-----SOMMET 14 --> unité lexicale = 'HERAUSFORDERN-V';
type de verbe = avec auxiliaire "haben"; dérivation = verbe-nom; fonction syntaxique = gouverneur; personne = 3;
nombre = singulier; catégorie = nom; type de nom = imparfait; personne du singulier = 3; genre = féminin;
typographie = majuscule.
-----SOMMET 15 --> unité lexicale = 'WERDEN-V';
type de verbe = avec auxiliaire sein; fonction syntaxique = gouverneur; désinence = non; catégorie = verbe;
conjugaison = participe passé; cas du complément indirect = datif; préposition régie = 'zu'.
-----SOMMET 16 --> unité lexicale = ','.
catégorie = ponctuation.
```

Lexical transfer

Lexical transfer is designed to establish a correspondence between the lexical elements of the input structure (source language) and the lexical elements of the output structure (target language), with possible changes to the nodes of the input structure. The sub-system that carries out lexical transfer is called TRANSF. The part of lexical transfer that involves tests on nodes in the tree other than the current lexical node is carried out in structural transfer. Such tests cannot be done in lexical transfer, which consists solely of a bilingual dictionary. In essence, the lexical transfer is based on consultation of a dictionary whose entries correspond to rules. TL transforms each node of the source language object tree into a target language subtree (which may consist of only one node), with the target language information obtained dynamically from the dictionary.

From an algorithmic point of view, the model is essentially a bilingual dictionary of simple transfer rules accessed by the source UL. Each rule is a sequence of 3-tuples (condition, image subtree, assignments), the last condition being empty. The automaton creates the object tree as follows. The UL of the current node is used to access the dictionary. The first triplet of the entry whose condition is verified is chosen. The image subtree (generally consisting of only one node) is added to the output, with the value of variables computed by the assignment part. The possibility of transforming one input node into an output subtree allows for the creation of compound words or tactic auxiliary nodes that will be used in the following phase (TS) in order to treat idioms and to carry out various disambiguation tests.

The linguistic data is listed in four files, plus an additional file per dictionary:

> declaration of variables;
> condition formats;
> condition functions and assignment procedures;
> dictionaries (from 1 to 7).

Only the data corresponding to the declaration of variables and the dictionaries are obligatory.

The German–French dictionary is currently limited to a small number of entries that are used to carry out tests to improve and develop the analysis modules. Each entry consists of: the name of a source lexical unit between single quotation marks; a separator '=='; and a list of triples, each containing:

(a) a condition: either a conditional expression or a call to a condition function;

(b) an image subtree of the current node, written in parenthesised form as in R O B R A. If the subtree consists of a single node, it need not be specified;

(c) an assignment part which contains, for each node of the image subtree: the name of the summit, the symbol ':', the name of the target U L assigned to this summit, optionally followed by a list of assigned values of target variables (expressed as formats, procedures or explicit statements).

For example,

```
+---------------------------------------------------------------------+
I                                                                     I
I '*ABLUFT'      ==      / O(1) /O :  'AIR'      ,+SM;                 I
I                            1 :  'éVACUER'  ,*X2DE,$B1,$G7.  I
I 'BRAUCHEN-V'   ==$POT /E(H) /E :  'UTILISER'  ,+VZI,$BLE,$NGPAI
I                          H :  'VIABLE'    ,*QL1,$QF5,$HM/  I
I                     /E(CI)/E :  'AVOIR.V'   ,+VIH;                 I
I                         CI:  'BESOIN'    ,*X3NM,               I
I                                          MANIP:=MANIPO.  I
I 'WARM-A'       ==      /     /    'CHAUD'     ,+QL1,$QF.           I
I                                                                     I
+---------------------------------------------------------------------+
```

Indexing in the dictionary is done with the help of an indexing manual. A decoded, intelligible version of this dictionary (as well as the other dictionaries in the system) can be obtained by means of VISULEX, a special program that operates on lexical data. Figure 16.9 (p.308) contains examples of VISULEX outputs.

An example of the results from lexical transfer is given in figure 16.10 (p.309).

Structural transfer

The main role of the structural transfer phase (TS) is to make the intermediate linguistic structure consistent with the grammar of the target language. At the beginning of structural transfer, the geometry and linguistic labels of the object tree still reflect the structure of the source language. Since analysis never achieves a fully universal interlingual representation, certain transformations and adaptations to the target language must be made. Structural transfer also serves to complete the work of lexical transfer, whenever the translation of a lexeme requires that several nodes of the tree structure be tested or modified. The model used to carry out these changes is the same as in structural analysis, i.e. R O B R A.

First, we complete the work of lexical transfer: ambiguities are

'BRAUCHEN-V'
                    --morphologie--

Valenz :
+ reiner Infinitiv
Akkusativobjekt
Praeposition ALS + Akkusativ
Praeposition FUER + Akkusativ
regiert einen Dass-Satz
regiert einen Infinitivsatz mit 'ZU'
    schwaches Verb :
    BRAUCH-

--si: Moeglichkeitsadjektiv oder -substantiv ?   --equivalents--
    --arbre: E(H)
E:'UTILISER'
    1ü comp qn,qch
    adj potentiel dérive du verbe: possib dériver nom fém ité
    négation possible sur 1' adjectif dérivé de potentialité ou non
H:'VIABLE'
    adjectif postposé
    adj: possib dériver nom fém
    autre équivalent possible
--sinon:
    --arbre: E(CI)
E:'AVOIR.V'
    1ü comp: de qch/de faire qch
CI:'BESOIN'
    lexème auxiliaire

'WARM-A'
                    --morphologie--

besondere Merkmale :
kann jedes Artikelwort entbehren
vom Adjektiv abgeleitetes Substantiv:
weibliches Substantiv, Flexionsmuster: E R D E :
    WAERM-
--cmt?--
    Komparativ- oder Superlativform :
    WAERM-
    Adjektiv, Positivform :
    WARM-
                    --equivalents--
'CHAUD'
adjectif postposé
adj: possib dériver adj et nom fém

Figure 16.9

```
                              ** TEXTE ORIGINE **

               Das Energieproblem ist zu einer grossen Herausforderung geworden.


                         *** GROUPE D'ETUDE POUR LA TRADUCTION AUTOMATIQUE ***

                                  --  ANALYSE STRUCTURALE  -- (cf. fig. 8)
                                  **************************
                                  --  TRANSFERT LEXICAL  --

RESULTAT DE L'EXECUTION, TEXTE : STREN   S1                              CODE LANGUE : ALY

                                 ULTXT
                                 ......1
                                   I
                                   I
                                   I
                                   I
                                   I
                                 *PHVB
                                 ......2
                                   I
                                   I
                                   I
     -----------------------------------------------------------------------
     I     I           I                  I                       I       I
     .     *GN         E3TRE              *GN                     DEVENIR .
     ......3 ......4   ......8            ......9                 .....15 .....16
           I                              I
           I                              I
           I                              I
        --------                   ------------------------
        I      I                   I       I       I      I
     LE        PROBLE2        ULCIBIN UN       <GA>   DE1FIER
     ......5 M.....6          .....10 .....11 .....12 .....14
               I                                I
               I                                I
               I                                I
               I                                I
               I                                I
               E1NERGI                          GRAND
               E.....7                          .....13
```

Figure 16.10

eliminated by conducting the required tests on the object tree. In lexical transfer, for each German word that requires contextual verification in order to be properly translated into French, a tactical image subtree is created which triggers a disambiguation procedure in TS.

The following contrastive linguistic operations are then carried out:

(a) treatment of the article;

(b) computation of French mood, tense, voice and aspect;

(c) contrastive evaluation of weight values, in preparation for ordering of French phrasal groups.

Finally, certain deletions and structural changes are carried out:

(d) deletion of superfluous function words and tactic nodes generated in lexical transfer;

(e) changes to argument structure for French predicates that differ from German predicates;

(f) placement of image subtrees.

An example of the results of structural transfer is shown in figure 16.11.

Syntactic generation

In addition to structural analysis and structural transfer, the syntactic generation phase (GS) also makes use of ROBRA. Much work remains to be done in GS; currently, only very basic operations are carried out, in order to allow morphological generation to produce the first German to French translations. The aim of GS is to generate all the lexical units of the target sentence in the correct order, and to assign to each the information that will enable morphological generation to determine its appropriate base form and affixes. Its principal operations are:

(a) The calculation of syntactic classes and functions. The algorithm traverses the object tree from top to bottom, and in recursive fashion, performing the following operations at each successive level:

(1) The syntactic functions of the dependents of a group are calculated on the basis of their logical relations *and* the actualisation features of their governor (e.g. conjugated verbs, non-conjugated, active vs. passive, etc.).

(2) The syntagmatic category of each of the groups whose syntactic function has just been determined is then calculated on the basis of the possible lexical category of the governor on its dependents *and* the valencies of that governor.

(b) The creation of function words. We then generate preposi-

Figure 16.11

```
                         -- TEXTE ORIGINE --

        Das Energieproblem ist zu einer grossen Herausforderung geworden.


              *** GROUPE D'ETUDE POUR LA TRADUCTION AUTOMATIQUE ***

                      --  TRANFERT LEXICAL  -- (cf. fig. 10)
                      ***************************
                    --  TRANSFERT STRUCTURAL  --

RESULTAT DE L'EXECUTION, TEXTE : STREN   S1                 CODE LANGUE : ALY

                                ULTXT
                                ......1
                                  I
                                  I
                                *PHVB
                                ......2
                                  I
                                  I
 --------------------------------------------------------------------
 I               I                        I                I       I
 .             *GN                      *GN              DEVENIR  .
 ......3       ......4                  ......9          ......14 ......15
                 I                        I
                 I                        I
         -----------------         -----------------
         I       I       I         I       I       I
   LE     PROBLE2 <JUXT>  <GA>      UN      DE1FIER
   ......5 M.....6 ......7 .....10 .....12 .....13
                     I       I
                   E1NERGI GRAND
                   E.....8 .....11

-----SOMMET 2  --> unité lexicale = '*PHVB';
classe syntagmatique = phrase verbale; forme morphologique du verbe = verbe conjugué;
temps morphologique francais simple = présent; fonction verbale = parfait; voix = non-réfléchie; mode = indicatif;
personne = 3; conjugaison = indicatif présent; temps universel = présent; aspect = acheve; nombre = singulier.
-----SOMMET 14  --> unité lexicale = 'DEVENIR';
fonction syntaxique = gouverneur; type d'auxiliaire exigé par le verbe = etre; argument 2 = 1; catégorie = verbe;
conjugaison = participe passé; complément direct = groupe adjectival.
```

tions and subordinate conjunctions, negation where it is required, conjugation auxiliaries and certain articles.

(c) The actualisation of pronominal and adjectival elements. Here, we calculate the number and gender of pronouns, adjectives and determiners that agree with their head noun or with the subject of the sentence.

(d) The ordering of the elements of the sentence in two stages:

(1) To each phrase, we assign a weight in the form of a relative numerical value. This value depends on the position that the group will occupy in relation to its governor and to other possible related elements to the left and the right. The elements that must appear to the left of the governor are assigned a negative value; those that must appear to the right, a positive value; and to the governor itself, the value zero is assigned. The further away the relative position of an element is from the governor, the greater the absolute value of the weight it is assigned. The weight of a given phrase is generally determined by its syntactic function, its category and the number of its constituents.

(2) Once all phrases have been assigned a weight, we can then order them and calculate number and gender agreement for adjectives and those past participles that agree with a preceding object (e.g. *l'essence* que le moteur a CONSOMMÉE).

An example of results after syntactic generation is given in figure 16.12.

Morphological generation

The morphological generation component (GM) we use is taken from the FRB model, which was developed by N. Nedobejkine for use in the Russian–French system. The model is complete, has been amply tested and is reliable. Moreover, it can be directly integrated without the slightest modification into any other bilingual application developed at GETA, and in this respect conforms to our laboratory's multilingualism policy.

The role of the algorithmic component SYGMOR is to produce, for each terminal node of the tree output by syntactic generation, the corresponding word in the target language. The input to morphological analysis was a text in the source language; the output of morphological generation will be an equivalent text in the target language.

SYGMOR transforms a tree into a string of masks of variables, and then transforms that string of masks into a string of characters. The latter process is conducted under the control of the following

-- TEXTE ORIGINE --

Das Energieproblem ist zu einer grossen Herausforderung geworden.

*** GROUPE D'ETUDE POUR LA TRADUCTION AUTOMATIQUE ***

-- TRANSFERT STRUCTURAL -- (cf. fig. 11)
****
-- GENERATION SYNTAXIQUE --

RESULTAT DE L'EXECUTION, TEXTE : STREN   S1   CODE LANGUE : ALY

```
                                              ULTXT
                                              ....1
                                              *PHVB
                                              ....2

                          E3TRE   DEVENIR                    *GN
                          .....8  ......9                    .....10

        *GN                                         UN   <GA>   DEIFIER
        ....3                                      ....11 ....12 ....14

 LE  PROBLE2 <JUXT>                                        GRAND
.....4 M.....5 .....6                                     .....13
                                                                        .....15
        EINERGI
        E.....7
```

-----SOMMET 7 --> unité lexicale = 'EINERGIE';
fonction syntaxique = gouverneur; catégorie = adjoint; type d'adjoint = adjectif; article exigé par le nom = pas d'article;
classe de provenance dérivationnelle = nom vers l'adjectif ou le nom de personne ou d'instrument;
code typographique pour les mots composés = constituant immédiat; genre = féminin; adjoint dérivable = adjectif de relation;
classe sémantique = objet concret; type de concret = matière; type du nom = nom commun; personne = 3; nombre = singulier;
rang en vue de la mise en ordre en génération = +51; nombre de constituants pour le groupe = +1.
-----SOMMET 14 --> unité lexicale = 'DEIFIER';
fonction syntaxique = gouverneur; catégorie = substantif; classe de provenance dérivationnelle = verbe vers le nom d'action;
complément direct = groupe nominal; genre = masculin; nom abstrait dérivable = masculin; type du nom = nom commun;
personne = 3; nombre = singulier; rang en vue de la mise en ordre en génération = +50;
nombre de constituants pour le groupe = +1.

Figure 16.12

linguistic data:

> declarations of variables, assignment formats and condition procedures;
>
> between 1 and 7 dictionaries;
>
> a grammar.

Each string is a processing unit, made up of a specific couple: name of Lexical Unit + mask of variable values. In processing each such couple, the system consults the grammar and has available registers:

> C the current mask;
>
> P the preceding mask;
>
> T the current string and the associated mask;
>
> S the previous output string and the associated mask.

While C is being processed, mask P is accessible. T changes at the same time as C and will be transferred to S at the end of the processing. The transfer of T to S empties T and C. As long as T is not empty, S contains the string of the preceding unit. The mask of the couple to be processed is placed in C, with T being the string to be constructed. The string created from the previous couple and the associated mask are found in S and P respectively. For each couple, the rules of the grammar are explored in the order of their enumeration, and the first rule whose condition is verified on the values of C and/or P applies.

A grammar rule carries out at least one of the following actions:

> (a) Concatenate to the right (D), to the left (G), or to the middle (M) of the current string (in T), the string indicated explicitly or by reference to a dictionary. M will always be the position in T corresponding to the point of the last concatenation.
>
> (b) Assign the given values P, and/or concatenate T to S (T: = S), or transfer T into S (S: = T) at the end of processing. Note that the concatenation of T to S is used to handle various cases of elision, contraction and the attachment of punctuation marks.
>
> (c) Transform all or part of string T, alone or concatenated to S.
>
> (d) Implement the rules indicated one after another. Those that are optional are placed in parentheses.

The dictionaries list the morphs of the language (as the right-hand member of the dictionary rules). The various concatenation operations transform these morphs into the words of the target language. Dictionary 1 can only be addressed by the lexical units (the left-hand member); the remaining dictionaries may be

accessed by the values of other variables. These values are automatically classified in alphabetical order, and it is in this order that
dictionary look-up proceeds. The first entry value corresponding to
the value contained in mask C is located, and within this entry, the
first condition/assignment/string triple whose condition is verified.
The assignments indicated (if any) are then carried out and the
morph (or character string) is inserted to the right, to the left or in
the middle of the current string.

G M uses six specialised dictionaries. The first contains the lexical
base forms of the U Ls of the language, and the five others list the
inflectional and derivational morphs according to their paradigms.
The first dictionary may be enlarged, but the other dictionaries
remain stable. (The various affixes and endings of the language
constitute a fixed, limited class of morphs, the complete list of
which results from the adopted strategy). The procedure for indexing the first dictionary is provided by an indexing manual that
N. Nedobejkine has entered into the computer. This manual indicates, for each U L to be indexed, the number and graphemic form
of its lexical base forms, the conditions for choosing among these
base forms and each base form's paradigm.

An excerpt from dictionary 1 is given below.

```
COBALT              ==          /MOT       /COBALT.
COEFFICIENT         ==          /MOT       /COEFFICIENT.
COEUR               ==PREF      /RIEN      /CARDIO-,
                    ==RELAT     /MOT       /CARDIAQUE,
                    ==          /MOT       /COEUR.
COEXISTER           ==NVBFEM    /MOT       /COEXISTENCE,
                    ==          /V1AZER3   /COEXIST.
COFACTEUR           ==          /MOT       /COFACTEUR.
COI"NCIDER          ==NVBDER    /URGENT    /COI"NCIDE,
                    ==          /V1AZER1   /COI"NCID.
COIFFER             ==NVBFEM    /MOT       /COIFFURE,
                    ==          /V1AZER1   /COIFF.
COKE                ==          /MOT       /COKE.
COL                 ==          /MOT       /COL.
COLLABORER          ==          /V1BION2   /COLLABOR.
COLLE               ==          /MOT       /COLLE.
COLLECTER.AGE       ==          /V1AAG2    /COLLECT.
COLLECTER.E         ==          /V1AFE2    /COLLECT.
COLLECTIF           ==          /ACTIF     /COLLECTI.
COLLECTIONNER       ==          /V1AZER2   /COLLECTIONN.
COLLECTIVISER       ==          /V1BION1   /COLLECTIVIS.
COLLECTIVISME       ==          /ISME      /COLLECTIV.
COLLER              ==          /V1AAG1    /COLL.
COLLIMATION         ==INSTR     /MOT       /COLLIMATEUR,
```

Dictionary 2 lists the forms of various verb stems and inflectional endings, and such regular derivational suffixes as -ABLE, -ABILITE. For example:

```
AIMER                    ==PRT      /PART1     /é,
                         ==INF      /          /ER,
                         ==FUT      /AUR1      /,
                         ==SAI      /PRES1     /,
                         ==PSS      /PSSMP1    /,
                         ==         /VERBE     /.
FUTUR1                   ==RIONS    /          /ERIONS,
                         ==RIEZ     /          /ERIEZ,
                         ==RAIENT   /          /ERAIENT,
                         ==RAIS     /          /ERAIS,
                         ==CDL      /          /ERAIT,
                         ==UNOPLU   /          /ERONS,
                         ==UNO      /          /ERAI,
                         ==DUEPLU   /          /EREZ,
                         ==DUE      /          /ERAS,
                         ==PLU      /          /ERONT,
```

Dictionary 3 is restricted to the derivational suffixes that form an action noun from a verb. For example:

```
AGE                      ==NVBPLU   /VALG      /AGES,
                         ==         /VALG      /AGE.
ANCE                     ==NVBPLU   /VALG      /ANCES,
                         ==         /VALG      /ANCE.
ATION                    ==NVBPLU   /          /ATIONS,
                         ==         /          /ATION.
CATION                   ==NVBPLU   /          /CATIONS,
                         ==         /          /CATION.
EMENT                    ==NVBPLU   /EMUET     /EMENTS,
                         ==         /EMUET     /EMENT.
```

Dictionary 4 includes all the affixes of the adjectival and nominal paradigms. For example:

```
CHIMIE                   ==AGPLU    /          /STES,
                         ==AGENT    /          /STE.
                         ==ADV      /          /QUEMENT,
                         ==ADJPLU   /          /QUES,
                         ==ADJ      /          /QUE,
                         ==PLU      /          /ES,
                         ==         /          /E.
EIEME                    ==CARD     /          /E,
                         ==ADV      /          /IèMEMENT,
                         ==ADJPLU   /          /IèMES,
                         ==         /          /IèME.
```

Dictionary 5 contains the prefixes of morphological negation.

Dictionary 6 contains the suffixes that derive an agentive or instrumental noun (or adjective) from a verb. For example:

```
EURRIC                   ==FEMPLU   /          /RICES,
                         ==FEM      /          /RICE,
                         ==PLU      /          /EURS,
                         ==         /          /EUR.
RICANT                   ==AGIFEMP  /          /RICES,
                         ==AGIFEM   /          /RICE,
                         ==AGIPLU   /          /EURS,
                         ==AGINST   /          /EUR,
                         ==FEMPLU   /          /ANTES,
                         ==FEM      /          /ANTE,
                         ==PLU      /          /ANTS,
                         ==         /          /ANT.
```

An example of a translation is given:

```
STREN     S1                       26 OCTOBRE 1983   21H 12MN 19S

              LANGUES DE TRAITEMENT: ALY-FRY

                   -- TEXTE ORIGINE --

Das Energieproblem ist zu einer grossen Herausforderung geworden.

STREN     S1

                   -- TEXTE TRADUIT --

----- ( TRADUCTION DU 26 OCTOBRE 1983      21H 12MN 02S ) -----
VERSIONS : ( A : 26/10/83 ; T : 26/10/83 ; G : 26/10/83 )

   Le problème énergétique est devenu un grand défi.
```

Conclusion

The experience acquired in the development of this model suggests that a second generation system like GETA's can certainly produce interesting results. However, important problems will remain unsolved until further research can address such questions as the following:

(a) economic strategies for pre-edition;

(b) methods for detecting and characterising well-defined and productive sublanguages;

(c) defining and developing expert systems to correct the output of analysis;

(d) defining formal models of linguistic representation that are linguistically more rigorous than the often too simplistic current models.

The immediate priority for the German–French system is to develop the transfer and syntactic phrases in order to be able to conduct further tests on the structural analysis component, which is already of respectable size but whose performance has not been properly evaluated. Once the structural analysis component has been tested on our working corpus (which is about thirty pages long), we will have to begin preparing for the conversion of the German–French system to the new version of ARIANE (ARIANE-78.5), which is now under development. This new version will allow us to handle some of the problems that are currently unresolved, such as the separable particles of German verbs.

Acknowledgements
I would like to thank Elliott Macklovitch for having helped me to translate this paper.

METAL : THE LRC
MACHINE TRANSLATION SYSTEM

The Linguistics Research Center of the University of Texas has been working on MT since its founding in 1961. Unlike some centres of MT research, its early focus was largely theoretical. Most US groups which developed first- and second-generation systems for MT were eliminated through the impact of the National Academy of Sciences report, *Languages and Machines,* released in 1966 ([1] ALPAC 1966). Limited funding continued, with some lapses, at the LRC. In 1978, Rome Air Development Center provided the means to apply the Center's theoretical findings; this was augmented in 1979 by support from Siemens AG, Munich. In 1980, Siemens became the sole project sponsor.

We will describe here the prototype MT system developed by the LRC. With the exception of the text-processing programs, which are written in SNOBOL, the LRC MT system is written in LISP and runs on a Symbolics Lisp Machine. The system includes a translation program, METAL, as part of a suite of programs designed to automate the complete process of translating technical texts.

The domain of application
Natural language texts range in 'complexity' from edited abstracts through technical documentation and scientific reports to newspaper articles and literary materials. As far as MT is concerned, the former portion of this spectrum is less complex than the latter because it is characterised by relatively less syntactic and semantic variety. Paradoxically, the order of complexity for human translators is essentially reversed due to a dramatic increase in the size of the vocabulary: no qualified human translator has much difficulty with straightforward syntax or normal idiomatic usage, but the prevalence and volatility of technical terms and jargon

poses a considerable problem. Time pressures impose another set of problems, for which there appear to be no solutions in the realm of unaided human translation, and few if any good ones in the realm of machine-aided translation as commonly interpreted (e.g., a text editor coupled with automated dictionary look-up).

In terms of the demand for MT, then, the market is in technical translation. There is no significant demand for machine translation of folklore, literary materials, and the like, but there is a substantial and growing demand for machine translation of technical texts.

There are several advantages in translating technical texts as opposed to more general texts. One of these concerns vocabulary: technical texts tend to concentrate on one subject area at a time, wherein the terminology (lexical semantics) is relatively consistent, and where the vocabulary is relatively unambiguous, even though it may be quite large. (This is not to say that lexical problems disappear!) Another advantage is that there is typically little problematic anaphora, and little or no 'discourse structure' as usually defined. Third, in accordance with current practice for high-quality human translation, revision is to be expected. That is, there is no a priori reason why machine translations must be 'perfect' when human translations are not expected to be so: it is sufficient that they be acceptable to the human beings who revise them, and that they prove cost-effective overall (including revision).

Notwithstanding the advantages of translating technical texts, there are definite problems to be confronted. First of all, the volume of such material is staggering: potentially tens or hundreds of millions of pages per year. Even ignoring all cost-effectiveness considerations, the existence of this much candidate material demands a serious concern for efficiency in the implementation. Second, the emphasis in MT is changing from information acquisition to information dissemination. The demand is not so much for loosely approximate translations from which someone knowledgeable about the subject can infer the import of the text (perhaps with a view toward determining whether a human translation is desired); rather, the real demand is for high-quality translations of, e.g., operating and/or maintenance manuals – for instructing someone not necessarily knowledgeable about the vendor's equipment in precisely what must (and must not) be done, in any given situation. Fidelity, therefore, is essential.

In addition to the problems of size and fidelity, there are problems regarding the text itself: the format and writing style. For example, it is not unusual to be confronted with a text which has been 'typeset' by a computer, but for which the typesetting commands are no longer available. This can be true even when the text was originally produced, or later transcribed, in machine-readable form. The format may include charts, diagrams, multi-column tables, section headings, paragraphs, etc. Misspellings, typographical errors, and grammatical errors can and do appear. Technical texts are notable for their frequency of 'unusual' syntax such as phrase- and sentence-fragments, a high incidence of acronyms and formulas, plus a plethora of parenthetical expressions. The 'discourse structure', if it can be argued to exist, may be decidedly unusual – as exemplified by a flowchart. Unknown words will appear in the text. Sentences can be long and complicated, notwithstanding the earlier statement about reduced complexity. Technically-oriented individuals are renowned for abusing natural language. The successful MT system will address these problems as well as those more commonly anticipated.

General system description

In this section we discuss the facilities of the LRC MT system which substantially automate the overall translation process, including the production and maintenance of the lexical databases along with several text-processing programs and METAL's place among them.

In any large software system, the problem of producing and maintaining the data sets on which the programs operate becomes important, if not critical. First of all, when there is a large volume of such material the data entry process itself can consume a significant amount of time; second, the task of insuring data integrity becomes an even larger time sink. We will briefly expand on these two problems, and indicate how the LRC MT system provides software tools to help cope with them.

There are two problems associated with data entry: creating the data in the first place, and getting it entered in machine-readable form. In most applications of database management systems, the creation of data is relatively straightforward: such data items as personal name, identification number(s), age, job title, salary, etc., serve as examples to illustrate the point that the data items usually pre-exist, thus data entry becomes relatively more important. In an application such as MT represents, however, creating the original data is the major bottleneck: one must decide, for each of thousands of words, many details of behaviour in a complicated linguistic environment. Certainly these details may be said to 'pre-exist', but a real problem arises when humans attempt to identify them. In general, the more sophisticated the MT system, the more of these details there are. Data entry, relatively speaking, becomes a small or insignificant issue – although it remains a significant issue in absolute terms.

In part of the LRC MT system, we have developed a sophisticated 'lexical default' program that accepts minimal information (the root form of the word, and its category) and automatically encodes almost all of the features and values that, for METAL, specify the details of linguistic behaviour. This is accomplished by a combination of morphological analysis of the root form of the input word, and search of the existing lexical database for 'similar' entries. Defaulted lexical entries are created in machine-readable form to begin with, and are available for human review/revision using standard on-line editing facilities. This greatly reduces both coding time and coding errors.

A potentially harder problem, however, is the maintenance of data integrity. Humans will make mental errors in creating lexical entries, and will aggravate these by making typographical errors during data entry. Even assuming a lexical default program (which, of course, does not make such mistakes), the process of human revision of the defaulted entries may introduce errors. Therefore, the LRC MT system includes a validation program that, working from a formal specification of what is legal in lexical entries, identifies any errors of format and/or syntax within each submitted entry. The formal specification is organised by language, by lexical category within language, and by feature within category. The incidence of semantic errors – which our validation program could not detect – has not been found to be significantly high. As a result, the use of this program has virtually eliminated incorrect coding of monolingual lexical entries, which used to be a major source of error in METAL translations. (Related programs employ similar techniques to identify errors in the phrase-structure grammar rules and transformations that constitute the remainder of our linguistic rule base.)

There is also the problem of maintaining an existing lexical database. In any MT system, there will be a need for changing existing entries in the light of experience; in a system like ours, which serves as a vehicle for research in MT, this problem is magnified by the occasional need for large-scale changes in lexical entries to accommodate new system features, or even theories of translation. As part of the LRC MT system, then, we have incorporated a general relational Data Base Management System (DBMS) along with a group of interface routines that transform, upon entry, both monolingual and transfer lexical entries from a format optimised for human use into a format more suitable for storage by the DBMS, and which reverse the transformation when retrieving the entries. Not only do the interface routines facilitate the entry, retrieval, and revision of lexical entries, but they also are integrated with the validation program so that a lexicographer may not, while using the DBMS, introduce errors in the format and/or syntax of a lexical entry. This same on-line DBMS is accessed directly by METAL during translation; therefore, changes made in the database are instantly reflected in translations, resulting in rapid turnaround that both encourages and enhances research and development.

Finally, there is the problem of tying all these modules together with a powerful, high-level user interface which optimises the task of entry acquisition and maintenance. The LRC has developed an Interactive Coder that uses a menu-selection scheme as the interface mechanism for acquiring and maintaining lexical entries. Starting with words typed in, or drawn from a concordance of unknown words resulting from the dictionary analysis of a new text, or appearing in an abbreviated transfer entry input file created by a text editor or perhaps a formatted dump from an existing on-line database, the Interactive Coder will use the database interface, the default program, the formal specifications governing lexical entries, and the validator, to facilitate the process of lexical entry acquisition and maintenance. The METAL system runs on a Lisp Machine with a high-resolution bit-mapped display and a 'mouse' pointing device, and the Interactive Coder takes advantage of this sophisticated, powerful menu interface medium to optimise the overall dictionary maintenance task. Figure 17.1 outlines the process of semi-automatic terminology acquisition and subsequent 'linguistic coding' supported by the lexical database management system.

As a glance at any technical manual will show, it is not always the case that all material in a document must or can be translated. Large portions of a text (up to fifty per cent of the characters, in our experience) may not be translatable material; the bulk of this may fall outside sentence boundaries, but some will fall within them. Thus it is necessary for a text to be marked, or annotated, to distinguish that which is to be translated (e.g. tables of contents, instructions, prose paragraphs) from that which is not (e.g. text formatting information, flowchart box boundaries, acronyms, and various attention-focusing devices). In the LRC system, a program determines which members of the machine's character set are unused in the text at hand, and a few of these are used to mark the text.

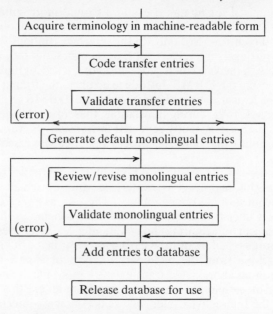

Figure 17.1 Coding Lexical Entries using the Interactive Coder.

Within multi-column tables, different sentences would inter-mingle if the MT system were to read the text in the usual computer fashion (i.e. horizontally across the line). Our text-processing component currently detects most such situations, and makes annotations that, after further automatic processing, will allow the translation component to 'see' the sentences as human beings would: reading vertically down the page. Another annotation program attempts to identify translatable units (isolated words, phrases, and sentences) and brackets them. Some untranslatable material (e.g. flowchart box labels) may appear to fall within sentence boundaries, but play no grammatical role in a sentence: therefore we introduced a 'toggle' convention for excluding such material so that it is invisible to the analyser. On the other hand, some untranslatable material within a sentence (e.g. equations and formulas) may indeed play a grammatical role: the annotation program marks such strings to be analysed, but carried through to the target language without translation. Human verification and emendation of the annotation component's output is necessary, as might be expected, but this task does not require significant training, and in particular it requires no knowledge of the target language(s) or the

MT system; it does require knowledge of what is to be translated.

Once the text has been annotated, another program extracts the sentences to be translated and prepares them for input to the MT system. Essentially, this entails copying any bracketed units into a new file, excluding toggled-out material. Anything outside of brackets is ignored completely. Complicating this process is the fact that, in order to minimise human intervention in reformatting the text after translation, the extraction program must record the location of the original sentences copied into the MT input file, so that the results of translation may be formatted just like the original document.

The next (optional) step is dictionary pre-analysis of the text. The MT input file is scanned, and the unique word occurrences are identified; each is then submitted to lexical analysis. This step offers the potential advantage of detecting dictionary shortages (i.e. unknown words) before they have a chance to affect the translation process; also, METAL's proposed spelling corrections can be noted for human review, and suspected acronyms can be verified. A concordance of the unknown and misspelled words is constructed; this may be accessed, for example, by the menu-driven Interactive Coder, if the user decides to create dictionary entries for unknown words before proceeding with the translation. After the user has checked (and, if desired, corrected) any textual errors noted during dictionary pre-analysis, translation may proceed.

Translation, briefly stated, involves reading in the source language (SL) file of sentences and printing out the corresponding target language (TL) file. This step – performed by METAL itself – is detailed below. The process of text reformatting is an important component of the overall process of translation. If an MT system provides only a sequence of translated sentences, then a human revisor must manually reformat the entire document, producing charts, figures, etc. This can be a source of substantial and unnecessary frustration, and can constitute a significant cost factor in translation. The alternative of providing unformatted translations to the end-user is distinctly unattractive. In the LRC MT system, a special reconstitution program automatically formats translated text as much like the original text as is possible: margins, indentation, and blank lines are all observed, and special forms (e.g. multi-column tables) are all reconstructed.

The last step involves human revision of the machine's translations. (This step may actually take place before reconstitution, using the MT input and MT output files separately, or an interlinear version of these two that is also available; but some small text format problems may remain that require human attention after reconstitution.) The linguistic and computational research behind METAL has always had as its goal a comprehensive analysis and translation of whole sentences in their context; the work of the revisor, then, should be mostly a matter of review. Where METAL fails to achieve a unified analysis of a sentence, a translation by phrases is provided which the revisor must render into a well-formed translation. Figure 17.2 depicts the steps in 'production' machine translation using the LRC MT system.

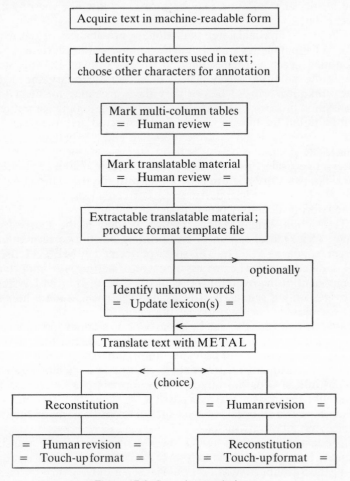

Figure 17.2. Steps in translation.

Linguistic techniques employed

Our distinction between 'linguistic techniques' and 'computational techniques' (discussed in the next major section) is somewhat artificial, but it has some validity in a broad sense, as should become clear from an overview of the points considered. In this section we present the reasons for our use of the following linguistic techniques: (a) a phrase-structure grammar; (b) syntactic features; (c) semantic features; (d) scored interpretations; (e) transformations indexed to specific rules; (f) a transfer component; and (g) attached procedures to effect translation.

In the LRC MT system we employ a phrase-structure grammar, augmented by sufficient lexical controls to make it resemble lexical-functional grammar ([566] Bresnan 1977). Of all our linguistic decisions, this is surely the most controversial, and consequently will receive the most attention. Generally speaking, there are two competing claims; first, that syntax rules per se are inadequate and wasteful (e.g. [636] Cullingford 1978); and second, that other forms of grammar (ATNs [626] Woods 1970; transformational [612] Petrick 1973; procedural [664] Winograd 1972; word-experts [652] Small 1980; etc.) are superior. We will deal with these in turn.

There are schools of thought that claim that, syntax rules per se are inappropriate models of language. Language should, according to this notion, be treated (almost) entirely on the basis of semantics, guided by a strong underlying model of the current situational context, and the expectations that may be derived therefrom. We cannot argue against the claim that semantics is of critical concern in Natural Language Processing. However, as yet no strong case has been advanced for the abandonment of syntax. Moreover, no system has been developed by any of the adherents of the 'semantics only' school of thought that has more-or-less successfully dealt with all of a wide range – or at least large volume – of material. A more damaging argument against this school is that every NLP system to date that has been applied to large volumes of text (in the attempt to process all of it in some significant sense) has been based on a strong syntactic model of language (see e.g. [90] Boitet et al. 1980; [579] Damerau 1981; [594] Hendrix et al. 1978; [259] Lehmann et al. 1981; [545] Martin et al. 1981; [615] Robinson 1982; and [617] Sager 1981).

There are other schools of thought that hold context-free phrase-structure (PS) rules in disrespect, while admitting the utility (necessity) of syntax. It is claimed that the phrase-structure formalism is inadequate, and that other forms of grammar are necessary. (This has been a long-standing position in the linguistics community; but this view is now being challenged by some, who are once again supporting PS rules as a model of natural language use ([584, 586] Gazdar 1981, 1983). The anti-PS positions in the Natural Language Processing community are all, of necessity, based on practical considerations, since the models advanced to replace PS rules are formally equivalent in generative power (assuming the PS rules to be augmented, which is always the case in modern NLP systems employing them). But cascaded ATNs ([628] Woods 1980), for example, are only marginally different from PS rule systems. It is curious to note that only one of the remaining contenders (a transformational grammar [579] Damerau 1981) has been demonstrated in large-scale application – and even this system employs PS rules in the initial stages of parsing. Other formal systems (e.g. procedural grammars [664] Winograd 1972) have been applied to semantically deep (but linguistically impoverished) domains – or to excessively limited domains (e.g. [652] Small's 1980 'word expert' parser seems to have encompassed a vocabulary of well under twenty items).

For practical application, it is necessary that a system be able to accumulate grammar rules, and especially lexical items, at a prodigious rate by current NLP standards. The formalisms competing with PS rules and dictionary entries of modest size seem to be universally characterisable as requiring enormous human resources for their implementation in even a moderately large environment. This should not be surprising: it is precisely the claim of these competing methodologies (those that are other than slight variations on PS rules) that language is an exceedingly complex phenomenon, requiring correspondingly complex techniques to model. For 'deep understanding' applications, we do not contest this claim. But we do maintain that there are some applications that do not seem to require this level of effort for adequate results in a practical setting. Our particular application – automated translation of technical texts – seems to fall in this category, as does, e.g., the EPISTLE text-critiquing system ([593] Heidorn et al. 1982).

The LRC MT system is currently equipped with approximately 550 PS rules describing the best-developed source language (German), and around 10000 lexical entries in each of the two main languages (German, and the best-developed target langage: English). The current state of our coverage of the SL is that the system is able to parse and acceptably translate the majority of sentences in previously-unseen texts, within the subject areas bounded by our dictionaries. We have recently begun the process of adding to the system an analysis grammar of the current TL (English), so that the direction of translation may be reversed; we anticipate bringing the English grammar up to the level of the German grammar in about two years' time. Our expectations for eventual coverage are that, for each SL, around 1000 PS rules will be adequate to account for almost all sentence forms actually encountered in technical texts. We do not feel constrained to account for every possible sentence form in such texts – nor for sentence forms not found in such texts (as in the case of poetry) – since the required effort would not be cost-effective whether measured in financial or human terms, even if it were possible using current techniques (which we doubt).

Our use of syntactic features is relatively noncontroversial, given our choice of the PS rule formalism. We employ syntactic features for two purposes. One is the usual practice of using such features to restrict the application of PS rules (e.g. by enforcing subject-verb number agreement). The other use is perhaps peculiar to our type of application: once an analysis is achieved, certain syntactic features are employed to control the course (and outcome) of translation – i.e. generation of the TL sentence. The 'augmentations' to our PS rules include procedures written in a formal language (so that our linguists do not have to learn LISP) that manipulate features by restricting their presence, their values if present, etc., and by moving them from node to node in the 'parse tree' during the course of the analysis. As is the case with other researchers employing such techniques, we have found this to be an extremely powerful (and of course necessary) means of restricting the activities of the parser.

We employ simple semantic features, as opposed to complex models of the domain. Our reasons are primarily practical. First, features seem sufficient for at least the initial stage of our application. Second, the thought of writing complex models of even one complete technical domain is staggering: the operation and maintenance manuals we have worked with (describing a digital telephone switching system) are part of a document collection that is expected to comprise some 100000 pages of text when complete. A research group the size of ours would not even be able to read that volume of material, much less write the 'necessary' semantic models subsumed by it, in any reasonable amount of time. (The group members would also have to become electronics engineers, in all likelihood.) If such models are indeed required for our application, we will never succeed.

As it turns out, we are doing surprisingly well without such models. In fact, our semantic feature system is not yet being employed to restrict the analysis effort at all; instead, it is used at 'transfer time' (described later) to improve the quality of the translations, primarily of prepositions. We look forward to extending the use of semantic features to other parts of speech, and to substantive utilisation during analysis; but even we were pleased at the results we achieved using only syntactic features.

It is a well-known fact that NLP systems tend to produce many readings of their input sentences (unless, of course, constrained to produce the first reading only – which can result in the 'right' interpretation being overlooked). The LRC MT system may produce multiple interpretations of the input 'sentence', assigning each of them a score, or plausibility factor ([615] Robinson 1982). This technique can be used, in theory, to select a 'best' interpretation from the available readings of an ambiguous sentence. We base our scores on both lexical and grammatical phenomena – plus the types of any spelling/typographical errors, which can sometimes be 'corrected' in more than one way.

Our experiences relating to the reliability and stability of heuristics based on this technique are decidedly positive: we employ only the (or a) highest-scoring reading for translation (the others being discarded), and our informal experiments indicate that it is rarely true that a better translation results from a lower-scoring analysis. (Surprisingly often, a number of the higher-scoring interpretations will be translated identically. But poorer translations are frequently seen from the lower-scoring interpretations, demonstrating that the technique is indeed effective.)

We employ a transformational component, during both the analysis phase and the translation phase. The transformations, however, are indexed to specific syntax rules rather than loosely keyed to syntactic constructs. (Actually, both styles are available, but our linguists have never seen the need or practicality of employing the open-ended variety). It is clearly more efficient to index transformations to specific rules when possible; the import of our findings is that it seems to be unnecessary to have open-ended transformations – even during analysis, when one might intuitively expect them to be useful.

It is frequently argued that translation should be a process of analysing the source language into a 'deep representation' of some sort, then directly synthesising the target language (e.g. [632] Carbonell et al. 1978). We and others ([227] King 1981) contest this claim – especially with regard to 'similar languages' (e.g. those in the Indo-European family). One objection is based on large-scale, long-term trials of the 'deep representation' (in MT, called the 'pivot language') technique by the MT group at Grenoble ([86] Boitet et al. 1980). After an enormous investment in time and energy, including experiments with massive amounts of text, it was decided that the development of a suitable pivot language (for use in Russian–French translation) was probably impossible. Another objection is based on practical considerations: since it is not likely that any NLP system will in the foreseeable future become capable of handling unrestricted input – even in the technical area(s) for which it might be designed – it is clear that a 'fail-soft' technique is necessary. It is not obvious that such is possible in a system based solely on a pivot language; a hybrid system capable of dealing with shallower levels of understanding is necessary in a practical setting. This being the case, it seems better in near-term applications to start off with a system employing a 'shallow' but usable level of analysis, and deepen the level of analysis as experience dictates, and theory plus project resources permit.

Our alternative is to have a 'transfer' component which maps 'shallow analyses of sentences' in the SL into 'shallow analyses of equivalent sentences' in the TL, from which synthesis then takes place. While we and the rest of the NLP community continue to explore the nature of an adequate pivot language (i.e. the nature of deep semantic models and the processing they entail), we can hopefully proceed to construct a usable system capable of progressive enhancement as linguistic theory becomes able to support deeper models.

Our transfer procedures (which effect the actual translation of SL into TL) are tightly bound to nodes in the analysis (parse tree) structure ([611] Paxton 1977). They are, in effect, suspended procedures – parts of the same procedures that constructed the corresponding parse tree nodes to begin with. We prefer this over a more general, loose association based on, e.g.,syntactic structure because, aside from its advantage in sheer, computational efficiency (search for structure transfer rules is eliminated), it eliminates the possibility that the 'wrong' procedure can be applied to a construct. The only real argument against this technique, as we see it, is based on space considerations: to the extent that different constructs share the same transfer operations, wasteful replication of the procedures that implement said operations (and editing effort to modify them) is possible. We have not noticed this to be a problem. For a while, our system load-up procedure searched for duplicates of this nature and automatically eliminated them; however, the gains turned out to be minimal: different structures typically do require different operations.

Computational techniques employed

Again, our separation of 'linguistic' from 'computational' techniques is somewhat artificial, but nevertheless useful. In this section we present the reasons for our use of the following computational techniques: (a) a 'some-paths', parallel, bottom-up parser; (b) associated rule-body procedures; (c) spelling correction; (d) another fail-soft analysis technique; and (e) recursive parsing of parenthetical expressions.

Among all our choices of computational techniques, the use of a 'some-paths', parallel, bottom-up parser is probably the most controversial. Our current parser operates on the sentence in a well-understood parallel, bottom-up fashion; however, the notion of 'some-paths' will require some explanation. In the METAL system, the grammar rules are grouped into 'levels' indexed numerically (0, 1, 2 . . .), and the parser always applies rules at a lower level(e.g. 0) before applying any rules at a higher level (e.g. 1). Thus, the application of rules is partially ordered. Furthermore, once the parser has applied all rules at a given level it halts if there exist one or more 'sentence' interpretations of the input; only if there are none does it apply more rules – and then, it always starts back at level 0 (in case any rules at that level have been activated through the application of rules at a higher level, as can happen with a recursive grammar). Thus, the rule-application algorithm is Markov-like, and the system will not necessarily produce all interpretations of an input possible with the given rule base. Generally speaking, the lower-level rules are those most likely to lead to readings of an input sentence, and the higher-level rules are these least likely to be relevant (though they may be necessary for particular input sentences, in which case they will eventually be applied). As a result, the readings derived by our parser are the 'most likely' readings (as judged by the linguists, who assign the rules to levels). This works very well in practice.

Our evolving choices of parsing methodologies have received our greatest experimental scrutiny. We have collected a substantial body of empirical evidence relating to parsing techniques and strategy variations. Since our evidence and conclusions would require lengthy discussion, and have received some attention elsewhere ([408] Slocum 1981), we will only state for the record that our use of a some-paths, parallel, bottom-up parser is justified based on our findings. First of all, all-paths parsers have certain desirable advantages over first-path parsers (discussed below);

second, our some-paths parser (which is a variation on an all-paths technique) has displayed clear performance advantages over its predecessor technique: doubling the throughput rate while increasing the accuracy of the resulting translations. We justify our choice of technique as follows: first, the dreaded 'exponential explosion' of processing time has not appeared, on the average (and our grammar and test texts are among the largest in the world), but instead processing time appears to be linear with sentence length – even though our system may produce all possible readings; second, top-down parsing methods suffer inherent disadvantages in efficiency, and bottom-up parsers can be and have been augmented with 'top-down filtering' to restrict the syntax rules applied to those that an all-paths top-down parser would apply; third, it is difficult to persuade a top-down parser to continue the analysis effort to the end of the sentence, when it blocks somewhere in the middle – which makes the implementation of 'fail-soft' techniques having production utility that much more difficult; and lastly, the lack of any strong notion of how to construct a 'best-path' parser, coupled with the raw speed of well-implemented parsers, implies that a some-paths parser which scores interpretations and can continue the analysis to the end of the sentence, come what may, may be best in a contemporary application such as ours.

We associate a procedure directly with each individual syntax rule, and evaluate it as soon as the parser determines the rule to be (seemingly) applicable ([553] Pratt 1973; [594] Hendrix 1978) – hence the term 'rule-body procedure'. This practice is equivalent to what is done in ATN systems. From the linguist's point of view, the contents of our rule-body procedures appear to constitute a formal language dealing with syntactic and semantic features/values of nodes in the tree – i.e. no knowledge of LISP is necessary to code effective procedures. Since these procedures are compiled into LISP, all the power of LISP is available as necessary. The chief linguist on our project, who has a vague knowledge of LISP, has employed OR and AND operators to a significant extent (we didn't bother to include them in the specifications of the formal language, though we obviously could have), and on rare occasions has resorted to using COND. No other calls to true LISP functions (as opposed to our formal operators, which are few and typically quite primitive) have seemed necessary, nor has this capability been requested, to date. The power of our rule-body procedures seems to lie in the choice of features/values that decorate the nodes, rather than the processing capabilities of the procedures themselves.

There are limitations and dangers to spelling correction in general, but we have found it to be an indispensable component of an applied system. People do make spelling and typographical errors, as is well known; even in 'polished' documents they appear with surprising frequency (about every page or two, in our experience). Arguments by LISP programmers [re: INTERLISP's DWIM] aside, users of applied NLP systems distinctly dislike being confronted with requests for clarification – or, worse, unnecessary failure – in lieu of automated spelling correction. Spelling correction, therefore, is necessary.

Luckily, almost all such errors are treatable with simple techniques: single-letter additions, omissions, and mistakes, plus two- or three-letter transpositions account for almost all mistakes. Unfortunately, it is not infrequently the case that there is more than one way to 'correct' a mistake (i.e. resulting in different corrected versions). Even a human cannot always determine the correct form in isolation, and for NLP systems it is even more difficult. There is yet another problem with automatic spelling correction: how much to correct. Given unlimited rein, any word can be 'corrected' to any other. Clearly there must be limits, but what are they?

Our informal findings concerning how much one may safely 'correct' in an application such as ours are these: the few errors that simple techniques have not handled are almost always bizarre (e.g. repeated syllables or larger portions of words) or highly unusual (e.g. blanks inserted within words); correction of more than a single error in a word is dangerous (it is better to treat the word as unknown hence a noun); and 'correction' of errors which have converted one word into another (valid in isolation) should not be tried.

In the event of failure to achieve a comprehensive analysis of the sentence, a system such as ours – which is to be applied to hundreds of thousands of pages of text – cannot indulge in the luxury of simply replying with an error message stating that the sentence cannot be interpreted. Such behaviour is a significant problem, one which the NLP community has failed to come to grips with in any coherent fashion. There have, at least, been some forays. Weischedel and Black ([622] 1980) discuss techniques for interacting with the linguist/developer to identify insufficiencies in the grammar. This is fine for system development purposes. But, of course, in an applied system the user will be neither the developer nor a linguist, so this approach has no value in the field. Hayes and Mouradian ([592] 1981) discuss ways of allowing the parser to cope with ungrammatical utterances; such work is in its infancy, but it is stimulating nonetheless. We look forward to experimenting with similar techniques in our system.

What we require now, however, is a means of dealing with 'ungrammatical' input (whether through the human's error or the shortcomings of our own rules) that is highly efficient, sufficiently general to account for a large, unknown range of such errors on its first and subsequent outings, and which can be implemented in a short period of time. We found just such a technique several years ago: a special procedure (invoked when the analysis effort has been carried through to the end of the sentence) searches through the parser's chart to find the shortest path from one end to the other; this path represents the fewest, longest-spanning phrases which were constructed during the analysis. Ties are broken by use of the standard scoring mechanism that provides each phrase in the analysis with a score, or plausibility measure (discussed earlier). We call this procedure 'phrasal analysis'.

Our phrasal analysis technique has proven to be useful for both the developers and the end-users, in our application: the system translates each phrase individually, when a comprehensive sentence analysis is not available. The linguists use the results to pin-point missing (or faulty) rules. The users (who are professional translators, editing the MT system's output) have available the best translation possible under the circumstances, rather than no usable output of any kind. Phrasal analysis – which is simple and independent of both language and grammar – should prove useful in other applications of NLP technology; indeed, IBM's EPISTLE system ([608] Miller et al. 1980) employs an almost identical technique ([598] Jensen and Heidorn 1982).

Few NLP systems have ever dealt with parenthetical expressions; but MT researchers know well that these constructs appear in abundance in technical texts. We deal with this phenomenon in the following way: rather than treating parentheses as lexical items, we make use of LISP's natural treatment of them as list delimiters, and treat the resulting sublists as individual 'words' in the sentence; these 'words' are 'lexically analysed' via recursive calls to the parser. Aside from the elegance of the treatment, this has the advantage that 'ungrammatical' parenthetical expressions may undergo phrasal analysis and thus become single-phrase entities as far as the analysis of the encompassing sentence is concerned; thus, ungrammatical parenthetical expressions need not result in ungrammatical (hence poorly handled) sentences.

The translation component: METAL

This section presents a more detailed description of METAL – the actual translation component of the LRC MT system. The top-level control structure is quite simple: the function TRANS-LATE is invoked with a sentence in the SL (currently, German) and returns as its value an equivalent sentence in the TL (currently, English). TRANSLATE invokes three functions in succession: PARSE (for sentence analysis), TRANSFER (for structural translation), and GENERATE (for sentence synthesis). After sketching the format and content of dictionary entries, we will briefly discuss how the linguistic rules (lexicons and grammars) govern analysis, transfer, and synthesis, illustrating this three-step process using example sentences.

METAL lexicons are divided into two types: monolingual, and bilingual (called 'transfer'). A monolingual lexicon must be created for each of the languages involved in the translation process; transfer lexicons link the SL and TL monolingual lexicons. Monolingual lexicons consist of entries for each lexical item. Each entry begins with a left parenthesis followed immediately by the canonical or 'dictionary' form of the entry, then a series of feature labels, each with a sequence of zero or more values enclosed within parentheses. The entry is terminated by a right parenthesis. The entries for the German noun stem (NST) *Ausgabe* and the corresponding English NST 'output' will serve as examples (see figure 17.3).

Space constraints do not allow a full analysis of the entries; simply stated, each monolingual entry provides METAL with the information necessary for analysis and synthesis of the lexical items. In addition to entries for distinct word stems, the METAL monolingual lexicons contain separate lexical entries for such morphemes as prefixes, infixes, suffixes, and punctuation.

Transfer lexicons consist essentially of canonical word pairs which indicate the many-many correspondence between the SL and TL word stems. Each pair may be augmented by an arbitrary collection of context restrictions that must be met in order for the indicated translation to take place. A sample transfer entry for the pair *Ausgabe* – 'output' is included in figure 17.3; there are no restrictions (conditions) placed on this transfer (indicating the translation of *Ausgabe* into 'output', or vice versa], other than the Subject Area tag [DP = Data Processing).

As an example of transfer restriction, it is possible to specify that a given German preposition corresponds to any of several English prepositions depending on the semantic type of its object noun.

German monolingual entry:
```
(Ausgabe              CAT (NST)
     ALO      (Ausgabe)
     PLC      (WI)
     SNS      (1)
     TAG      (DP)
     CL       (P-N S-0)
     DR       (NP RD)
     FC       (PP)
     GD       (F)
     SX       (N)
     TY       (ABS DUR)
)
```

English monolingual entry:
```
(output               CAT (NST)
     ALO      (output)
     PLC      (WI)
     SNS      (1)
     TAG      (DP)
     CL       (P-S S-01)
     DR       (NP RD)
     FC       (PP)
     ON       (VO)
     SX       (N)
)
```

German-English Transfer entry:
```
(Ausgabe (NST DP) 0              ! output (NST DP) 0              )
(                                    +                            )
```

Figure 17.3. German monolingual, English monolingual, and Transfer entries for *Ausgabe* = output.

Four entries for the German preposition *vor,* shown in figure 17.4, will illustrate this. In these entries the appropriate English translation is defined by a restriction on semantic type (TY) and sometimes grammatical case (GC). These transfer entries are valid for all subject areas, but must be tried in a particular order (as evidenced by numeric 'preference factors' in the entries). Thus the presence in context of an object noun of semantic TYpe other than ABStract, DURative, or PuNcTual results in the English translation 'in front of'; else the presence of a Dative object noun of type ABStract or PuNcTual will result in the English translation 'before'; else the presence of a Dative object noun of type DURative will result in the English translation 'ago' [which will later be postponed]; otherwise, the translation 'in front of' is chosen.

```
(vor(PREP ALL) 30              ! in_front_of(PREP ALL) 0     )
(      OPT TY * ABS DUR PNT !                                )
(                         +                                  )


(vor(PREP ALL) 20              ! before(PREP ALL) 0          )
(      GC D                    !                             )
(      TY ABS PNT              !                             )
(                         +                                  )


(vor(PREP ALL) 10              ! ago(PREP ALL) 0             )
(      GC D                    !                             )
(      TY DUR                  !                             )
)                              +                             )


(vor(PREP ALL) 0               ! in_front_of(PREP ALL) 0     )
(                         +                                  )
```

Figure 17.4. German-English Transfer entries for *vor* = in front of, before, or ago.

Analysis

For human-engineering reasons, one of the most convenient forms for expressing a grammar is via context-free phrase-structure rules. Context-free rules alone may or may not fully describe human language (see [586] Gazdar 1983 for arguments that CF grammars are indeed sufficient), but, in any case, more general phrase-structure rules preclude efficient computational treatment, and CF rule-based systems seem to function as well as or better than any other technique, in practice. It has become traditional to augment the context-free rules by associating with them procedures in some programming language in order to provide more generative power, while maintaining computational tractability. In METAL, these 'rule-body procedures' are invoked as soon as the parser finds a phrase matching their constituent phrase structure.

The traditional purpose of such procedures is to restrict the application of a rule by tests on syntax (e.g. number agreement between noun and verb) and/or semantics (e.g. whether the proposed syntactic subject can be interpreted as an agent); if such tests fail, the syntactic phrase is not built. In METAL, these procedures not only accept or reject rule application, but they also construct an interpretation of the phrase. Traditional parsers automatically build a 'parse tree' and may add the output of such procedures as semantic information; in METAL, the parser (i.e. the LISP program) makes no commitment to a syntactic structure, but instead, linguistic procedures construct the interpretation (phrase) and compute its weight, or plausibility measure. The weight of a phrase is used when comparing it with any others that span the same sequence of words in order to identify the most likely reading.

A rule-body procedure in our system has several components: a constituent test part that checks the sons to ensure their utility in the current rule; an agreement TEST part to enforce syntactic and/or semantic correspondence among constituents; a phrase CONSTRuctor, which formulates the interpretation (phrase) defined by the current rule; and one or more target-language-specific transfer parts which operate during the second stage of translation (following complete sentence analysis). The inter-constituent test, the phrase constructor, and the transfer procedures may include calls to case frame procedures and/or transformations, as well as simpler routines to test and set syntactic and semantic features/ values.

Case frames may apply semantic and syntactic agreement restrictions to the predicate (verb structure) and its arguments (noun and prepositional phrases) when constructing a clause. Each predicate's lexical entry specifies its possible 'central arguments'. For German, the case frame will identify the case role-players according to voice (e.g. active) and mood (e.g. indicative) of the clause, and information about each potential argument such as its semantic type, form (noun phrase or prepositional phrase), and grammatical case (e.g. accusative) or prepositional marker. The restrictions can be general, or specific to the individual verb, preposition, and/or noun. The frame will fail, causing application of the clause rule to be rejected, if any of the restrictions are not met. Otherwise, case roles are assigned to the central arguments and the 'peripheral arguments' are then identified.

The geometry of interpretations typically (though not always) parallels their original phrase structure. In other words, they are usually topologically equivalent to what the parser would produce if it were automatically constructing a tree. Some rules, however, incorporate transformations which may arbitrarily alter the phrase being constructed. The transformation module allows a linguist to specify a structural descriptor to any depth, to perform syntactic and/or semantic tests as in rule body procedures, and to specify a new structure into which the old is transformed. The transformation program attempts to match the 'old' pattern descriptor with the currently instantiated phrase. If the match is successful, and the specified conditions are met, a new phrase is constructed using the 'new' pattern descriptor, with the (old) matched phrase usually

providing (most of) the structural contents, and constructor operations may further annotate the phrase with new features and/or
values. The transformation module can have no effect on the parsing algorithm, whatever the outcome of its application, unless the
rule is written so that failure to complete a transformation causes
the interpretation to be rejected; in such a case, only the fact of the
rejection has an effect on the parser: it abandons that search path,
just as it would if any other condition in the rule-body procedure
were unsatisfied.

NN	NST	N-FLEX
0	1	2
(LVL 0)	(REQ WI)	(REQ WF)
TEST	(INT 1 CL 2 CL)	
CONSTR	(CPX 1 ALO CL)	
	(CPY 2 NU CA)	
	(CPY 1 WI)	
ENGLISH	(XFR 1)	
	(ADF 1 ON)	
	(CPY 1 MC DR)	
SPANISH	...	

Figure 17.5. A German Context-free PS rule for building a Noun
STem + an inflectional ending into a NouN.

A grammar in METAL consists of a number of partially-ordered
(LeVeLled), augmented phrase-structure syntax rules, plus a collection of indexed transformations. A relatively simple PS rule for
building nouns will be used to illustrate the parts and format of
METAL grammar rules (see figure 17.5).

The first line consists of a left-hand element, the 'father' node
(here, NN), and one or more right-hand elements – the 'sons'
(here, NST and N-FLEX). In the example rule, the left-hand
element is the noun (NN) node and the right-hand elements are
the noun stem (NST) and the nominal ending (N-FLEX) nodes.
The second line enumerates the elements (from 0 to n) for reference in the rule-body procedure. Each constituent may have individual conditions, called 'column tests', to restrict exactly what
elements fit the rule. If any column test fails, the grammar rule will
fail — i.e. the parser will abandon its attempt to apply this rule. In
this example, the column test for the first element (NST) requires

it to be word-initial (WI) — i.e. preceded by a blank space in the matrix sentence; the column test for the second element (N-FLEX) requires it to be word-final (WF) – i.e. followed by a blank space.

In addition to the column tests, which apply only to single elements, each rule has a TEST part that states agreement restrictions between the right-hand elements. Failure of any agreement test will also result in failure of the entire rule. In the example rule, the single agreement test states that there must be an intersection (INT) of the inflectional class (CL) values for the two constituents; i.e. the values for the feature CL coded on the NST and the N-FLEX are compared to insure that they have at least one value in common.

Only after all conditions have been satisfied is it possible for METAL to build the appropriate syntax tree. This is done in the CONSTR part of the rule, which can also add or copy information in the form of features and values from the sons to the father. In the example rule, the CONSTRuctor (by not applying a transformation) would produce the tree represented below:

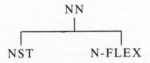

In the example rule-body procedure, the CONSTRuctor will copy all features with their associated values from the first element (i.e. the NST), except for the allomorph (ALO) and inflectional class (CL) features, using the operation CPX. CONSTR in this rule will also copy (CPY) the grammatical number (NU) and case (CA) features from the second constituent (the N-FLEX), and the word initial (WI) feature from the first constituent (the NST).

Transformations may be applied in the TEST, CONSTR, and/or Transfer portions of grammar rules. These range from simple movement and deletion operations to highly complex transformations which add structure, perform tests, etc. The following exemplifies a simple movement transformation:
```
(XFM (&:1 (&:2 &:3))
     (&:1 (&:3 &:2)))
```
This transformation simply exchanges the two sons (no.2 and no.3) of the current node (no.1): each ampersand represents one and only one constituent, or node.

Determining whether a sequence of words constitutes a clause is handled by a case frame, which is invoked in the TEST portion of clause-level rules. Simply stated, the case frame uses the argument information coded on the verb stem's lexical entry to identify its arguments, perform agreement tests, and label those arguments. In METAL, an argument may be a noun phrase, prepositional phrase, or adverbial phrase, depending on the verb. For a more detailed discussion of the grammar or lexicon, see [73] Bennett 1982.

Transfer

The purpose of the TRANSFER module is to restructure the most plausible interpretation of the SL sentence into an interpretation of an equivalent sentence in the TL(s). Every non-terminal node (phrase) in every sentence interpretation has attached to it the 'suspended' rule-body procedure that originally created it; this eliminates the need to search through a monolithic 'transfer grammar' for a matching pattern or routine – and also eliminates the danger of inadvertently applying an inappropriate pattern or routine that happened to match (part of) the same structure. The suspended procedure associated with the root phrase in the most plausible interpretation is (re)invoked by TRANSFER. The appropriate target-language-specific transfer part of a rule-body procedure can recursively transfer all or some of the node's sons (i.e. its non-terminal constituents) in any order, apply transformations, and/or lexically transfer a terminal son. Lexical transfer replaces a SL canonical form with a TL canonical form using the appropriate transfer lexicon; this process may be sensitive to sentential context. The TL stem is created and appropriate suffixes are added to create the proper TL word. Features in TL lexical entries may be used to help select the proper sense (i.e. word).

The final parts of a grammar rule are the Transfer sections [in figure 17.5, ENGLISH and SPANISH]. In the multi-lingual METAL system, there is a separate transfer section for each target language. (In our English analysis grammar, there are, e.g. GERMAN and CHINESE sections, which means that METAL can translate bi-directionally as well as into multiple languages.) The appropriate transfer section(s) are individually invoked only after a sentence [S] has been analysed, at which point the system will perform the transfer operations specified, generally moving down the tree to the terminal nodes where lexical substitution takes place. In our example rule (figure 17.5), the first operation is

(XFR 1)

which causes the system to recursively invoke TRANSFER on

the first son (i.e., the NST). Because the NST happens to be a terminal (lexical) node, it will be translated using the appropriate transfer entry. The remaining two operations (ADF and CPY) are performed as the system ascends the tree. Thus, while analysis generally proceeds bottom-up, transfer proceeds top-down. At each node in the tree, all nodes below are accessible for reading (to determine context) and writing (to pass down information neces- sary for proper transfer).

Transfer in METAL is not a particularly simple process. Con- sider the following sentence pair:

> German: *die auszugebenden Resultate*
> gloss: the to-be-output results
> English: the results to be output

Here, the German participial verb form must be post-posed in English; a transformation (conditioned on the form of the partici- pial phrase) must be employed in cases like these. Prepositions present notorious problems; they must be translated and positioned with respect to their object NPs at least:

> German: *vor diesem Haus*
> English: in front of this house

> German: *vor dieser Woche*
> English: before this week

> German: *vor einer Woche*
> gloss: ago one week
> English: one week ago

Clearly the relationship is complex: both the German noun (i.e. its semantic type) and its determiner (if any) influence the selection of a suitable English translation, as well as its position in the phrase.

A TL case frame, when applied during the transfer phase, will order the case role-fillers as required by the verb based on voice, mood, etc. The syntactic form of the central arguments is chosen and, if necessary, prepositions are introduced as specified in the transfer verb entry. Consider the following examples:

> German: *aus Gold besteht die Tuer*
> gloss: of gold consists the door
> English: the door consists of gold

> German: *auf Gold besteht der Mann*
> gloss: on gold insists the man
> English: the man insists on gold

Here, it is not only true that the complements must be re-ordered in English, but it is also necessary to translate the verb–preposition combination as a unit. This, in turn, may reflect on (help disambiguate) the semantic type of the matrix-subject, as the following examples illustrate:

> German: *aus Gold besteht er*
> gloss: of gold consists [it]
> English: it consists of gold

> German: *auf Gold besteht er*
> gloss: on gold insists [he]
> English: he insists on gold

Various of these factors can and do interact, as illustrated by the following example:

> German: *die aus Gold bestehende Tuer*
> gloss: the of gold consisting door
> English: the door consisting of gold

In the METAL system, the transfer procedures attached to analysis rules interact with complex transfer lexical entries to determine the proper form and wording of the target-language structures. Generally speaking, each node appearing in an analysis tree is responsible for producing its appropriate translation, in context. (This is not always true, since a higher-level node can usurp the function of one or more of its sons – either performing transfer directly, or assigning a new transfer procedure to be executed in place of the original.) We have found this combination of techniques (lexical transfer interacting with grammatical structural transfer procedures) to be a flexible and powerful tool that facilitates high-quality translation. The top-level node (phrase) in the newly constructed TL tree is eventually returned by TRANSFER as its functional value, and this in turn is used for synthesis.

Synthesis

The GENERATE function synthesises the translation by simply taking the TL tree produced by TRANSFER, and inflecting and appending together all of the lexical allomorphs (words and their inflections) located in its terminal nodes. The value of the function GENERATE is a sentence; it is returned to the function TRANSLATE, which returns that sentence as its functional value. For synthesis into multiple target languages, transfer and synthesis (but not analysis) may be invoked multiple times.

The METAL parser

The parser – the LISP program that interprets a sentence according to the linguistic rules – is the heart of METAL. If the parser is inefficient, the analysis effort will consume far too much space and time to be of practical benefit. The current METAL parser is a variation on the left-corner parallel, bottom-up parsing algorithm. During an extensive series of experiments comparing a dozen parsers on the basis of their practical performance characteristics ([408] Slocum 1981), a left-corner parser augmented by top-down filtering ([553] Pratt 1973), which closely resembles an Earley parser, was determined to be the most efficient, and soon replaced the previous METAL parser based on the Cocke–Kasam–Younger algorithm ([519] Aho and Ullman 1972). However, the newest, 'some-paths' implementation of the left-corner parser has since proven to be even more efficient than the fast left-corner parser.

The LRC parsers work with a special data structure called a 'chart' that records the complete state of an analysis at every point in that analysis. Roughly speaking, the parser starts by adding every word in the sentence to the chart; it then draws grammatical inferences from those additions. The grammatical inferences are of two varieties: (1) interpretations of the syntax rules found to apply to the current portion of the input, referred to as instantiated phrases; and (2) predictions about what types of phrases (what grammatical categories) may appear next in the input. After the parser has drawn all possible grammatical inferences on a given 'level', the chart is examined for phrases which span the sentence, and whose syntactic categories appear in the user-definable list of acceptable ROOTCATEGORIES: usually '(S)'. These phrases constitute the interpretations of the input sentence; if there are none, the parser reverts to the lowest LeVeL for which there are pending rules, and continues.

In order for the parser to add a word to the chart, the word must be lexically analysed. There are three ways to do this: (1) the word may appear in the dictionary as an entry; (2) the word may be decomposed into a sequence of morphemes, each of which appears in the dictionary; or (3) the word may have a lexical entry generated on-the-fly. In METAL, any combination of the three is possible. Words, or sequences of letters that appear to be words, are looked up in the dictionary; independently, an attempt is made to decompose each word into an acceptable sequence of morphemes, each of which appears in the dictionary. (In METAL, the dictionary is composed of lexical entries in the usual sense, plus any literals appearing in phrase-structure grammar rules.) Lexical entries for numbers are automatically generated. Definitions for unknown words and non-words are also generated. Parenthetical expressions are 'lexically analysed' via a recursive call to the parser: each is parsed as if it were a complete sentence, then its interpretations are automatically transformed into 'lexical entries' for incorporation into the analysis of the encompassing sentence. Unknown words may be decomposed into sequences of known words and other lexical items (numbers, acronyms, etc.), especially if these are flagged by punctuation marks (e.g. hyphens, slashes) in the input. For example, the German 'word' '10*mal*' may be decomposed into '10' and *mal* [in English, 'times'].

Morphological analysis is a relatively simple process, relying on a letter tree [discrimination net] to indicate the legal transitions from character to character in known morphemes (defined via lexical entries or as literals in syntax rules). For a highly synthetic language like German, this tree is searched recursively to discover successive morphemes in a word. As a bonus, METAL includes a program capable of correcting the most common spelling and typographic errors (deletion, substitution, addition, wrong case, and transposition); thus typical transcription errors pose no problem. The results of morphological analysis can be ambiguous in many ways: morph sense, morph category, and even morph boundaries may be indeterminate. The parser (using the phrase structure rules) sorts out the ambiguities according to word and sentential context, as a natural part of its operation – i.e. lexical disambiguation (including homograph resolution) is not a process distinct from other forms of disambiguation (e.g. syntactic).

As PARSE finds phrase-structure rules that are applicable to a current sequence of morphs, words, and/or phrases in the ongoing analysis, it does not automatically build a syntactic structure expressing this fact; instead, it invokes a special routine which is responsible for determining (through the invocation of the rule-body procedure) the applicability of the rule, and for constructing and scoring the interpretation. This special routine constructs a preliminary parse tree, invokes the rule-body procedure to determine if the rule is applicable (and possibly to annotate the tree and/or transform it), and if the resulting interpretation is acceptable it scores the phrase based on the scores of its constituents and any preference assigned by the rule-body procedure; it rejects the interpretation if its score falls below cutoff, else attaches the 'suspended' rule-body procedure to the phrase (important in the transfer phrase, as explained elsewhere). The scores of the root nodes in the sentence analyses will be used later to determine the 'most plausible analysis' for transfer and synthesis.

If the METAL parser fails to achieve a unified interpretation of a sentence (or of a parenthetical expression, which is recursively parsed as if it were a sentence), it attempts to 'fake' an analysis of the sentence. A phrasal analysis is constructed from the fewest, largest, highest-scoring phrases that together span the input sentence. An S phrase is built just as if there were a grammar rule with the discovered phrases listed in its right-hand side. A default rule-body procedure is attached, to be invoked during the transfer phase; this procedure will simply use the (XFR) operator to invoke TRANSFER on the constituent phrases of the dummy S phrase just built.

Recent experimental results

In the last four years, METAL has been applied to the translation into English of over 1000 pages of German telecommunication and data processing texts. To date, no definitive comparisons of METAL translations with human translations have been attempted; this situation will soon be remedied. However, some stimulating quantitative and qualitative statistics have been gathered.

On our Symbolics LM-2 Lisp Machine, with 256K words of physical memory, preliminary measurements indicate an average performance of 5–6 seconds (real time) per input word; this is already six times the speed of a human translator, for like material. The paging rate indicates that, with added memory, we could expect a significant boost in the performance ratio. With a faster, second-generation Lisp Machine, we would expect another substantial reduction of real-time processing requirements: preliminary measurements (before system tuning) show a doubling of the throughput rate; further speed increases are anticipated.

Measuring translation quality is a vexing problem – a problem not exclusive to machine translation or technical texts, to be sure. In evaluating claims of 'high-quality' MT, one must carefully consider how 'quality' is defined; 'percentage of words [or sentence] correct [or acceptable]', for example, requires definition of the operative word 'correct'. A closely related question is that of who determines correctness. Acceptability is ultimately defined by the user, according to his particular needs: what is acceptable to one user in one situation may be quite unacceptable in another situation, or to another user in the same situation. For example, some professional post-editors have candidly informed us that they actually look forward to editing MT output because they 'can have more control over the result'. For sociological reasons, there seems to be only so much that they dare change in human translations; but as everyone knows (and our informants pointed out), 'the machine doesn't care'. The clear implication here is that 'correctness' has traditionally suffered where human translation is concerned; or, alternately, that 'acceptability' depends in part on the relationship between the translator and the revisor. Either way, judgements of 'correctness' or 'acceptability' by translators and editors is likely to be more harsh when directed toward MT than when directed toward human translation. It is not yet clear what the full implications of this situation are, but the general import should be of some concern to the MT community.

For different (and obvious) reasons, qualitative assessments by MT system vendors are subject to bias – generally unintentional – and must be treated with caution. But one must also consider other circumstances under which the measurement experiment is conducted: whether (and for how long, and in what form) the text being translated, and/or its vocabulary, was made available to the vendor before the experiment; whether the MT system was previously exercised on that text, or similar texts; etc. At the LRC, we conduct two kinds of measurement experiments: 'blind', and 'follow-up'. When a new text is acquired from the project sponsor, its vocabulary is extracted by various lexical analysis procedures and given to the lexicographers. This may include a partial or full concordance of the text, in which each word is displayed in a context that includes the full matrix sentence. The lexicographers then write ('code') entries for any novel words discovered in the text. The linguistic staff never sees the text prior to a blind experiment. Once the results of the blind translation are in, the project staff are free to update the grammar rules and lexical entries according to what is learned from the test, and may try out their revisions on sample sentences from the text. Some time later the same text is translated again, so that some idea of the amount of improvement can be obtained.

In addition to collecting some machine performance statistics, we count the number of 'correct' sentence translations and divide by the total number of sentence units in the text, in order to arrive at a 'correctness' figure. (For our purposes, 'correct' is defined as 'noted to be unchanged for morphological, syntactic, or semantic reasons, with respect to the original machine translation, after revision [by professional post-editors] is complete'. (Non-essential stylistic changes are not considered to be errors.) In the course of experimenting with over 1000 pages of text in the last four years, our 'correctness' figures have varied from 45 per cent to 85 per cent (of full-sentence units) depending on the individual text and whether the experiment was of the 'blind' or 'follow-up' variety.

The single numerical assessment of greatest interest in the MT community is almost certainly cost-effectiveness. Until METAL is evaluated by unbiased third parties, taking into account the full costs of translation and revision using METAL versus conventional (human) techniques, the question of METAL's cost-effectiveness cannot adequately be addressed. However, we have identified some performance parameters that are interesting. Our sponsor has calculated that METAL should prove cost-effective if it can be implemented on a second-generation Lisp Machine supporting 4–6 post-editors who can sustain an average total output of about sixty revised pages/day. At 275 words/page, and 8 hours/day, this works out to 1.7 seconds/word, minimum real-time machine performance. If the new second-generation Lisp Machines are, as generally claimed, three times as fast as the current generation (represented by our LM-2), then our immediate target is 5.2 seconds/word, minimum real-time performance – about what we now experience on our LM-2. If this level of performance can be sustained while maintaining a high enough standard of quality that an individual revisor can handle 10–15 pages/day, METAL will have achieved cost-effectiveness.

We have also measured revision performance: the amount of time required to edit texts translated by METAL. In the first such experiment, conducted late in 1982, two Siemens post-editors revised METAL's translations at the rate of 15–17 pages/day (depending on the particular editor). In a second experiment, conducted in mid-1983, the rates were only slightly higher (15–20 pages/day), but the revisors nevertheless reported a significant improvement in their subjective impression of the quality of the output. In a third experiment, conducted in early 1983, the revisors reported further improvement in their subjective impression of the quality of the output, and their revision rates were much higher: around 29 pages/day. Thus, our experimental performance figures indicate that we may already have reached the goal of cost-effectiveness.

Conclusions
MT research at the Linguistics Research Center involves the selection and testing of Natural Language Processing techniques in a real-world environment. With an efficient computational component such as we now have, it becomes possible to empirically validate new linguistic theories as they are proposed. Our research can therefore answer questions about their extensibility, and the limits of their application.

Some 'old' Natural Language Processing techniques are producing surprisingly good results, and new ones are being developed. Others have not proven to be effective, and have been abandoned. METAL is capable of producing useful English translations for a wide variety of German sentences; translation from English into German has recently begun. However, further development is currently underway to resolve a number of remaining problems. Two areas in which the LRC is making improvements in the METAL linguistic component are the treatment and placement of adverbials, and more extensive use of semantics.

In the future, we look forward to the development, phased introduction, and empirical assessment of more advanced NLP techniques – especially for anaphora resolution and the use of semantic models. We see no evidence that today's advanced but experimental NLP techniques will soon [in this decade, or even century] be able to supplant the more primitive techniques that are currently effective in a large-scale application such as ours. But we nevertheless hope that such techniques can, even if in elementary form, be effectively utilised within practical applications of current techniques to further improve the overall quality and cost-effectiveness of translation. We have tried to anticipate this eventuality in the design of the LRC MT system by allowing for future evolution, possibly even revolution. In the process of applying advanced NLP techniques to practical problems in real-world settings, such as translation, we fully expect the feedback of experience to substantially influence both the form and content of linguistic theories.

ISOMORPHIC GRAMMARS AND THEIR USE
IN THE ROSETTA TRANSLATION SYSTEM

Assume that we wish to design a system that translates sentences from some natural language L into a natural language L'. Assume further that this system translates sentences in isolation, without regard to their context, that it uses information about the languages L and L', in the form of grammars and dictionaries, but that it does not use extra-linguistic information. Such a system will in general not be able to translate a sentence of L unambiguously into L', but will define a set of 'possible translations'. It seems reasonable to assume that for the large majority of sentences the 'correct translation' or the 'best translation' is a member of this set, given correct grammars and dictionaries. From a research point of view it may be interesting to study these possible translations or more precisely the relation possible-translation, which can be regarded as a symmetric relation, while the relation best-translation obviously is not symmetric. From an application point of view the relation possible-translation is interesting as well, because it may be the core of a realistic translation system, which is able to disambiguate on the basis of additional information: knowledge of the world and understanding of the context.

With regard to the way in which the relation between the grammars G and G' of the languages L and L' is defined, I will distinguish three kinds of systems.

1. The first kind operates as follows. On the basis of a grammar G of the source language L an analysis component constructs a syntactic structure (a surface structure or some type of deep structure) of the input sentence. This syntactic structure is transformed into a syntactic structure of the target language L' by a transfer component. A generation component converts this syntactic structure into a sentence of L' (during every stage of this process

ambiguities may arise). This is the well-known design of a transfer system, with structural transfer. The systems METAL ([417] Slocum et al. 1982 and chapter 17) and GETA ([89] Boitet et al. 1982 and chapters 10, 16) are examples of this (cf. also [496] Whitelock et al. 1983). This approach may be the most feasible one for translation between two languages in one direction, but it has the theoretical disadvantage that there is not a separate grammar G' of L' in the system (or only a partial one). G' is in fact defined by G and the transfer component.

2. In the second kind of system a grammar G of L defines an analysis component which translates from L into an intermediate language IL. A grammar G' of L' defines a generation component which translates from IL into L'. This is the approach of interlingual systems (like CETA, cf. [210] Hutchins 1978) and of transfer systems where the transfer component performs only lexical transfer, as is the plan in EUROTRA ([229] King 1982 and chapter 19). It is a nice approach if the goal is to translate between a number of languages. An attractive property of these systems seems to be that the grammars G and G' can be developed more or less independently. However, there are two kinds of problems.

(i) It is difficult to define the right IL. Its expressions must contain all necessary information for the translation, which may involve both 'deep semantic' and 'surface structure' information, but they must be as independent as possible of any specific language. Furthermore, translation into and from IL must be essentially easier than into and from a natural language.

(ii) The applicability of the generation component to the IL expressions may be a problem. The grammar G' can be regarded as a generative grammar where IL expressions function as deep structures. It seems impossible to define the IL independently in such a way that each IL expression corresponds to a well-formed sentence, therefore the generation component must have a filtering effect. But then it becomes hard to guarantee that the generation component is applicable to all IL expressions that may be constructed by any analysis component. There is a conflict between the idea that G' is developed independently and the requirement that the generation component is applicable to the output of any analysis component.

3. In the third kind of system there are explicit grammars G and G', which are not developed independently but are attuned to each other, for instance in the following way.

Let G and G' be 'compositional' grammars, where sentences are built up starting from basic expressions by applying syntactic rules, which prescribe how bigger expressions can be constructed from smaller ones. Basic expressions are the smallest meaningful units, e.g. stems of content words. Each syntactic rule corresponds to a meaning operation. The tuning of the grammars comes down to ensuring that (i) for each basic expression of G there is at least one basic expression of G' with the same meaning, and (ii) for each syntactic rule of G there is at least one syntactic rule of G' corresponding to the same meaning operation.

A sentence s' of L' is considered a possible translation of a sentence s of L if s and s' are derived from corresponding basic expressions by applying corresponding syntactic rules. In other words: sentences are translations of each other if they can be derived 'in parallel'.

This approach can be used for bilingual as well as multilingual translation, but here only the latter will be discussed. The intermediate language in this case consists of 'neutral' representations of derivation processes. If we make a comparison with the interlingual approach of (2) the advantage is that the intermediate language and the translation into and from IL follow immediately from the grammars. A disadvantage is that writing corresponding grammars for a set of languages is more difficult than writing a grammar for each language separately. This may lead to more complex grammars.

This chapter is devoted to the third approach: the isomorphic grammar approach, first described in ([248] Landsbergen 1982). It underlies the ROSETTA system, which is being developed at Philips Research Laboratories. The chapter discusses the way in which isomorphic grammars define the relation possible-translation; it does not discuss methods for choosing the best translation which are currently used in the ROSETTA system or planned to be used in the future. In the second section the kind of grammars used in ROSETTA are discussed: M-grammars. In the third section it is shown how each grammar defines an analysis and a generation component and how an analysis component and a generation component for different languages together form a translation system. In the fourth section isomorphic grammars are discussed. Then it is shown how a well-known translation problem can be solved in the framework. In the concluding section the results are summarised.

M-grammars

The ROSETTA system is based on Montague Grammar ([619] Thomason 1974 and [581] Dowty et al. 1981). For those who are familiar with Montague Grammar I want to point out immediately that the system does not use intensional logic as an intermediate language and that its syntax is not categorial. But the grammars used in the system, called M-grammars, do obey the Compositionality Principle of Montague Grammar, i.e. there is a close relation between syntactic structure and semantic structure. Furthermore, derivation trees, whch reflect the syntactic and semantic structure of sentences play an important role in the system. I will not go into a detailed comparison between M-grammars and Montague's original proposals. M-grammars can be seen as a computationally viable variant of Montague grammars, which is in accordance with the extensions proposed by Partee ([610] Partee 1976).

An M-grammar consists of three components: a syntactic component, a semantic component and a morphological component.

The syntactic component of an M-grammar defines a set of objects called S-trees (surface trees). This is done in two steps: (i) the definition of the set of possible S-trees: the domain T; (ii) the definition of the set of syntactically well-formed S-trees: the subset T_M of T. First, I will describe what an S-tree is and how the domain T is defined.

An *S-tree* t is an object of the form

$$(1) \quad N[r_1/t_1, \ldots, r_n/t_n] \quad (n \geq 0)$$

where N is a node, the r_i are syntactic relations and the t_i are S-trees.

If $n = 0$ we speak of a terminal S-tree and write N[ ] or just N.

The syntactic relations are defined by enumeration. Examples are subject, object, head, modifier, etc.

A node is defined as a syntactic category followed by a tuple of attribute–value pairs, for example:

SENTENCE { mood: declarative, voice: active }

The syntactic categories are defined by enumeration.

For each syntactic category the corresponding attributes are defined, for each attribute the set of possible values is defined.

So, given a set of syntactic relations and a set of syntactic categories with the corresponding attributes and values, the set of possible S-trees is defined. I will call this the domain (T) of S-trees.

An example of an S-tree is (2):

SENTENCE { mood: declarative, voice: active }
        [subject/NP { number: singular, person: 3 }
          [det/ART { sort: definite },
          head/NOUN { stem: 'boy',
                number: singular }],
      aux/VERB { stem: 'be', form: sing 3 },
      head/VERB { stem: 'talk', form: ing }]

A more familiar representation of the above (2) is (3):

(3)

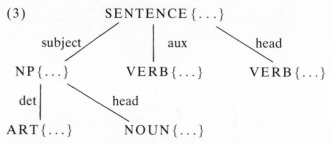

As it is not the purpose of this chapter to indulge in detailed notations, I will in the sequel often use an abbreviated notation, for instance

SENTENCE (the boy is talking)

for (2) and (3).

The trees are called surface trees, because the leaves of a tree for a complete sentence correspond to words in the right order. The correspondence between these leaves, the terminal S-trees, and actual words is defined by the morphological component.

The syntactic component of an M-grammar defines the domain T of S-trees by specifying the syntactic relations, the syntactic categories and the corresponding attributes and values.

The set T_M of well-formed S-trees, a subset of T, is defined by specifying (1) a set of basic S-trees, (2) a set of M-rules, (3) a set of final syntactic categories.

(1) The basic S-trees are a subset of T, in most cases they correspond to stems of content words, but not necessarily.

(2) An M-rule R_i defines a function F_i from n-tuples of S-trees (for n-ary rules) to finite sets of S-trees. So application of R_i to a tuple $t_1, \ldots, t_n$ results in a set $F_i(t_1, \ldots, t_n)$. The set is empty if the rule is not applicable. In most cases an applicable rule will result in a set with one element.

The basic S-trees and the M-rules together define a subset T_M^+ of T, consisting of the basic S-trees and all S-trees that can be formed by applying M-rules to S-trees already in T_M^+.

(3) The set of final syntactic categories defines a subset T_M of T_M^+, consisting of the S-trees with a final syntactic category at the top.

An example of the syntactic component of an M-grammar, G_E;
   (i) the definition of the domain T.
The syntactic relations are {head, det, aux, subject}.
The syntactic categories and the relevant attributes and values are:

SENTENCE with attributes mood, values: {declarative, interrogative}
and voice, values: {active, passive}

NP   with attributes number, values: {singular, plural, unspecified}
and person, values: {1, 2, 3}

ART   with attribute sort, values: {definite, indefinite}

NOUN   with attributes stem, values: {'boy', ...}
number, values: {singular, plural, unspecified}

VERB   with attributes stem, values: {'be', 'talk', ...}
form, values: {sing3, ing, stemform}

(ii) the definition of the set T_M of well-formed S-trees.

(1) The basic S-trees are

b_1: NOUN{stem: 'boy', number: unspecified}

b_2: VERB{stem: 'talk', form: stemform}

(2) The M-rules are:

R_1: if t_1 is an S-tree with category NOUN,
then $F_1(t_1)$ consists of NP{number: singular, person: 3}
[det/ART{sort: definite},
head/t_1']
where t_1' is t_1 with number: singular.
else $F_1(t_1)$ is empty.

R_2: if t_1 is an S-tree with category NP,
and t_2 is an S-tree with category VERB,
then $F_2(t_1, t_2)$ consists of
SENTENCE{mood: declarative, voice: active}
[subject/t_1,
aux/VERB{stem: 'be', form: sing3},
head/t_2']
where t_2' is t_2 with form: ing.
else $F_2(t_1, t_2)$ is empty.

(The precise notation for M-rules is not discussed here.)

(3) The set of final categories is {SENTENCE}.

G_E defines the S-tree (2) of 'the boy is talking'. Application of R_1 to basic expression b_1 leads to NP (the boy). Application of R_2 to this NP and b_2 leads to S-tree (2).

The derivation process can be displayed in a derivation tree. This is defined as follows:

A *derivation tree* is either (the name of) a basic expression b_i or an expression of the form $R_i\langle d_1, \ldots, d_n \rangle$, where R_i is (the name of) an M-rule and $d_1, \ldots, d_n$ are derivation trees; (4) is the graphical representation.

(4)

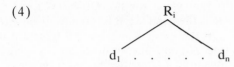

$$R_i$$
$$d_1 \quad . \quad . \quad . \quad . \quad d_n$$

A derivation tree 'generates' a set of S-trees: the S-trees that can be derived by applying the rules in the way the tree suggests. It is important to notice the difference with Montague's original definition. In an M-grammar a derivation tree is an independently defined object, that may or may not correspond with the successful derivation of an actual expression. So a derivation tree may define an empty set of S-trees. On the other hand an S-tree t from the domain T defines a set of derivation trees; the derivation trees that are able to generate t.

We define the relation $R_{G,syn}^+$ ('generates') between derivation trees (d) and S-trees (t) recursively.

$dR_{G,syn}^+t$ iff d and t are basic expressions and d = t.
        or d has the form $R_i\langle d_1, \ldots, d_n\rangle$
        and $\exists t_1, \ldots, t_n$ such that $d_1R_{G,syn}^+t_1$
                 . . .
                 . . .
                $d_nR_{G,syn}^+t_n$
        and $t \in F_i(t_1, \ldots, t_n)$

$dR_{G,syn}t$ iff $dR_{G,syn}^+t$ and t has a final category.

It is easy to establish that $T_M = \{t \mid \exists d: dR_{G,syn}\, t\}$.

The derivation tree for SENTENCE (the boy is talking) is (5); (6) shows the tree in graphical notation, and for each node of the tree the corresponding S-tree.

(5) $R_2\langle R_1\langle b_1\rangle, b_2\rangle$

(6)

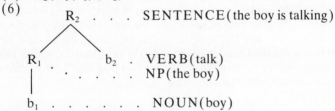

        R_2 . . . SENTENCE (the boy is talking)

R_1 . b_2 .   VERB (talk)
| · . . . . NP (the boy)

b_1 . . . . . . NOUN (boy)

So, according to example grammar G_E:

    $d_6R_{G,syn}t_3$

where d_6 is the derivation tree (6) and t_3 the S-tree (3).

It is important to notice that the leaves of an S-tree like (3), corresponding to words, are not necessarily basic S-trees.

VERB(talking) and NOUN(boy) have been derived from basic S-trees, but have a different value for the attributes form and number respectively. ART(the) and VERB(is) have been introduced by the M-rules. On the other hand, basic S-trees are not necessarily terminal S-trees. Idioms like 'to make up one's mind', which are basic expressions from a semantic, but not from a syntactic point of view, may be represented as compound basic S-trees.

The syntactic component defines besides $R_{G,syn}$ a second, rather trivial relation: the relation $R_{G,leaves}$ between S-trees and the sequences of leaves of these S-trees.

In example grammar G_E:

$$t_3 R_{G,leaves} s_L$$

where t_3 is the S-tree of (3) and s_L is ART(the) + NOUN(boy) + VERB(is) + VERB(talking).

The basic expressions and rules of an M-grammar must be chosen in accordance with the Compositionality Principle. Each basic S-tree must have a well-defined meaning (or set of meanings), each syntactic rule must correspond to a meaning rule. Where a syntactic rule defines a function that operates on its argument expressions (S-trees) and forms a new expression, the corresponding meaning rule defines a function that operates on the meanings of the argument expressions and delivers the meaning of the new expression. So there is a close correspondence between the way in which the form of an expression is derived from basic expressions and the way in which its meaning is derived from basic meanings. The derivation process of the meaning can be represented in a similar way as the derivation process of the form. I will illustrate this for example grammar G_E. Assume that the semantic component of G_E assigns to the syntactic rules R_1 and R_2 the respective meaning rules M_1 and M_2, and to the basic expressions b_1 and b_2 the respective basic meanings B_1 and B_2 (a basic expression may have more than one meaning, but in the example we assume that there is no such ambiguity). Then the way in which the meaning of 'the boy is talking' is derived can be represented in the tree (7), which has the same geometry as the derivation tree (6), but which is labelled with names of meaning rules and basic meanings.

(7)

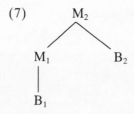

I will call this tree a *logical derivation tree*. Ordinary derivation trees will from now on be called syntactic derivation trees in contexts where misunderstandings might arise.

In Montague Grammar the meaning rules and the basic meanings are expressed in intensional logic. In that case recursive application of the meaning rules in the way indicated by the logical derivation tree results in an expression of intensional logic that represents the meaning of the sentence. As we will see in the next section, in the ROSETTA system the logical derivation trees themselves are used as meaning representations. Therefore in an M-grammar only the unique names of the meaning rules and the basic meanings have to be specified.

More formally, the semantic component of an M-grammar specifies for each M-rule R_i a meaning rule M_i and for each basic expression a set of basic meanings B_i. A logical derivation tree is either (the name of) a basic meaning B_i or an expression of the form $M_i\langle e_1, \ldots, e_n \rangle$, where M_i is (the name of) a meaning rule and $e_1, \ldots, e_n$ are logical derivation trees.

The semantic component defines a relation $R_{G,sem}$ between logical derivation trees (e) and syntactic derivation trees (d).

$e R_{G,sem} d$ holds iff d is a basic expression and e a corresponding
basic meaning
or d has the form $R_i\langle d_1, \ldots, d_n \rangle$,
e has the form $M_i\langle e_1, \ldots, e_n \rangle$,
M_i is the meaning rule of R_i
and $e_1 R_{G,sem} d_1$
$\cdots$
$\cdots$
$e_n R_{G,sem} d_n$

In example grammar G_E:

$e_7 R_{G,sem} d_6$

where e_7 is the logical derivation tree (7) and d_6 is the syntactic derivation tree (6).

As we have seen, the syntactic component of an M-grammar defines sequences of terminal S-trees, not sentences. It is the task of the morphological component to define the relation $R_{G,morph}$ between the these sequences of terminal S-trees and sequences of actual words (strings of symbols). $R_{G,morph}$ is defined by means of a monolingual dictionary and morphological rules, not to be discussed in this chapter. If s_L is a sequence of terminal S-trees and s is a sequence of words corresponding to s_L according to the morphological component, we write

$s_L R_{G,morph} s$

In example grammar G_E s_L is $ART(the) + NOUN(boy) + VERB(is) + VERB(talking)$ and s is 'the boy is talking'.

(If the morphological rules are not only used for inflection, but also for derivation, e.g. for composite words, words do not always correspond to terminal S-trees. In that case the definition of the relation $R_{G,leaves}$ has to be adjusted.)

Summarising, we see that an M-grammar defines a relation between logical derivation trees and sentences: R_G.

(8) $eR_G s$ iff $\exists d, t, s_L : eR_{G,sem}d$ and $dR_{G,syn}t$ and
$$tR_{G,leaves}s_L \text{ and } s_L R_{G,morph}s$$

(e ranges over logical derivation trees, d over syntactic derivation trees, t over S-trees, s_L over sequences of leaves, s over sentences)

In example grammar G_E:

$e_7 R_G$ 'the boy is talking'

*Analysis and generation modules for M-grammars
and their use in ROSETTA*

In the previous section we have seen how an M-grammar defines a relation R_G between logical derivation trees and sentences. In this section I will briefly indicate how this relation can be expressed in terms of functions, called $ANALYSIS_G$ and $GENERATION_G$. Here $ANALYSIS_G$ is a function from sentences to sets of logical derivation trees and $GENERATION_G$ is a function from logical derivation trees to sets of sentences. We require that for each logical derivation tree e and for each sentence s holds:

(9) $eR_G s \leftrightarrow e \in ANALYSIS_G(s) \leftrightarrow$
$$s \in GENERATION_G(e)$$

The problem is of course that $ANALYSIS_G$ and $GENERATION_G$ must be defined constructively, in such a way that they can be effectively computed. This is achieved by defining computable functions for each of the four relations used in the definition of R_G: $R_{G,morph}$, $R_{G,leaves}$, $R_{G,syn}$ and $R_{G,sem}$.

From the definition of $R_{G,sem}$ (the relation between logical derivation trees e and syntactic derivation trees d) follows immediately that it can be expressed as a computable function in both directions: A-TRANSFER for analysis and G-TRANSFER for generation.

(10) $eR_{G,sem}d \leftrightarrow e \in A\text{-}TRANSFER_G(d) \leftrightarrow$
$$d \in G\text{-}TRANSFER_G(e)$$

For $R_{G,morph}$ (the relation between sequences of leaves s_L and sentences s, not discussed here), this is possible too. The functions are called A-MORPH for analysis and G-MORPH for generation.

(11) $s_L R_{G,morph} s \leftrightarrow s_L \in A\text{-}MORPH_G(s) \leftrightarrow$
$$s \in G\text{-}MORPH_G(s_L)$$

For $R_{G,syn}$ (the relation between syntactic derivation trees d and S-trees t) the situation is more complicated. The function for analysis is called M-PARSER, for generation M-GENER-ATOR.

(12) $d R_{G,syn} t \leftrightarrow d \in M\text{-}PARSER_G(t) \leftrightarrow$
$$t \in M\text{-}GENERATOR_G(d)$$

That M-GENERATOR is a computable function is easy to establish, if the functions F_i corresponding to the M-rules R_i are computable. M-GENERATOR has to apply the functions for the rules in the derivation tree recursively and this recursion is finite, because the derivation tree has a finite depth.

M-PARSER is a computable function as well, but only thanks to two conditions that M-grammars have to obey. The first condition is the reversibility condition on M-rules. It says that each rule R_i defines not only the 'compositional' function F_i, but also an 'analytical' function F'_i, the 'reverse' of F_i. F'_i is a function from S-trees into finite sets of n-tuples of S-trees, for an n-ary M-rule. The condition says:

$t \in F_i(t_1, \ldots, t_n)$ iff $\langle t_1, \ldots, t_n \rangle \in F'_i(t)$.

The second condition (the measure condition) is that the compositional function F_i delivers a result t which is 'bigger', according to some measure of complexity, than each of the arguments $t_1, \ldots, t_n$. Thanks to these conditions M-PARSER can be defined as an effective procedure which applies recursively the analytic M-rules to an S-tree in a top-down fashion. For more details, cf. [605] Landsbergen 1981. In that paper a somewhat simpler version of M-grammar is discussed, but for the control structure of M-PARSER this does not make any difference.

The relation $R_{G,leaves}$ can be expressed as a computable function $LEAVES_G$ from S-trees (t) to their sequence of leaves (s_L) in a trivial way.

(13) $t R_{G,leaves} s_L \leftrightarrow s_L = LEAVES_G(t)$

$LEAVES_G$ is the function we need for generation. The definition of a computable function for analysis, which operates on a sentence s_L (a sequence of terminal S-trees) and delivers the S-trees of which s_L is the sequence of leaves, is not trivial. To make this possible, M-grammars must obey a third condition, the surface syntax condition, which requires that for each M-grammar a set of 'surface rules' exists. The surface rules, in combination with the morphological component must define a set of S-trees which includes all correct S-trees but may also include incorrect ones (correct or incorrect with respect to the syntactic component of the M-grammar). So the 'surface-syntax' is weaker than the 'M-syntax'; it defines a superset of the set of correct S-trees. The essential difference between surface rules and M-rules is that application of a surface rule to a tuple of S-trees $t_1, \ldots, t_n$ leads to an S-tree of the form $N[r_1/t_1, \ldots, r_n/t_n]$; i.e. a surface rule does not make any change in its arguments, nor in their order, whereas M-rules have the power to do so. Thanks to this property of surface rules an effective parser (called S-PARSER) can be defined. S-PARSER constructs for any sentence a finite set of S-trees, in accordance with the surface rules, such that the leaves of each S-tree constitute this sentence.

S-PARSER does not express the relation $R_{G,\text{leaves}}$ for all t in T (the set of possible S-trees), but it does for all t in T_M (the set of correct S-trees of G).

(14) $\forall t \in T_M: tR_{G,\text{leaves}}s_L \leftrightarrow t \in \text{S-PARSER}(s_L)$

We are now able to define the functions ANALYSIS_G and GENERATION_G.

(15) $e \in \text{ANALYSIS}_G(s)$ iff $\exists s_L, t, d$:
$s_L \in \text{A-MORPH}_G(s)$ and $t \in \text{S-PARSER}_G(s_L)$ and
$d \in \text{M-PARSER}_G(t)$ and $e \in \text{A-TRANSFER}_G(d)$

(16) $s \in \text{GENERATION}_G(e)$ iff $\exists d, t, s_L$:
$d \in \text{G-TRANSFER}_G(e)$ and
$t \in \text{M-GENERATOR}_G(d)$ and
$s_L = \text{LEAVES}_G(t)$ and $s \in \text{G-MORPH}_G(s_L)$

The correctness of (9) can easily be proved, on the basis of (8) and (10)–(16).

Before discussing the way in which the functions ANALYSIS and GENERATION define a translation function, let us consider another example grammar, G_D, for a fragment of Dutch corresponding to the English fragment defined by G_E in section 2. I will present G_D in an abbreviated form:

1. Basic S-trees.
   b'_1: NOUN(*jongen*), basic meaning B_1
   b'_2: VERB(*praat*), basic meaning B_2
2. M-rules.
   R'_1: NOUN(*jongen*)→NP(*de jongen*)
   R'_2: NP(*de jongen*)+VERB(*praat*)→
   SENTENCE(*de jongen praat*)

R'_1 corresponds to meaning rule M_1, R'_2 with M_2, as defined in the section on M-grammars.

3. Final categories: {SENTENCE}.

G_D generates the Dutch sentence *de jongen praat,* of which (17) is the syntactic derivation tree.

(17) R_2' . . .   SENTENCE(*de jongen praat*)

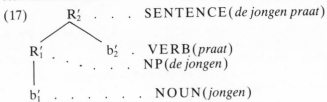

R_1' . b_2' .   VERB(*praat*)

| . . . . .   NP(*de jongen*)

b_1' . . . . . .   NOUN(*jongen*)

The basic expressions and rules of G_D are chosen in such a way that they correspond in meaning to the basic expressions and rules of G_D. The result is that the logical derivation tree of *de jongen praat* is identical to that of 'the boy is talking': (7). This suggests that logical derivation trees may play the role of intermediate expressions in translation, if the grammars are attuned to each other in this way. Curry ([578] 1961) already proposed that a distinction be made between the structure of the derivational history and the structure of the produced expressions and expected that languages would differ less in the former than in the latter. Dowty ([580] 1982) gives some interesting examples of similarity in derivational history, for English, Japanese, Breton and Latin. In the next section the tuning of grammars will be discussed in more detail.

I will now define the relation possible-translation (TR-REL) as follows. Let L be a language defined by grammar G, L' defined by G'. Assume that G and G' use the same set of meaning rules and basic meanings. Let IL be the set of possible logical derivation trees. G defines a relation R_G between IL and L, G' defines a relation $R_{G'}$ between IL and L'.

We define, for each s in L, s' in L':

sTR-REL$_{L-L'}s'$ iff $\exists e \in IL$: eR_Gs and $eR_{G'}s'$

As we have seen, relations R_G and $R_{G'}$ can be expressed by means of computable functions ANALYSIS$_G$ and GENERATION$_G$, resp. ANALYSIS$_{G'}$ and GENERATION$_{G'}$. With the help of these functions translation functions (TR-F) from L into L' and vice versa can be easily defined.

TR-F$_{L-L'}$(s) =$_{def}$ {s' | $\exists e \in IL$:
        $e \in$ ANALYSIS$_G$(s) and s' $\in$ GENERATION$_{G'}$(e)}

TR-F$_{L'-L}$(s') =$_{def}$ {s | $\exists e \in IL$:
        $e \in$ ANALYSIS$_{G'}$(s') and s $\in$ GENERATION$_G$(e)}

From the definitions it follows immediately that

sTR-REL$_{L-L'}s' \leftrightarrow s' \in$ TR-F$_{L-L'}$(s)$\leftrightarrow s \in$ TR-F$_{L'-L}$(s')

The definition of TR-F$_{L-L'}$ leads to the design of the ROSETTA system according to diagram (18). It should be kept in mind that the system as it is defined here delivers a *set* of translations. For disambiguation additional knowledge is needed. Possible methods for providing this knowledge are outside the scope of this paper, but I want to mention two of these methods, which are currently studied and have already been partially implemented in ROSETTA. The first is to enrich IL with a many-sorted type system and to prefer readings in which the IL functions are applied to arguments that match with the types these functions expect. The second method is to disambiguate by means of an interactive dialogue with the user (as in the ITS system [325] Melby et al. 1980 and chapter 9).

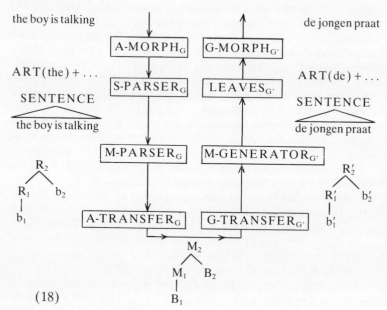

(18)

Isomorphic M-grammars

In the previous section we have seen that logical derivation trees can serve as intermediate expressions in a translation system, if the grammars of the languages are attuned to each other in such a way that they assign indeed the same logical derivation tree to sentences that we wish to consider possible translations of each other. So these sentences have not only the same meaning, but their meaning is also derived in the same way from the same basic meanings and the sentences have a similar derivational history. In this section this 'tuning' will be defined precisely, as the 'isomorphy relation' between grammars. I will first give a few auxiliary definitions.

A *syntactic derivation tree* d of G is called *well-formed* iff it defines at least one S-tree t, i.e. iff

$\exists t : d R_{G,syn} t.$

Thanks to the surface syntax condition (previous section), the morphological component does not function as a filter, therefore a well-formed syntactic derivation tree defines a sentence too.

A *logical derivation tree* e is *well-formed* with respect to G iff there is at least one corresponding syntactic derivation tree of G that is well-formed, i.e. iff

$\exists d : e R_{G,sem} d$ and d well-formed.

Now we are able to define the isomorphy relation as follows.

Two grammars are *isomorphic* iff each logical derivation tree which is well-formed with respect to one grammar is also well-formed with respect to the other grammar.

More formally:

$(19) \quad G \sim G'$ iff $\forall e : (\exists s : e R_G s \leftrightarrow \exists s' : e R_{G'} s')$

Isomorphy is an equivalence relation, therefore it makes sense to speak of a set of isomorphic grammars. In the previous section it was shown that for each grammar G the relation R_G can be expressed as a function $ANALYSIS_G$ and as a function $GENERATION_G$. It was also shown that for each pair of M-grammars G, G' a translation function can be defined as the composition of $ANALYSIS_G$ and $GENERATION_{G'}$. From (19) it follows that for isomorphic G and G', for some sentence s, for some logical derivation tree e:

if e ϵ $ANALYSIS_G(s)$ holds, then

$GENERATION_{G'}(e)$ is not empty.

So every sentence of the source language that is correct according to G will be translated.

A prerequisite for the isomorphy of two grammars is that for each basic expression in one grammar there is a basic expression in the other grammar with the same meaning and for each rule of one grammar there is a rule in the other grammar with the same meaning rule. But this is not sufficient; there are also demands concerning the applicability of the rules. The situation is sketched in figures (20), (21) and (22).

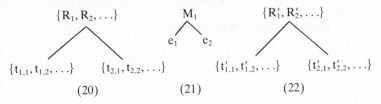

$\{R_1, R_2, \ldots\} \qquad M_1 \qquad \{R'_1, R'_2, \ldots\}$

$e_1 \qquad e_2$

$\{t_{1,1}, t_{1,2}, \ldots\} \quad \{t_{2,1}, t_{2,2}, \ldots\} \qquad \{t'_{1,1}, t'_{1,2}, \ldots\} \quad \{t'_{2,1}, t'_{2,2}, \ldots\}$

$(20) \qquad\qquad (21) \qquad\qquad (22)$

(21) shows the logical derivation tree $e = M_1\langle e_1, e_2 \rangle$. Meaning rule M_1 corresponds to a set of rules $\{R_1, R_2, \ldots\}$ of G and to a set of rules $\{R'_1, R'_2, \ldots\}$ of G'. Assume that the subtrees e_1 and e_2 are well-formed with respect to G and G' and correspond each to sets of S-trees of G and of G' in the way the figures indicate. So e_1 corresponds to S-trees $t_{1,1}$, $t_{1,2}$, etc. of G and to S-trees $t'_{1,1}$, $t'_{1,2}$, etc. of G'.

Now e is well-formed with respect to G if at least one of the rules R_k is applicable to at least one of the possible pairs $t_{1,i}$, $t_{2,j}$. If G and G' are isomorphic this implies that e is well-formed with respect to G', i.e. that at least one of the rules of R'_k is applicable to one of the possible pairs $t'_{1,i}$, $t'_{2,j}$.

Isomorphy of grammars is not always easy to prove. I will describe four cases in which such a proof is relatively easy.

I. Let G and G' be context-free grammars. Even without going into formal detail it may be easy to see that a context-free grammar is a special case of an M-grammar. For example, a context-free rule $S \rightarrow AB$ can be formulated as an M-rule R_1 which is applicable to a pair of S-trees t_1, t_2 with the respective categories A and B at the top. The rule creates a new top node with category S and t_1 and t_2 as immediate subtrees.

Assume a one-to-one relation between the non-terminal symbols (the syntactic categories) of G and G': $A_i \leftrightarrow A'_i$. Assume a similar one-to-one relation between the terminal symbols (the basic expressions) of G and G': $a_i \leftrightarrow a'_i$, where a_i and a'_i have the same meaning. Assume further a one-to-one relation between production rules of G and G' with the same meaning operation. This relation must be in accordance with the other relations.

So, if $P \rightarrow Q_1 \ldots Q_n$ is a production rule of G, then $P' \rightarrow Q'_1 \ldots Q'_n$ is a production rule of G', where $P \leftrightarrow P'$ and $Q_i \leftrightarrow Q'_i$ $(1 \leqslant i \leqslant n)$.

Then G and G' are isomorphic grammars. This is not a very interesting case of isomorphy: the relation possible-translation is only possible between expressions with the same surface structure.

II. Let G and G' be context-free grammars. Assume that there is a one-to-one relation between the non-terminal symbols of G and G' and between the terminal symbols of G and G', as in case I. Assume that for each production rule
$$P \rightarrow Q_1 \ldots Q_n$$
of one grammar there is a production rule
$$P' \rightarrow Q'_{i_1} \ldots Q'_{i_n}$$
of the other grammar, with the same meaning operation, where $Q'_{i_1} \ldots Q'_{i_n}$ is a permutation of $Q'_1 \ldots Q'_n$, and $P \leftrightarrow P'$, $Q_i \leftrightarrow Q'_i$ $(1 \leqslant i \leqslant n)$.

G and G' are isomorphic grammars, which can define the relation possible-translation between expressions with different surface structures.

More complicated cases of isomorphy between context-free grammars are possible, but I will continue with two cases of isomorphy between more powerful M-grammars.

III. Let G and G' be M-grammars. Assume a one-to-one relation between the categories of G and G'. Assume a one-to-one relation between the basic expressions. Two corresponding basic expressions have corresponding categories and the same basic meaning. Assume a one-to-one relation between the M-rules of G and G', where corresponding rules have the same meaning rule. Corresponding rules operate on S-trees of corresponding categories and deliver S-trees of corresponding categories. (The category of an S-tree is the category at the top node of the S-tree.) The rules are complete, i.e. they are applicable to all expressions with the required categories.

Then G and G' are isomorphic. The grammars G_E and G_D described earlier are examples of this kind of isomorphy.

IV. Let G and G' be M-grammars. The relation between G and G' in this case is similar to III, except that instead of one-to-one relations we have relations between sets.

There is a correspondence between sets of categories, e.g.

$\{A\} \leftrightarrow \{D', E'\}$

$\{B, C\} \leftrightarrow \{F'\}$

There is a correspondence between sets of basic expressions with the same meaning, which is in accordance with the relation between the categories.

E.g. $\{A(\ )\} \leftrightarrow \{D'(\ )\}$

       $\{B(\ ), C(\ )\} \leftrightarrow \{F'(\ )\}$

There is a correspondence between sets of M-rules, with the same meaning, in such a way that corresponding sets of rules are (together) complete for corresponding sets of categories and deliver expressions with categories in corresponding sets.

E.g. $\{R_1, R_2\} \leftrightarrow \{R_1', R_2'\}$

where $R_1: A(\ldots) + B(\ldots) \rightarrow A(\ldots)$

        $R_2: A(\ldots) + C(\ldots) \rightarrow A(\ldots)$

and $R_1': D'(\ldots) + F'(\ldots) \rightarrow D'(\ldots)$

        $R_2': E'(\ldots) + F'(\ldots) \rightarrow E'(\ldots)$

In that case G and G' are isomorphic.

Case IV is close to what is really needed. The main remaining problem is to have sets of M-rules that are complete for the categories at the top node. Often the applicability of a rule is dependent on other information in the S-tree as well. For example, in the next section we will see a rule which substitutes an NP for a syntactic variable x_j in a clause with category CL1. The rule is not applicable to all expressions with category CL1, but only to those which contain an instance of x_j. Apparently the notion of syntactic category must be extended for cases like this to something like 'CL1 with variable x_j'. In the context of Montague grammar, compound categories like this have been proposed by Janssen ([597] 1980). There may also be cases where the sets of rules are intended to be incomplete, because they are assumed to be applied to a set of expressions (caused by an ambiguity earlier in the derivation process) and they are assumed to filter out some of them. In complicated grammars filters like this may lead to situations where isomorphy is hard to prove.

Isomorphic grammars for a classic translation problem

In this section I will show how a 'classic' translation problem can be solved in the framework of isomorphic grammars. It concerns the translation of the English sentence 'Peter likes to sleep' into the Dutch sentence *Peter slaapt graag*. The problem is that in sentences like this the verb 'to like' must be translated into the Dutch *graag* (German: *gern*), which is an adverb. Because of this the English and the Dutch sentences have quite different surface structures. The example grammars I will describe here are over-simplified and have no other aim than to illustrate that isomorphic grammars are a more powerful tool for translation than might be expected. I will first sketch two grammars G_1 and G_1' for which the isomorphy is straightforward and which relate the English sentence 'Peter desires to sleep' to the Dutch *Peter wenst te slapen*. Then I will describe the extension that is necessary for the above-mentioned problem sentences. In the example grammars the abbreviated notation for S-trees will be used. The definition of the domain of S-trees is omitted.

The grammar of the English fragment, G_1:
1. basic S-trees:

$VAR(x_1), VAR(x_2), \ldots$ (syntactic variables, corresponding
to logical variables $X_1, X_2, \ldots$)
    VERB1(sleep),   (basic meaning B_1)
    VERB2(desire), (B_2)
    NP(Peter),   (B_3)

2. M-rules:

The functioning of the rules will be sketched by illustrative examples and some additional comments.

R_1: $VERB1(sleep) + VAR(x_i) \rightarrow CL1(x_i \text{ sleep})$

The rule operates on an intransitive verb and a variable and constructs a clause in which the variable is the subject.

R_2: $VERB2(desire) + VAR(x_i) + VAR(x_j) \rightarrow$
$$CL2(x_i \text{ desire } x_j)$$

The rule operates on a verb that expects a to-infinitive phrase, and two variables. It constructs a clause with the variables as the subject and the complement.

$R_{3,j}$: $CL1(x_i \text{ sleep}) + CL2(x_i \text{ desire } x_j) \rightarrow$
$$CL1(x_i \text{ desire (to sleep)})$$

This is a rule scheme with an instance for every variable index j. The rule substitutes a clause for the complement variable in another clause. A condition for the applicability of the rule is that the subject variable of the complement and the subject variable of the main clause have the same index.

$R_{4,j}$: $NP(Peter) + CL1(x_j \text{ desire (to sleep)}) \rightarrow$
$$SENTENCE(Peter \text{ desires (to sleep)})$$

The rule (scheme) substitutes an NP for the subject variable of a clause, assigns the simple present tense to the verb and adjusts its form to number and person of the subject.

The meaning rules corresponding to these rules are named respectively: M_1, M_2, $M_{3,j}$, $M_{4,j}$.

3. Final categories: $\{SENTENCE\}$

G_1 generates the S-tree of the sentence 'Peter desires to sleep', with derivation tree (23). Its logical derivation tree is (24).

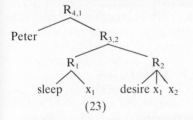

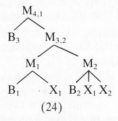

(23)   (24)

The grammar of the corresponding Dutch fragment, G_1':

1. basic S-trees:

$VAR(x_1)$, $VAR(x_2)$, . . ., corresponding to logical variables
$$X_1, X_2, . . .$$

$VERB1(slaap)$, (basic meaning B_1)
$VERB2(wens)$, (B_2)
$NP(Peter)$ (B_3)

2. M-rules

$R'_1: \text{VERB}1(slaap) + \text{VAR}(x_i) \rightarrow \text{CL}1(x_i\ slaap)$

$R'_2: \text{VERB}2(wens) + \text{VAR}(x_i) + \text{VAR}(x_j) \rightarrow$
$$\text{CL}2(x_i\ wens\ x_j)$$

$R'_{3,j}: \text{CL}1(x_i\ slaap) + \text{CL}2(x_i\ wens\ x_j) \rightarrow$
$$\text{CL}(x_i\ wens\ (te\ slapen))$$

$R'_{4,j}: \text{NP}(Peter) + \text{CL}1(x_j\ wens\ (te\ slapen)) \rightarrow$
$$\text{SENTENCE}(Peter\ wenst\ (te\ slapen))$$

The meaning operations corresponding to these rules are named respectively: M_1, M_2, $M_{3,j}$, $M_{4,j}$.

3. Final categories: {SENTENCE}

G'_1 generates the S-tree for *Peter wenst te slapen,* with a derivation tree similar to (23), and a logical derivation tree identical with (24). G_1 and G'_1 are isomorphic grammars. This is easy to verify: there is a one-to-one relation between the syntactic categories and an according one-to-one relation between the basic expressions with the same meaning and between the rules with the same meaning of the two grammars. It is easy to verify that if a rule of one grammar is applicable to some type of expressions, the corresponding rule of the other grammar is applicable to the corresponding expressions.

In order to generate the sentence 'Peter likes to sleep' grammar G_1 has to be extended to G_2. The only difference is the addition of the basic expression $\text{VERB}2(like)$, with basic meaning B_4.

For the generation of *Peter slaapt graag* the following extensions of G'_1 to G'_2 are needed. Two syntactic categories are added: ADV2 and ADVP. The basic expression $\text{ADV}2(graag)$ is added, with basic meaning B_4, the same two-place function as for VERB2 (like) in the English fragment.

Two M-rules are added:

$R'_5: \text{ADV}(graag) + \text{VAR}(x_i) + \text{VAR}(x_j) \rightarrow$
$$\text{ADVP}(x_i\ graag\ x_j)$$

The rule operates on a two-place adverb and two variables and constructs an adverbial phrase with the adverb as the head and the variables as arguments.

The corresponding meaning operation is M_2, the same as for rule R_2 and R'_2.

Though the rules differ syntactically, they have the same semantics: the application of a two-place function to its arguments.

$R'_{6,j}: \text{CL}1(x_i\ slaap) + \text{ADVP}(x_i\ graag\ x_j) \rightarrow$
$$\text{CL}1(x_i\ slaap\ graag)$$

The rule (scheme) substitutes a clause for the second argument variable of the ADVP. The result is a clause in which the main verb of the substituted clause becomes the main verb and the ADVP a modifier. The rule is only applicable if the subject variable of the clause and the first argument variable of the ADVP have the same index.

The isomorphy of G_2 and G_2' can be established as follows. The relation between the syntactic categories is no longer one-to-one. VERB2 of G_2 corresponds to VERB2 and ADV2 of G_2'. CL2 of G_2 corresponds to CL2 and ADVP of G_2'.

The relation between the basic S-trees is the same as for G_1 and G_1', except for the addition of VERB2(like)$\leftrightarrow$ADV2(graag).

The relation between the rules is extended as follows:

$$R_1 \leftrightarrow R_1'$$
$$R_2 \leftrightarrow \{R_2', R_5'\}$$
$$R_{3,j} \leftrightarrow \{R_{3,j}', R_{6,j}'\}$$
$$R_4 \leftrightarrow R_4'$$

The derivation tree of 'Peter likes to sleep' according to G_2 is (25).

The derivation tree of *Peter slaapt graag* according to G_2' is (26).

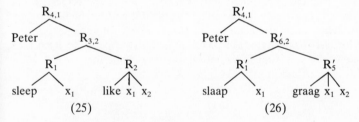

(25) and (26) are isomorphic derivation trees, corresponding to the same logical derivation tree.

So we have reached our goal: the grammars G_2 and G_2' define the relation possible-translation not only between 'Peter desires to sleep' and *Peter wenst te slapen,* but also between 'Peter likes to sleep' and *Peter slaapt graag.*

Conclusion

In this paper I have presented a compositional approach to machine translation. According to this approach a sentence s' is considered a possible translation of a sentence s if

(1) s' and s have the same meaning.

(2) the meanings of s' and s are derived in the same way from the same basic meanings.

(3) the surface structures of s' and s (which may differ) have similar derivational histories.

I have shown in some detail how a translation system can be defined in accordance with this approach. If isomorphic M-grammars are written for the set of languages involved, computable analysis and generation functions can be derived from these grammars. The composition of an analysis function for a language L_1 and a generation function for a language L_2 constitutes a translation function from L_1 to L_2. The intermediate language of this translation system follows directly from the grammars. If the grammars are strictly isomorphic it is guaranteed that the generation component is applicable to any intermediate expression that is delivered by an analysis component. In the fourth section I have given guidelines for achieving this isomorphy.

As for the power of this approach, it should be noted that the M-rules have in principle transformational power (the three conditions on M-grammars are the only formal restriction on their power). Furthermore there is the possibility to have compound S-trees as basic expressions, which enables to handle complex idioms as well as the translation of words into complex expressions.

The practical experience with this approach in the ROSETTA system has been very favourable thus far. Of course the isomorphy approach is not a panacea for all translation problems (especially not for the disambiguation problem), but it appears to be a nice framework for a clear formulation of these problems.

EUROTRA

EUROTRA planning started in late 1977, and in early 1978 a permanent working group – the Eurotra Co-ordination Group – was set up, drawing its membership from European Universities and Research Units working on machine translation and related areas. This group met at roughly monthly intervals over the next four and a half years, with the aim of establishing the technical basis of a machine translation system which was to be developed collaboratively in Europe and was to be designed to meet the specific needs of the European Community. (The working group operated under the aegis of the Commission of the European Communities.) Over this planning period it was possible to finance a relatively small amount of contract work aimed at helping in laying down the linguistic and software basis for the project, so that around a hundred or so people were eventually involved in some way with EUROTRA planning, but it is perhaps worth emphasising that this was almost entirely on a part-time or even spare-time basis, so that it is only recently that anyone has begun to work on the project full time.

Of the specific needs of the Community, one is primordial and has affected the whole of the system design. This is the requirement of multilinguality. The Community has, at the moment, seven official languages, with the prospect of perhaps adding at least another two. The system therefore had to be designed from the start to deal with a large number of language pairs. (Initially, Greece was not yet a member, so there were thirty pairs. Now, with the addition of Greek, there are forty-two.)

Any possibility of gathering a large enough team of qualified people into one central place was also ruled out from the beginning, partly on practical and social grounds, partly because one important subsidiary aim in the proposed project was to stimulate and encour-

373

age European expertise in machine translation and in computational linguistics in general. This could hardly be achieved by removing those who had the necessary expertise from their own countries and concentrating them in one geographic location.

These two requirements – multilinguality and de-centralised development – are fundamental criteria for design decisions.

The planning work came to fruition with a decision of the Council of Ministers of the Community, which became effective on 13 November 1982, to set up a research and development programme whose objective is 'the creation of a machine translation system of advanced design (EUROTRA) capable of dealing with all the official languages of the Community'.

Objectives of the programme

The research and development programme lasts for five and a half years from the date of the Council decision. It is broken up into three separate phases, each with its own subsidiary objectives, and with a check-point at the end of each phase.

The first phase, whose end is now (late 1984) approaching, involves carrying out several tasks in parallel. First, an organisational structure must be set up. This involves the individual Member States of the Community agreeing to participate in the programme (the Council decision only commits them to that part of the work which is to be paid for out of central funds) and setting up research units to carry out the work to be done in the Member States. Simultaneously, linguistic specifications defining the framework for linguistic work are being drawn up as are software specifications defining a core software which is to be used by all the groups working in the project.

During the second phase, which also lasts two years, a first small system is to be built, which will cover approximately 2500 lexical entries per language and will work on a pre-chosen corpus enlarged to account for common linguistic phenomena not included in the corpus. This system must meet several criteria which will be examined in more detail below.

The third phase, lasting a year and a half, involves enlarging the basic system by a factor of ten, to give a system, still working in a specified subject area and over specified text-types, but dealing with more complex text and a vocabulary of approximately 20000 lexical entries per language.

It should be emphasised that this final system is not intended to be an industrial system. It will be a pre-industrial prototype, intended to confirm the validity of the approach taken and as a basis for possible industrial development after the end of the research and development programme.

Thus, the overall development strategy is to use the first, preparatory phase, to set up specifications, in the full knowledge, in the case of the linguistic specifications, that they cannot in the nature of things be definitive. The only real test of specifications in a computer application is the application itself, and no attempt will have been made before the second phase to construct any kind of realistic system.

The construction of a small system during the second phase gives the first real opportunity for extensive testing and feed-back, and can therefore be regarded as supplying evidence for the modification and stabilisation of the linguistic specifications, which then serve as the basis for development of the pre-industrial prototype. It is hard to over-emphasise the importance of this: of the seven languages to be treated by EUROTRA, some have received quite extensive attention at the hands of linguists, some have been almost completely neglected. And of the language pairs to be dealt with, very few have received any kind of intensive contrastive study even in theoretical terms, much less in computational terms. Thus the possibility of modifying and refining the underlying linguistics during the second phase of the project is crucial to the success of the whole enterprise.

This leads to a further general and extremely important point. EUROTRA is essentially a linguist's project. It is anticipated that something close to a hundred people will be involved full-time during the last two phases of the project on nothing but linguistic work. Furthermore, they will be working in decentralised teams, with all the communication difficulties automatically thereby implied. For this reason, it is part of the EUROTRA credo that everything possible must be done to facilitate the life of the linguist.

The first basic option

Given the requirement of multi-linguality, the basic choice of a system design is severely constrained, in that it is clearly quite impracticable to develop forty-two separate bi-lingual systems, one for each language pair. Thus, some kind of neutral switching point must be defined to serve as an interface between one language and another. Here two plausible candidates exist. The first involves defining an interlingua, which would be independent of any specific language. The source-language text would then be mapped on to an interlingual representation and the target-language translation generated from the same interlingual representation. However, no such interlingua as yet exists, there are some who believe on theoretical grounds that it would be impossible to define one, and, even if it were practicable, the definition would, in itself, constitute a very long term research project.

EUROTRA has therefore adopted the alternative possibility, whereby two switching points are defined rather than one, the first a representation on to which the source language text is mapped, the second a representation from which the target translation is generated, with a separate system component mapping the first representation on to the second. This is, of course, the familiar transfer-based design.

EUROTRA imposes two further constraints on this design. First that it must be possible for the analysis and generation modules to be written by linguists who know nothing of the languages involved other than the language for which they are constructing the analysis or generation module. Secondly, and primarily for economic reasons, the size of transfer modules must be reduced as much as possible.

These two constraints, together with an obvious desire to produce high-quality translation, have led to a systematic attempt to identify linguistic phenomena which *can* be treated interlingually and to represent them as such in the structure serving as the interface between analysis and transfer and between transfer and generation. Most of the description of linguistic representation given in a later section will concentrate on these phenomena.

A note of caution should be added, however. No pretension is made to having identified linguistic universals in any sense: the most that can be claimed is 'Euroversality' and even that only after the extensive testing to be carried out during the second phase. Furthermore, no attempt is made to work with language independent canonical structures: the output from analysis will reflect the structure of the source language and the input to generation will, at least to some extent, be tailored to make sure that a mono-lingual generation writer can carry out his task.

It is perhaps appropriate to make another point here. Because of the attempt to identify and use interlingual information, which can be presumed to be unchanged by transfer, the generation component of EUROTRA is intended to be rather more powerful than is usual in most machine translation systems. It is not restricted to generating the correct morphological form and producing the correct word order, but can carry out transformations of arbitrary complexity, in the limit freely choosing an appropriate surface syntactic form in the target language.

Project organisation

The main organisational lines of the technical work are predicated on the preceding sections: a transfer based design and a requirement for decentralised collaborative development can be put together to give an organisational scheme whereby the monolingual components of the system, analysis and generation, are developed by language groups working in the Member States independently of one another, and transfer components are developed by language groups working in collaboration. Of course, the various linguistic modules must ultimately fit together to make a single multi-lingual translation system. To facilitate this, quite a lot of work is carried out by a central team. This includes the drawing up and maintaining of linguistic specifications, which provide a framework for the work to be done by the language groups and a definition of the crucial interface structure. From the software end, specifications are being drawn up for a core software which will be used by all the groups (Johnson and Rosner discuss this more fully in chapter 11), and ample software documentation aimed primarily at use by linguists is being prepared. This documentation defines and describes the use of specially defined high-level languages to be used by all language groups for the description of linguistic data and the expression of linguistic strategies.

More basic options

A number of other critical choices have had heavy influence on EUROTRA's design. The first of these is modularity. It is by now banal to say that a good system design must be modular; but EUROTRA has made a conscious decision to push modularity to its limits. Thus, machine translation is seen as involving a large number of sub-tasks, each of which can be defined linguistically, and which can be quite diverse in their nature. It is potentially feasible for different modes of description to be used for different kinds of linguistic tasks, so that the linguist can choose the mode of expression which is the most natural for the task in hand. Thus, the main modules of analysis, transfer and generation are in their turn broken down into smaller modules, each with a well-defined input and output and each performing a specific task.

Heavy modularity in its turn serves the ends of two other basic options. The first of these is system robustness. EUROTRA, in its final form, is not intended as an interactive system, although some interaction can be foreseen during development. This means that, in quite a strong sense, the system is not allowed to fail. It must always produce *some* output, even if the input text is defective in some way or if the system itself malfunctions in some way (for example, if analysis fails to compute a full interface structure or if a lexical entry is missing). One way foreseen to help in achieving this is that any module in the system may, if the linguist so desires, be accompanied by a fall-back module, or even a set of fall-back modules, whose job is to identify what has gone wrong and take emergency action to repair the situation.

Modularity also helps to ensure that the system is easy to improve and to repair.

Then, too, the system is intended to be extremely open-ended. Its design makes it possible to add new subject fields and new text-types very easily, just as it makes it possible to add new languages. Furthermore, although no extensive use of heavy inferencing or artificial intelligence techniques is planned for the system to be developed over the next four and a half years, it is hoped that the extreme modularity of the system together with the possibility that different modules may behave in different ways will make it possible to add such techniques relatively easily.

The last paragraph should not be read as meaning that E U R O T R A intends to work primarily on a syntactic level. The analysis aimed at is quite 'deep' and involves extensive use of semantics both as a part of the representation aimed at and as a tool in achieving the analysis.

To summarise what has been said in these introductory sections: E U R O T R A is a multilingual transfer-based system, aiming at rather an ambitious level of analysis relying heavily on interlingual representation but recognising that a true full interlingua is not currently feasible, highly modular and open-ended, intended to be robust in operation and to be easily improvable.

EUROTRA's linguistic framework

This and following sections will concentrate on a description of the linguistic representation aimed at as the output of analysis, and especially those parts of it which are intended to be interlingual and thereby both to improve the quality of the translation and to cut down the size of the transfer components. Very little will be said about linguistic strategies to be used to construct the representation during analysis, mainly because the decision about what strategy is appropriate to a particular language is the responsibility of the language group rather than imposed from outside. (The project contains linguists coming from many different backgrounds and with experience with a variety of earlier systems; leaving them freedom to develop their own strategies uses their background to maximum profit, allows freedom for experimentation and permits different languages to be treated differently.)

However, it should be noted that different kinds of information may be considered as defining different dimensions of representation (e.g. surface syntax, deep syntax, semantic). Separating out the different dimensions provides the EUROTRA linguist with an important conceptual tool for the investigation of the language he is working on, and also can be used to provide a first simple stratificational strategy. Much current EUROTRA work exploits the possibility of such a separation.

Apart from the requirements given in earlier sections, several other criteria have been used in drawing up the definition of the linguistic representation. The first is a check on the desire to extract and use as much interlingual information as possible. Clearly this could be pushed to an extreme where, for example, a pure logical form was aimed at. But EUROTRA is first and foremost a translation system; any information included in the linguistic representation must be justifiable in terms of its being needed in order to translate. In other words, the level of analysis aimed at is what is necessary for translation and no more.

Since analysis and generation are to be entirely monolingual, in the limit being developed by linguists who know nothing of any other language than the language they are working on, this imposes another constraint on the representation. It cannot, for example, reflect all the lexical distinctions made by all the languages. A linguist working on the analysis of English simply does not know that there are two translations possible for 'wall' in Italian, so he cannot be expected to distinguish two senses of 'wall' in constructing analysis. The distinction must be made in transfer. This seems at first sight to go counter to the requirement that transfer be kept as small as possible, but it is a straightforward practical necessity. If analysis were to take account of all the distinctions in all the languages, the linguists would have to know all the languages very well indeed.

The same requirement of monolinguality also determines the nature of the representation input to generation. The linguist constructing it knows only one language, in theory, and yet he must obviously know what kinds of representation to expect. This is achieved by defining an equivalence relation between linguistic representations, stipulating that certain elements of the representation are critical for equivalence to hold. The starting point for generation is then defined as being the set of structures equivalent, in this sense, to the set of structures produceable by an analysis module working on the same language.

Other criteria, for example, the feasibility of computing a certain kind of information, have played a role in determining the content of the representation, but what has been said is sufficient to illustrate the grounds on which choices have been made.

Linguistic representation

The linguistic representation is, not very surprisingly, a tree structure, with some special characteristics. First, the nodes of the tree are not simple labels, but structures of some complexity. The easiest way to characterise these structures is to describe each node as carrying a set of named properties with associated values, with the further complication that properties may be inter-related in that any property may be declared to imply the existence of other specified properties or to be further sub-classified by other properties. (The resulting structure is conceptually quite similar to a LISP property list.) A definition of the set of legal complex labellings to be found on the nodes is communicated by the linguist to the system via a set of declarations which can then be used to check linguistic well-formedness of nodes created during processing.

Secondly, cross-references between nodes is possible: thus, a node which represents a pronoun, for example, can be related to the node representing its antecedent.

What has been described so far is a tree, in the conventional sense, with complex labelling on the nodes and with the possibility of co-indexing. The other peculiarities of the tree come from theoretical linguistic considerations. The tree structure incorporates a version of dependency, which is not quite the standard version. In a standard dependency representation (e.g. ([590] Hays 1964) there are no non-terminal nodes. Each node is labelled by a lexical item, which is the 'governor' of the constituent, that is, roughly, the most important item in the constituent, whose presence determines what other items may be present, their surface appearance in the case of, for example, agreement phenomena, and which shares most properties with the constituent as a whole. Thus the governor of a noun group is the noun, the governor of a verbal phrase is the verb and so on. A standard dependency representation of 'John likes foreign films' would be something like:

```
        likes
        /\
  John    films
            |
          foreign
```

This representation has the advantage that it makes explicit that the properties of a node are dependent on the properties of the dominating node. It has, however, the disadvantage that it is difficult to distinguish the properties of a complex expression as a whole from the properties of the governor. The following examples, taken from ([63] Arnold 1983) illustrate this.

(1) Jules drank a *bottle of wine*.

(2) *The good* die young.

(3) *Kim's quickly solving the problem* surprised me.

In (1) the syntactic governor is 'bottle'. However, semantically and pragmatically, it is clear that the object of drink should be something liquid. Thus, there is an obvious difference between the semantic properties of the constituent 'a bottle of wine' and the governor. In (2), the most plausible candidate for the governor of 'the good' is 'good', even though 'good' is an adjective and 'the good' is a nominal expression. Here there is a difference in the syntactic properties. Similarly, in (3), the expression in italics seems, considered as a whole, to be nominal, but its governor ('solving') is obviously verbal.

A further disadvantage of a standard dependency representation is that the position of the governor in the original surface string is not recoverable from the structure alone, simply because the governor is placed on the dominating node. This can sometimes be damaging. Consider the two Italian phrases 'Ho visto *il piccolo libro*' and 'Ho visto *il libro piccolo*': both these would get the same representation for the phrase in italic, along the lines of:

libro

/\

il   piccolo

despite the fact that the first sentence could be paraphrased as simply 'I saw the small book', remaining neutral on the question of whether any other books exist, whilst the second suggests quite strongly that it is the small book as opposed to some other book, and would be paraphrased 'I saw the *small* book', with heavy emphasis on the 'small'. To say this more generally, surface order carries information on scope and emphasis, and a representation which obscures surface order is for this reason not very satisfactory.

E U R O T R A's linguistic representation, therefore, is a modified version of a standard dependency representation. The governor is lowered into the sub-tree into a position which reflects its position in surface order, and is explicitly marked as the governor with the complex labelling on the node. This gives the tree structure for the earlier example:

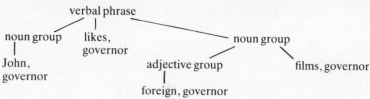

where surface order is preserved, and where it is possible to indicate the properties of any item in that group.

This representation has something in common with a constituency representation, although it is not a normal constituency representation, where the fundamental principle is that items should be represented as single units if they behave distributionally and semantically as single units (cf. [613] Radford 1981). In the EUROTRA representation, there is a requirement that each node has at least one terminal daughter. So the standard constituency representation

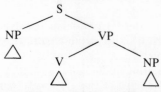

is not possible in EUROTRA, since the VP has no terminal daughter.

The main reason for this (for other reasons, see ([63] Arnold 1983) is that a standard constituency structure imposes rather 'deep' tree structures, which can complicate rule writing for the linguist. The simple fact that word order varies across the EUROTRA languages means that translation involves some re-ordering of nodes. In general, it is easier to write such rules if all that is involved is the re-ordering of sister nodes rather than dealing with more complex geometries. Choosing a representation which produces rather 'flat' trees minimises the amount of complex geometric manipulation that is required.

A further point should be made here. The complex labelling on the nodes is used to indicate different *kinds* of linguistic information, so the same linguistic representation simultaneously captures quite diverse dimensions of description. This means that the geometry of the tree structure must be weak enough to allow interpretation of all these different kinds of information. (So far, only two have been relevant, surface syntactic constituency and dependency relations: it will become clear later that other kinds of linguistic information are also involved.) This means that two demands on the geometry must be met:

(a) It must be capable of grouping any set of items together to which a collective property must be assigned (e.g. that a noun group has the property of being definite).

(b) Any items that are in relationship must be grouped together (e.g. the subject and object relations in the sentence 'John likes foreign films' are captured by representing the two noun groups and the verb as sister nodes).

This section so far has described the major choices made in deciding on the geometrical form of the tree structure which provides the skeleton of the linguistic representation. It can be summarised by saying that a geometry which is representationally weak has been adopted: its relative impoverishment will be compensated for by the richness of the labelling system.

Co-indexing

One problem with any representation in the form of a tree structure is that it is too weak to allow representation of some phenomena which must be captured if the translation is to be correct.

The most obvious of these is the antecedent relation. A simple example will illustrate the importance of this: translating from English into French, the 'they' in 'the girls were hungry, so *they* bought themselves some food' can only refer to 'the girls' and is translated by *elles*. Conversely, if the sentence is 'The boys were hungry, . . .' the 'they' is translated by *ils*. So the referent of the pronoun must be included in the linguistic representation.

A further example comes from constructions like 'Peter expects John to come' where 'John' is the agent of 'come', despite not being related directly to 'come' in the surface structure. There are, amongst our languages, languages which do not allow this kind of construction, using instead a construction similar to 'Peter expects that John will come', where 'John' is explicitly related to the verb of which he is agent. Again, in order to get the correct translation, the fact that 'John' is agent of 'come' must be captured in the linguistic representation.

In order to resolve problems of this sort, the representation includes a co-indexing mechanism, which makes it possible to state explicitly that a node refers to another node, even when this node is far removed in the tree, perhaps even at a different level. To illustrate this, the second, slightly more complex example, will be used. An empty element, i.e. a node carrying no lexical material, is introduced in the correct position to give a relationship with 'come'. This is then co-indexed with the antecedent, to give the (oversimplified) representation below, from which it can be seen that property and relational information is distributed over the co-indexed nodes, with the property information appearing in the surface structure location of the item itself, and the relational information appearing in the 'understood' position.

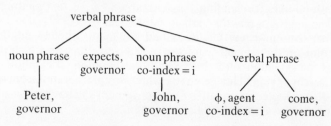

The co-indexing mechanism is very powerful, and the price paid for it is complication of the linguistic representation. Consequently, its use is in practice constrained to those cases were it is needed for correct translation, in conformity with the 'as much as necessary, and no more' principle stated earlier.

The representation presented in this and the preceding section can be viewed as a version of X-bar theory, with the maximal projection of X being 1. Jackendoff ([596] 1977) argues that X^3 is required for all major categories. In the case of EUROTRA, the richer labelling system is used to achieve what the Jackendoff richer geometry achieves. (Arnold ([63] 1983) discusses this issue in more detail.)

The complex labellings

An attempt has been made in the last two sections to describe the geometrical shape of the tree structure which is at the basis of EUROTRA's linguistic representation. In this and the following sections, attention is concentrated on the content of the representation, that is, the kinds of information represented in the complex labellings on the nodes of the tree via properties and values of properties. No attempt will be made to give an exhaustive description; in particular scant attention will be paid to language-specific information.

As has already been said, in defining the representation to be aimed at, much effort has gone into trying to identify interlingual information, that is, in terms of the complex labelling, properties and values which are the same for all languages and are consistently interpreted in all languages.

Semantic relations

The first kind of information of this type is semantic relations. A rather large set of semantic relations has been defined covering both valency-bound arguments to the predicate and circumstantials. The clearest example of the use of circumstantial semantic relations is in determining the correct choice of preposition in prepositional phrases. In a recent internal report ([170] ETL-4-GB final report) the English language group distinguishes twenty-five different uses of the English preposition 'on', using fourteen semantic relation assignments (often together with other information) to capture these uses. The following examples are extracted from this report:

'He arrived *on Thursday*'
'He lives *on bread*'
'He is *on the committee*'
'He is working *on a new project*'

It is easy to see that each of these requires a different translation, depending on the semantic relation involved (time location, means, membership, concerning). It is exactly this type of information which is captured by the values of the semantic relation property.

Semantic relation values indicating the type of relation entered into by valency bound arguments are used when different languages express the same semantic relations differently at the surface. The classic example of this is 'I like Paul' versus *Paul me plaît,* where the experiencer and patient roles are the same in both languages, but the surface syntactic structures are almost exact inverses of each other. Without semantic relations, the transfer writer has to map one surface structure on to the other. With semantic relations, the main burden can be left to generation.

At the moment, a fairly stable list of circumstantial relations has been drawn up and has proved satisfactory in simulated testing against the EUROTRA languages. The question of an appropriate set of relations for argument places is rather more open, as is the related question of establishing a classification of predicate types. It seems unlikely that predicate-typing can be used as a way of *determining* the number and the nature of the arguments to the predicate, but such a classification might well prove to give clues to the dictionary maker.

Because of space limitations, a number of interesting issues to do with semantic relations cannot be discussed here (for example, the

question of middle-valencies, the homogeneity of the argument/
circumstantial system and so on).

Before finishing this section though, it should be emphasised that
the list of possible semantic relations is, like all the other properties
and values, to be considered as preliminary. No definitive list can
be drawn up before the empirical test of constructing a system using
the tentative proposals has been carried out.

Time values

The use of semantic relations in EUROTRA cannot be considered an enormous innovation: several systems use deep case
systems (cf. [582] Fillmore 1968), and EUROTRA's main
difference from these systems lies in the number of relations distinguished rather than in any strong conceptual difference.

Much more ambitious is an attempt to distinguish a set of interlingual time values, intended to capture the 'time-meaning' of a
sentence and avoid the necessity of mapping morphological tense
on to morphological tense in transfer. The complexity of such a
mapping can be illustrated by considering the translation of Dutch
present tense into English; the following examples illustrate some
of the possible translations:

Ik lees een boek I am reading a book (progressive)
Ik lees vaak boeken I often read books (present)
Morgen lees il zes boeken Tomorrow I will read six books
    (future)

Setting up an appropriate list has, not surprisingly, proved to be
non-trivial, and work is still continuing in this area. However, if a
satisfactory set of values can be determined, the effort will have
been worthwhile.

Modality values

An even more open question is that of a set of modality values.
At first sight, it seems plausible that an interlingual representation
of modality would be extremely useful especially given that our
seven languages break down into two major families, Germanic
and Romance, the first of which tends to express modality lexically
(via, for example, the use of modal verbs), the second morpho-
syntactically, so that modality becomes a property of the sentence
as a whole rather than being identifiable with some particular lexical
item. And certainly, the translation of, for example, English 'can'
into French 'pouvoir' or 'savoir' depends on being able to make a
distinction between 'ability' and 'possibility'. Thus, quite a lot of
effort has gone into trying to define an interlingual representation

of modality. Rather surprisingly, however, some recent work ([171] ETL-4-DK Final Report 1984) offers evidence that a bi-lingual treatment of modality between two languages representative of the two families (Danish and French) might be less problematic than was originally thought. Although this work is based on too small a sample to be in any way definitive, it may turn out that the use of interlingual tools for the representation of modality is more restricted than seemed a priori likely.

Other interlingual features

Many other linguistic phenomena require interlingual treatment. For example, there is a clear distinction between semantic singularity/plurality and morphological number, just as there is a clear distinction between sex and morphological gender. These, and other similar phenomena, are represented by interlingual values in the linguistic representation. They will not be discussed in any detail here, however, where it seemed more interesting to pick out what might be called problem areas which are still the subject of intensive work.

Syntactic categorial information

Rather more conventional, but equally necessary for producing correct translation, are the syntactic categorial values which also appear in the complex labellings. These cover the normal syntactic categories (noun group, verb group, etc.). They are not interlingual in the sense given earlier, in that whilst all our languages share the same properties and values, the distribution of the values across the languages does not coincide. Their use needs no illustration.

The same is true for part of speech information, which is also included in the complex labelling.

Syntactic function information

Two kinds of syntactic function information may be included in the complex labelling, surface syntactic function and deep syntactic function (e.g. deep subject is distinguished from surface subject in passive sentences such as 'Rome was invaded by the barbarians').

Strictly speaking, this kind of information is redundant in the final linguistic representation aimed at as the output from analysis, providing that the correct semantic relation has been calculated and assigned. But its use during analysis and generation is clear, and even during transfer it may serve as confirmatory evidence for a choice primarily based on semantic information.

Syntactic information, both functional and categorial, can also be used to serve another important purpose. In the introductory sections, some emphasis was put on the idea that the system must always give some output, no matter what goes wrong, and the use of fall-back measures to ensure this was mentioned. Syntactic information is archetypically the sort of information which would be used by a fall-back module if semantic processing had been unsuccessful, just as tense information provides fall-back information for the interlingual time values.

Language specific information

A very brief description has been given of that part of the content of the linguistic representation which is presumed to be interlingual, either in the strong sense that properties and values are shared by all languages and distributed over the languages in the same way, or in the weaker sense that although properties and values are the same, the distribution of the values varies in different languages.

In addition to this information, the output from analysis will contain purely language specific information, for example morphological properties and values, which is also represented via the labelling system. This information too, can of course, be used in transfer to determine translational equivalents, just as it can be used by fall-back processes.

The linguistic representation summarised

The representation consists of a tree, allowing co-indexing between nodes, encapsulating a compromise between dependency theory and constituency theory. Each node contains a complex labelling, in the form of properties and associated values, which describes the relational and categorial properties of the constituent represented by the sub-tree dominated by that node. The relational properties always indicate relations to the governing node at the same level of the tree, which operates as governor no matter what the kind of relation described is. The kinds of relation indicated include semantic relation, deep syntactic function and surface syntactic function. Categorial information includes, at least, syntagmatic category or syntactic word class information, and, potentially, other categorial properties such as semantic features, time values, modality values, singularity/plurality features and so on.

The linguistic representation aimed at as the output from analysis contains all the interlingual information plus the language specific information which is then available for use, if necessary, in transfer. The output from transfer contains the interlingual information plus any information put there by transfer in order to make generation possible. The richness of the representation aimed at is intended to serve the double purpose of ensuring high-quality translation and reducing the size of transfer.

Dictionary information

Dictionary entries too may contain rather a lot of information. As well as the conventional morphological and part of speech information they may contain an indication of any valency bindings associated with the lexical entry, including semantic restrictions where these are relevant, co-occurrence and sub-categorisation restrictions, indications of possible surface structures relevant to the lexical entry, subject field and thesaurus type information, and, if desired, a semantic formula representing common-sense knowledge about the lexical unit and providing a basis for limited inferencing. (This is not an exhaustive list, but is merely indicative.)

The aim of all this information is to facilitate the construction of the linguistic representation and to make it possible to use, for example, case-frame-driven or semantically based linguistic strategies. The general principle here is that the precise content of a dictionary entry is dictated by the use to which the dictionary information will be put. Therefore a rigid format should not be decided in advance and imposed on the linguist; rather, when he decides he needs a particular type of information he should be able to add it easily.

It should be mentioned that dictionary information is not contained in a single monolithic block, with access to it possibly only at (a single) defined point during processing. In principle, dictionary information is no different from any other kind of information; it can be retrieved and used at any time.

A linguist's version of the core software

Johnson and Rosner (chapter 11) have given an account of the principles of EUROTRA's software design. Therefore, all that is given here is a brief outline of the most relevant aspects from a linguist's point of view.

The general architecture of the software is that of a controlled production system, whose basic building blocks are EUROTRA grammars. Each grammar is a primitive, consisting of a set of patterns and associated actions, where the patterns are matched against the data structure and trigger application of the associated action when a match is successful. Each grammar is accompanied by a precise definition of the class of structures it is intended to work on, and the underlying software ensures that no structure which does not meet the definition is passed to the grammar. Similarly, each grammar is accompanied by a definition of acceptable results, and no structure is passed on by the grammar which does not meet that definition. It is possible to give a ranked set of definitions of acceptable results, so that if the 'best' result is not achieved the next best will be passed on, and so on.

Much of the flexibility of EUROTRA from the point of view of the linguist comes from the fact that a grammar is primitive. This means that different grammars may have different natures. For example, one very natural way for a linguist to express himself is in the form of general re-write rules, and many grammars will be of this form. But there are linguistic tasks which are not most suitably described this way – morphological analysis aimed at identifying lexical units for example, or the expression of a preference based strategy where competing representations have to be compared and scored in order to choose the best. For these tasks, it is possible to stipulate that the underlying interpretation scheme of the grammar should be that appropriate to the task in hand, rather than a general re-write interpreter. The description of the task can then also be adapted to its nature.

The grammars are linked together into processes, the linking being specified by the linguist. For each process, once again, what the input must be and what constitutes an acceptable result is defined. Processes in their turn can be linked together, always with defined input and output if desired.

Processes making up larger processes can be executed in sequence or in parallel, according to a parameter whose value is given by the linguist. What processes are in sequence, what structure is passed from one process to the next is further governed by another parameter, which may stipulate, for example, that the second process is only interested in structure created by the first or in structure used by the first. This parameter actually involves two characterisations of the structure produced by the first, a distinction between new structure versus old structure, and a distinction between used structure and non-used structure.

If this description is now turned on its head, taking as starting point the whole system and working down to the grammars, the linguist proceeds as follows. It is given in the overall system design that there are three processes executed in sequence, analysis, transfer and generation. Within each of these major processes he determines linguistically coherent sub-processes, defines their required input and output, whether they should be executed in sequence or in parallel, and, if they are to be executed in sequence, what parts of the structure produced by each process in the sequence is to be visible to the next. This is repeated for each process in return, until he arrives at a task which can no longer be sensibly sub-divided. For this task, he defines a grammar, defining its input and output and specifying, in addition, what interpretation scheme is appropriate.

Thus, the system as a whole is declarative, since linguistic facts are expressed declaratively in the grammars, but the control provided by being able to link processes and grammars together provides an easy way to express the procedural knowledge inherent in the definition of a linguistic strategy. Specifying input and output helps to prevent undesired interactions between processes in a perspicuous way, whilst at the same time ensuring that the system is self-documenting.

Fall-back procedures are catered for by allowing the outputs of processes to be ranked in order of desirability, and by using the definition of the input to a process as a way of determining exactly what the situation is.

Conclusion

In conclusion, it is worth asking what exactly it is that is so special about E U R O T R A. Its software design contains several innovative features, all of them springing from a desire to make the linguist's life as easy as possible. Linguistically, it tends to reflect the state of art, which is only to be expected in a project which must produce a working system within four and a half years. Nonetheless, it contains quite a large research component, coming partly from an attempt to identify and use interlingual phenomena, but mainly from its multi-lingual nature, which implies an enormous amount of pure linguistic research to be done over the forty-two language pairs, most of which have up to now received little study and no attention at all in terms of computational linguistics.

Acknowledgement

Neither of the authors could claim ownership of the work and ideas presented here. They are the result of a collaborative effort by a very large number of people over several years, to the point where it is frequently impossible to identify who was the original author of an idea, or even, sometimes, of a particular phrase. Our thanks should go to all these people, and especially to Douglas Arnold, Louis Des Tombe and Lieven Jaspaert, whose work on the linguistic specifications for E U R O T R A has provided much of the material for this chapter. The authors are, of course, solely responsible for misreporting, inaccuracies or errors.

ABBREVIATIONS, TITLES AND ACRONYMS

AJCL	American Journal of Computational Linguistics
ALLC	Association for Literary and Linguistic Computing
ASIS	American Society for Information Science
ATA	American Translators Association, Ossining, New York, USA
Babel	International Journal of Translation
BCS	British Computer Society, London, England
CACM	Communications of the Association for Computing Machinery
CAIS	Canadian Association for Information Science, Ottawa, Ontario, Canada
Daedalus	Journal of the American Academy of Arts and Sciences
EDRS	ERIC Document Reproduction Service, Arlington, Virginia, USA
FBIS	Foreign Broadcast Information Service, Washington, DC
FTD	USAF Foreign Technology Division, Wright Patterson AFB, Ohio, USA
ICCL	International Conference on Computational Linguistics [COLING]
IEEE	Institute of Electronics and Electrical Engineers
IJCAI	International Joint Conference on Artificial Intelligence
ITL	Review of the Institute of Applied Linguistics, Louvain
JPRS	Joint Publications Research Service
LRC	Linguistics Research Center, University of Texas, Austin, Texas, USA
Meta	Journal des traducteurs (Translators' Journal)
NTIS	National Technical Information Service, Springfield, Virginia, USA
RADC	Rome Air Development Center, Griffiss Air Force Base, New York, USA

BIBLIOGRAPHY

MT up to 1973

1. ALPAC (1966) *Language and Machines; Computers in Translation and Linguistics,* Washington, DC: Publication 1416, National Academy of Sciences.

2. Bar-Hillel, Y. (1953) 'A Quasi Arithmetical Notation for Syntactic Representation', in *Language,* Vol.29.

3. —— (1960) 'The Present Status of Automatic Translation of Languages', in F. L. Alt, (ed.), *Advances in Computers,* vol.1, Academic Press, New York, pp.91-163.

4. —— (1971) 'Some Reflections on the Present Outlook for High-Quality Machine Translation', in W. P. Lehmann and R. Stachowitz (eds), *Feasibility Study on Fully Automatic High Quality Translation.* Final technical report RADC-TR-71-295. Linguistics Research Center, University of Texas at Austin.

5. Booth, A. D., ed. (1967) *Machine Translation,* North Holland, Amsterdam.

6. Ceccato, S. (1961) *Linguistic Analysis and Programming for Mechanical Translation,* Gordon and Breach, New York.

7. Cleave, J. P. (1957) 'A Type of Program for Mechanical Translation', *MT* 4:3, pp.54-8.

8. Colmerauer, A. et al. (1971) *TAUM-71,* TAUM Group, University of Montreal.

9. Delavenay, E. (1963) *La Machine à Traduire,* Presses Universitaires de France, Paris.

10. Dostert, B. H. (1973) *Users' Evaluation of Machine Translation,* Georgetown MT System, 1963-73. Final technical report RADC-TR-73-239. RADC.

11. Dostert, L. E. (1957) 'Brief History of Machine Translation Research', in L. E. Dostert (ed.), *Research in Machine Translation,* Georgetown University Press, pp.3-10.

12. Edmundson, H. P., ed. (1961) *Proceedings of the National Symposium on Machine Translation,* Prentice-Hall, Englewood.

13. Edmundson, H. P. and Hays, D. G. (1958) 'Research Methodology for Machine Translation', *MT* 5:1, pp.8-15.

14. Garvin, P. L. (1967) 'The Georgetown-IBM Experiment of 1954: An Evaluation in Retrospect', in W. M. Austin (ed.), *Papers in Linguistics in Honor of Leon Dostert,* Mouton, The Hague, pp.46-56.

393

15. Ghizetti, A., ed. (1966) *Automatic Translation of Languages. Papers presented at the NATO Summer School held in Venice, July 1962*, Pergamon Press, Oxford.

16. Gross, M. (1973) *Notes on the Feasibility of High Quality Mechanical Translation*. Final scientific report. Laboratoire D'Automatique Documentaire et Linguistique, University of Paris.

17. Harper, K. E. (1963) 'Machine Translation', in *Current Trends in Linguistics*, vol.1, Mouton, The Hague, pp.133-42.

18. Hays, D. G. (1961) 'Linguistic Research at the Rand Corporation', in H. P. Edmundson (ed.), *Proceedings of the National Symposium on Machine Translation*, Prentice-Hall, Englewood Cliffs.

19. —— (1961) 'Grouping and Dependency Theories', in H. P. Edmundson (ed.), *Proceedings of the National Symposium on Machine Translation*, Prentice-Hall, Englewood Cliffs.

20. Hofmann, T. (1971) 'REF-Bug', in R. Kittredge (ed.), *Working Papers in the Linguistics of Machine Translation*, TAUM Group, University of Montreal.

21. Hoof, H. van (1973) 'Machine Translation', in *International Bibliography of Translation*, Verlag Dokumentation, Munich, pp.464-504.

22. Josselson, H. H. (1971) 'Automatic Translation of Languages since 1960: A Linguist's View', in *Advances in Computers*, vol.11, Academic Press, New York, pp.1-53.

23. Kulagina, O. S. and Mel'čuk, I. A. (1967) 'Automatic Translation: some Theoretical Aspects and the Design of a Translation System', in A. D. Booth (ed.), *Machine Translation*, Mouton, The Hague.

24. Lamb, S. (1961) 'MT at the University of California, Berkeley', in H. P. Edmundson (ed.), *Proceedings of the National Symposium on Machine Translation*, Prentice-Hall, Englewood Cliffs.

25. Lehmann, W. P. and Stachowitz, R. A. (1972) 'Machine Translation in Western Europe: a Survey', *Current Trends in Linguistics*, vol.9, Mouton, The Hague, pp.688-701.

26. —— (1972-75) *Development of German-English Machine Translation System*, Final (annual) report(s), University of Texas at Austin, Linguistics Research Center.

27. Locke, W. N. and Booth, A. D., eds (1956) *Machine Translation of Languages,* Wiley, New York.

28. Macdonald, R. R. (1963) *General Report, 1952-1963,* Occasional Papers on Machine Translation, No.30, Georgetown University Press.

29. Matthews, G. H. (1961) 'The Use of Grammars within the Mechanical Translation Routine', in H. P. Edmundson (ed.), *Proceedings of the National Symposium on Machine Translation*, Prentice-Hall, Englewood Cliffs.

30. Mounin, G. (1964) *La Machine à Traduire Histoire des Problèmes Linguistiques,* Mouton, The Hague.

31. Nikolaeva, T. M. (1958) 'Soviet Developments in Machine Translation: Russian Sentence Analysis', *MT* 5:2, pp.51-9.

32. Oettinger, A. G. (1958) 'A Survey of Soviet Work on Automatic Translation', *MT* 5:3, pp.101-8.

33. Panevova, J. and Sgall, P. (1973) review of P. Garvin, 'On Machine Translation: Selected Papers', *The Prague Bulletin of Mathematical Linguistics* 19, pp.59-64.

34. Pankowicz, Z. L. (1967) *Commentary on ALPAC Report,* RADC, Rome.

35. —— (1973) *Users' Evaluation of Machine Translation,* Georgetown MT System (1963-73). Technical Report RADC-TR-73-239, University of Texas at Austin.

36. Panov, D. J., Ljapunov, A. A. and Muxin, I. S. (1956) *Avtomatizacija perevoda s odnogo jazyka na drugoj,* Izdatel'stvo AN SSSR, Moscow.

37. Panov, D. J. (1960) *Automatic Translation,* Pergamon Press, Oxford.

38. Pendergraft, E. D. (1967) 'Translating Languages', in H. Borko (ed.), *Automated Language Processing; the State of the Art,* Wiley, New York.

39. Reifler, E. (1954) 'The First Conference on Mechanical Translation', *MT* 1:2, pp.23-33.

40. —— (1958) 'The Machine Translation Project at the University of Washington, Seattle', in *Proc. 8th Intern. Congr. Linguists, Oslo,* Washington, pp.514-18.

41. Reynolds, A. C. (1954) 'The Conference on Mechanical Translation', *MT* 1:3, pp.47-55.

42. Rozentsveig, V. Y. (1958) 'The Work on Machine Translation in the Soviet Union, Fourth International Congress of Slavicists Reports, Sept. 1958', *MT* 5:3, pp.95-100.

43. Sakamoto, Y. (1970) 'Automatic Language Processing System – Natural Language Dictionary for Machine Translation', Res. Electrotech. Lab. no.707, pp.1-132.

44. Sinaiko, H. W. and Brislin, R. W. (1973) 'Evaluating Language Translations: Experiments of Three Assessment Methods', *Journal of Applied Psychology* 57, 3, pp.328-34.

45. Sinaiko, H. W. and Klare, G. R. (1972) 'Further Experiments in Language Translation: Readability of Computer Translations', *ITL* 15, pp.1-29.

46. —— (1973) 'Further Experiments in Language Translation: A Second Evaluation of the Readability of Computer Translations', *ITL* 19, pp.29-52.

47. Stout, T. M. (1954) 'Computing Machines for Language Translation', *MT* 1:3, pp.41-6.

48. Vauquois, B. (1966) 'Langages Artificiels, Systèmes Formels et Traduction Automatique', in A. Ghizetti (ed.), *Automatic Translation of Languages. Papers presented at the NATO Summer School held in Venice, July 1962*, Pergamon Press, Oxford.

49. —— (1968) 'Structures Profondes et Traduction Automatique. Le Système du CETA', in *Revue Roumaine de Linguistique*, Bucharest, Tôme XIII, No.2.

50. Vauquois, B., Veillon, G. and Veyrunes, J. (1966) 'Syntax and Interpretation', in *Mechanical Translation*, vol.9, part II.

51. Weaver, W. (1949) *Translation*, New York (mimeographed).

52. Yngve, V. H. 1955) 'Sentence-for-Sentence Translation', *MT* 2:2, pp.29-37.

53. —— (1957) 'A Framework for Syntactic Translation', *MT* 4:3, pp.59-65.

54. —— (1961) 'MT at the Massachussetts Institute of Technology', in H. P. Edmundson (ed.), *Proceedings of the National Symposium on Machine Translation*, Prentice-Hall, Englewood Cliffs.

55. —— (1964) 'Implications of Mechanical Translation Research', *Proc. Amer. Phil. Soc.* 108, 275.

56. —— (1967) 'MT at MIT in 1965', in A. D. Booth (ed.), *Machine Translation*, Mouton, The Hague.

57. Zemanek, H. (1961) 'Methoden der automatischen Sprachübersetzung', *Sprache im technischen Zeitalter*, Heft 2, pp.87-110.

MT 1973 onwards

58. Addis, T. R. (1977) 'Machine Undertanding of Natural Language', *International Journal for Man–Machine Studies* 9, 2, pp.207-22.

59. Amano, S. (1982) *Machine Translation Project at Toshiba Corporation*. Technical note, Toshiba Corporation, R&D Center, Information Systems Laboratory, Kawasaki, Japan.

60. Ambrosi, A., Ayotte, R., Bellert, I., Dansereau, J., Kittredge, R., Poulin, G. and Stewart, G. (1973) 'Le système de traduction automatique de l'Université de Montreal (TAUM)', *Meta* 18, 1-2, pp.277-89.

61. Andreyewski, A. (1980) 'Whither Automation and the Translator', *ATA Chronicle* 9, 2, pp.3-4.

62. —— (1981) 'Translation: Aids, Robots, and Automation', *Meta* 26, 1, pp.57-66.

63. Arnold, D. (1983) 'The Eurotra Interface Structure: A Linguistic View'. Unpublished paper.

64. Arthern, P. J. (1979) 'Machine Translation and Computerized Terminology Systems: A Translator's Viewpoint', in B. M. Snell (ed.), *Translating and the Computer,* North-Holland, Amsterdam.

65. —— (1981) 'Aids Unlimited: The Scope for Machine Aids in a Large Organization', *ASLIIB Proceedings* 33, 7-8.

66. Asakura, S. and Nakano, S. (1974) 'An Experimental System of Machine Translation by the Programmed Sentence Scan Method', *Mem. Chubu Institute of Technology,* vol.10A, pp.73-84.

67. Association for Computing Machinery (1973) 'Natural Language Translation by Computer Emerges from Government Research Stage', *CACM* 16, 3, p.196.

68. Back, M-C. (1980) 'Die Disambiguierung franzoesischer Verben', *LA Neue Folge,* Heft 3.2.

69. Back, T. (1983) 'Der Operator DIHOM: Linguistische Grundlagen, Algorithmus und Leistungstest'. Arbeitspapier c17. sfb-100-c.

70. Barnes, A. M. N. (1983) 'An Investigation into the Syntactic Structures of Abstracts, and the Definition of an "Interlingua" for their Translation by Machine'. msc thesis, Centre for Computational Linguistics, University of Manchester Institute of Science and Technology.

71. Baudot, Jean et al. (1981) 'Un modèle de mini-banque de terminologie bilingue', *Meta* 26, 4, pp.315-31.

72. Bektaev, K. B., Kenesbaev, S. K. and Piotrovskii, R. G. (1973) 'Discussion of Engineering Linguistics'. Joint Publications Research Service, Arlington, Virginia.

73. Bennett, W. S. (1982) 'The Linguistic Component of METAL'. Working paper lrc-82-2, Linguistics Research Center, University of Texas at Austin.

74. Berg, H-H. (1973) 'Uebersicht zur maschinellen Verarbeitung von Texten des "neueren Deutsch"', in H-H. Berg and W. Lenders (eds), *Bericht ueber die Arbeiten der Clearingstellen zum aelteren und neueren Deutsch,* Institut fuer deutsche Sprache, Abteilung Linguistische Datenverarbeitung, Mannheim, pp.8-69.

75. Bertsch, E. (1977) 'Automatische Lexikographie, Analyse und Uebersetzung: Resume einer Fachtagung (Automatic Lexicography, Analysis, and Translation: Summary of a Symposium)', *Sprache und Datenverarbeitung* 1, 1, pp.74-5.

76. Bevan, N. (1980) 'Human Factors in the Use of Eurodicautom and Systran'. Report on the CETIL workshop for the Commission of European Communities, Luxembourg.

77. —— (1982) 'Psychological and Ergonomic Factors in Machine Translation', in V. Lawson (ed.), *Practical Experience of Machine Translation,* North-Holland, Amsterdam, pp.75-8.

78. Blatt, A. (1983) 'Nominalanalyse des Englischen im Uebersetzungs-system SUSY', *Dokumentation K7*, SFB-100-K.

79. Boitet, C. (1976) 'Towards an Adaptive and Interactive System in Automatic Translation', in A. Zampolli (ed.), *Computational and Mathematical Linguistics: Proceedings of the ICCL, Pisa, 1973*, Florence.

80. —— (1976) 'Méthodes sémantiques en traduction automatique', *TA Informations* 1, pp.3-42.

81. —— (1976) 'Problèmes actuels en traduction automatique: un essai de reponse', *AJCL* 2, microfiche 48, p.68. From *Proceedings of the Sixth ICCL [COLING 76]*, Ottawa.

82. —— (1976) 'Un essai de reponse à quelques questions théoriques et pratiques liées à la traduction automatique: Définition d'un système prototype', doctoral thesis, University of Grenoble.

83. —— (1977) 'Wo steht die GETA Anfang 1977?', in *Dritter Europaeischer Kongress ueber Dokumentationssysteme und -netze, Luxembourg*, vol.2, Munich.

84. —— (1977) 'Où en est le GETA debut 1977?', *TA Informations* 18, pp.3-20.

85. —— (1977) 'Mechanical Translation and the Problem of Under-standing Natural Languages', *Table-Ronde IFIP*, Toronto, and *Colloque Franco-Sovietique sur la TA*, Moscow.

86. Boitet, C., Chatelin, P. and Daun Fraga, P. (1980) 'Present and Future Paradigms in the Automatized Translation of Natural Languages', *Proceedings of the Eighth ICCL [COLING 80]*, Tokyo, pp.430-6.

87. Boitet, C. and Gerber (1984) 'Expert Systems and Other New Techniques in MT', *Proceedings of COLING 84*, Stanford.

88. Boitet, C., Guillaume, P. and Quezel-Ambrunaz, M. (1978) 'Manipulation d'arborescences et parallélisme: le système ROBRA', *Proceedings of COLING 80*, Bergen.

89. —— (1982) 'Implementation and Conversational Environment of ARIANE-78.4, an Integrated System for Automated Translation and Human Revision', *Proceedings of the Ninth ICCL [COLING 82]*, Prague, pp.19-27.

90. Boitet, C. and Nedobejkine, N. (1980) 'Russian–French at GETA: Outline of the Method and Detailed Example', *Proceedings of the Eighth ICCL [COLING 80]*, Tokyo.

91. —— (1981) 'Recent Developments in Russian-French Machine Translation at Grenoble', *Linguistics* vol.19-3/4.

92. Bonner, M. (1981) 'Report of the Saarbruecken Colloquium, 2. Internationales Kolloquium Maschinelle Uebersetzung, Lexiko-graphie und Analyse, Saarbruecken', *ALLC Bulletin* 9, 3, pp.11-12.

93. Bosted, D. A. (1982) 'Quality Control Procedures in Modification of the Air Force Russian-English MT System', in V. Lawson (ed.), *Practical Experience of Machine Translation*, North-Holland, Amsterdam, pp.129-33.

94. Bourbeau,·L. (1974) *La nominalisation*. T A U M Group, University of Montreal.

95. —— ed. (1981) *Linguistic Documentation of the TAUM-AVIA-TION Translation System*, T A U M Group, University of Montreal.

96. Brandtner, K. (1980) 'Realisierung der Analyse direkter Komplemente', *LA Neue Folge*, Heft 3.2.

97. Brinkmann, K-H. (1974) 'On the Development and Operation of Terminology Data Banks as a Requirement for Machine-Aided Translation', *Nachrichten duer Dokumentation* 25, 3, pp.99-104.

98. —— (1979) 'Ueberlegungen zum Aufbau und Betrieb von Terminologie-Datenbanken als Voraussetzung der maschinenunterstuetzten Uebersetzung', *Nachrichten fuer Dokumentation* 25, pp.99-105.

99. —— (1974) 'Terminologists, Lexicographers, and Computers', *Philips Terminology Bulletin* 3, 3/4, pp.8-9.

100. —— (1975) 'The T E A M Program System', *Philips Terminology Bulletin* 4, 2/3, pp.20-37.

101. —— (1978) 'Perspectives of Machine Translation', *Nachrichten fuer Dokumentation* 29, 3, pp.104-8.

102. —— (1979) 'Perspectives d'avenir de la traduction automatique (Future Prospects for Machine Translation)', *Meta* 24, 3, pp.315-25.

103. —— (1980) 'Terminology Data Banks as a Basis for High-Quality Translation', *Proceedings of the Eighth ICCL [COLING 80]*, Tokyo.

104. Brinkmann, K-H. and Tanke, E. (1976) 'The T E A M Program System and International Co-Operation in Terminology/Le système des programmes T E A M et la cooperation internationale dans le domaine de la terminologie', *International Co-Operation in Terminology, First Infoterm Symposium, Vienna, 1975*, Infoterm Series, vol.3, Munich, pp.180-92.

105. Brockhaus, K. (1976) 'Zum formalen Aufbau einer Grammatik fuer automatische Uebersetzungsverfahren (unter Beruecksichtigung der Programmierung)', paper presented at the Internationales Kolloquium *Automatische Lexigraphie, Analyse und Uebersetzung*, Saarbruecken.

106. Brown, G. P. (1974) 'Some Problems in German to English Machine Translation', master's thesis, MIT.

107. Bruderer, H. E. (1975) 'Maschinelle Sprachuebersetzung fuer den Apollo-Sojus-Flug', *Neue Zuercher Zeitung* 156, p.42.

108. —— (1975) 'The Present State of Machine Translation and Machine-Aided Translation', *ALLC Bulletin* 3, 3, pp.258-61.

109. —— (1975) 'Control of an Automatic Translation System in the Computer Centre of Zurich University', *ALLC Bulletin* 3, 3, pp.197-200.

110. —— (1975) 'Vorfuehrung eines automatischen Uebersetzungsverfahrens im Rechenzentrum der Universitaet Zuerich (Developing an Automatic Translation System at the Computing Centre of the University of Zurich)', *ALLC Bulletin* 3, pp.197-200.

111. —— (1976) 'A Register of Machine Translation and Machine-Aided Translation Projects', *Computers and the Humanities* 10, 1, p.45.

112. —— (1977) 'Automatic Language Processing in Switzerland: A State of the Art Report (Nov. 1976)', *Sprache und Datenverarbeitung* 1, 1, pp.76-77.

113. —— (1977) 'The Present State of Machine and Machine-Assisted Translation', in *Commission of the European Communities, Third European Congress on Information Systems and Networks: Overcoming the Language Barrier,* vol.1, Verlag Dokumentation, Munich, pp.529-56.

114. —— (1977) *Handbook of Machine Translation and Machine-Aided Translation: Automatic Translation of Natural Languages and Multilingual Terminology Data Banks,* North-Holland, Amsterdam.

115. —— (1928) *Handbuch der maschi:nellen und maschinenuntersteutzten Sprachuebersetzung: Automatische Uebersetzung natuerlicher Sprachen und mehrsprachige Terminologiedatenbanken.* Verlag Dokumentation, Munich.

116. Bruderer, H. E. and Hays, D. G. (1975) 'LATSEC Shows MT in Zuerich', *AJCL* 4, microfiche 29, p.9.

117. Buchmann, B., Warwick, S. and Shann, P. (1984) 'Design of a Machine Translation System for a Sublanguage', *Proceedings of COLING 84,* Stanford.

118. Burden, D. (1981) 'Natural Human Languages Automatically Translated by Computer: The SYSTRAN II System', *Computers and People* 30, 5-6, pp.12-15, 24.

119. Burge, J. (1978) 'The TARGET Project's Interactive Computerized Multilingual Dictionary', *AJCL* and the Translation and English Language Center, Carnegie-Mellon University, Pittsburgh.

120. Bush, C. (1974) 'Structural Passives and the "Massive Passive" Transfer', Master's thesis, Brigham Young University.

121. Buttelmann, H. W. (1974) *Semantic Directed Translation of Context Free Languages,* technical report, Computer and Information Science Research Center, Ohio State University, Columbus.

122. Chafe, W. L. (1974) 'An Approach to Verbalization and Translation by Machine', final report RADC-TR-74-271, Department of Linguistics, University of California at Berkeley. Also *AJCL* (1974), microfiche 20, p.10.

123. —— (1976) 'Foundations of Machine Translation, Linguistics', *AJCL* 2, microfiche 46, pp.8-9.

124. Chaloupka, B. (1976) 'Xonics MT System', *AJCL* 2, microfiche 46, pp.37-9.

125. Chandioux, J. (1976) 'Leibniz, A Multilingual System', *AJCL* 2, microfiche 46, p.25.

126. —— (1976) 'METEO: An Operational System for the Translation of Public Weather Forecasts', *AJCL* 2, microfiche 46, pp.27-36.

127. —— (1976) 'METEO: un système operationnel pour la traduction automatique des bulletins meteorologiques destinés au grand public', *Meta* 21, 2, June, pp.127-33.

128. Chandioux, J. and Guéraud, M. (1981) 'METEO: un système à l'épreuve du temps', *Meta* 26, 1, pp.18-22.

129. Chauché, J. (1975) 'Les systèmes ATEF et CETA', *TA Informations* 16, 2, pp.27-38.

130. —— (1975) 'The ATEF and CETA Systems', *AJCL* 2, microfiche 17, pp.21-40.

131. Chauché, J., Guillaume, P. and Quezel-Ambrunaz, M. (1973) 'Le système ATEF', GETA report G-2600-A, University of Grenoble.

132. Chaumier, J. et al. (1977) 'Evaluation of the SYSTRAN Automatic Translation System', Report no.5, study carried out for the Commission of the European Communities, Bureau Marcel van Dijk, Paris.

133. Chaumier, J., Mallen, M. C. and Van Slype, G. (1977) 'Evaluation du système de traduction automatique SYSTRAN', Commission of the European Communities, Luxembourg.

134. Chevalier, M., Dansereau, J. and Poulin, G. (1978) 'TAUM-METEO: description du système', TAUM Group, University of Montreal.

135. Chevalier, M., Isabelle, P., Labelle, F. and Laîné, C. (1981) 'La traductologie appliquée à la traduction automatique', *Meta* 26, 1, pp.35-47.

136. Chinese-English Translation Assistance (1974) 'Translation Project for Chinese Terms', *The Linguistic Reporter* 16, 2, p.2.

137. Clark, R. (1981) 'Machine Aids: A Small User's Reaction', *Machine Aids for Translators: ASLIB Proceedings* 33, p.7-8.

138. Commission of the European Communities (1976) *Plan of Action for the Improvement of the Transfer of Information between European Languages,* Brussels.

139. —— (1977-8) *Overcoming the Language Barrier: Proceedings of the Third European Congress on Information Systems and Networks, Luxembourg, 1977,* Verlag Dokumentation, Munich, 1977 (vol.1)-1978 (vol.2).

140. —— (1977) *Evaluation of the SYSTRAN Automatic Translation System,* Report no.5: *Synthesis of the Economic and Qualitative Evaluations,* Bureau Marcel van Dijk, Brussels.

141. Couture, B. (1976) 'La Banque de terminologie au service de l'entreprise', *Meta* 21, 1, pp.100-9.

142. Cox, H. N., Dunfee, W. P., Schaad, T. A. and Vogel, L. A. (1979) 'Programmer's Panel Address Translation', *IBM Tech. Disclosure Bulletin* 22, 3, pp.652-3.

143. Crawford, T. D. (1974) 'A Review of the Cardiff Machine Translation Project', in A. Zampolli (ed.), *Computational and Mathematical Linguistics: Proceedings of the Third Symposium on the Use of Computers in Literary and Linguistic Research,* Cardiff.

144. —— (1974) 'Machine Translation: The Development of BABEL Program', *Lettera* 8, pp.54-60.

145. —— (1974) 'Project BABEL: Machine Translation with English as the Target Language', *Proceedings of the Third Symposium on the Use of Computers in Literary and Linguistic Research,* Cardiff.

146. Crespi Reghizzi, S. and Mandrioli, D. (1978) 'Basic Problems of Perspective of Automatic Elaboration of Natural Languages', *Riv. Inf.* 8, 1, pp.5-48.

147. Culhane, P. T. (1977) 'Semantics via Machine Translation', *Russian Language Journal* 31, 108, pp.35-42.

148. Daley, D. H. and Vachino, R. F. (1973) 'The West German Federal Bureau of Languages and Machine-Aided Translation in Germany', *Federal Linguist* 5, 3-4, pp.14-18.

149. Dansereau, J. (1978) 'Traduction automatique et terminologie automatique', *Meta* 23, 2, pp.132-40.

150. Desroches, S. and Bourbeau, L. (1980) *TAUM-BESCH,* TAUM Group, University of Montreal.

151. *Dokumentations-System TITUS II* (1974) Zentralstelle fuer Textildokumentation und Textilinformation, Duesseldorf.

152. Du Feu, V. M. (1975) 'Acta Mathematica Sinica (review)', *ALLC Bulletin* 3, 3, p.277.

153. Dubuc, R. and Gregoire, J-F. (1974) 'Banque de terminologie et traduction', *Babel* 20, 4, pp.180-4.

154. Duckitt, P. (1981) 'Translating and Online', *Machine Aids for Translators: ASLIB Proceedings* 33, 7-8.

155. Ducrot, J-M. (1974) 'Perspectives et avantages offerts par la traduction automatique des analyses de documents selon la mèthode TITUS', in *1er Congrès national francais sur l'information et la documentation, Communications,* Paris, pp.321-33.

156. —— 'The TITUS II System', undated technical report.

157. —— 'TITUS IV: General Description of the Method', pp.1-39, undated technical report.

158. —— 'TITUS IV: Système de traduction automatique et simultanée en 4 langues', Institut Textile de France, undated technical report.

159. Earp, R. and Cherinka, J. (1978) 'Getting Started in a Natural Language Translation Project', *Journal of Educational Data Processing* 15, 2, pp.25-34.

160. Eggers, H. (1976) 'Probleme der Identifikationsgrammatik und ihrer Anwendung', *Vorabdrucke des Internationalen Kolloquiums 'Automatische Lexikographie, Analyse und Uebersetzung',* Saarbruecken, pp.24-9.

161. —— (1980) *Maschinelle Uebersetzung, Lexikographie und Analyse,* 2 vols, Saarland University, Saarbruecken.

162. —— (1981) 'Das Lemmatisierungssystem SALEM des Sonderforschungsbereichs 90 "Elektronische Sprachforschung" in Saarbruecken', *ALLC Bulletin* 9, 2, pp.9-15.

163. Eggers, H., Luckhardt, H-D., Maas, H. D. and Weissgerber, M. (1980) *SALEM – ein Verfahren zur automatischen Lemmatisierung deutscher Texte,* Max Niemeyer Verlag, Tuebingen.

164. Elliston, J. S. G. (1979) 'Computer Aided Translation: A Business Viewpoint', in B. M. Snell (ed.), *Translating and the Computer,* North-Holland, Amsterdam, pp.149-58.

165. Engelberg, K-J., Hauenschild, C., Knöpfler, S. and Pause, P.E. (1984) 'CON³TRA. Ein prozedurales Modell des Textverstehens für die Uebersetzung', *Sonderforschungsbereich* 99, Bd.93.

166. EUROTRA Council Decision (1982) *Official Journal of the European Communities,* No.L317/21.

167. EUROTRA Reports ET-9 *Final Report* (1982) (Johnson, Krauwer, Maas, Maegaard).

168. —— ETS-1 *Final Report* (1983) (Johnson, Krauwer, Rosner, Varile).

169. —— *Syntax and Semantics of the Eurotra Formalism: Preliminary Version* (1984) (Maas and Maegaard).

170. —— ETL-4-GB *Final Report* (1984) (the English Language Group).

171. —— ETL-4-DK *Final Report* (1984) (the Danish Language Group).

172. —— ETL-3 *Final Report* (1984) (Arnold, Des Tombe, Jaspaert).

173. —— *EUROTRA Linguistic Specifications: Version 3* (1985) (Arnold, Des Tombe, Jaspaert).

174. Faiss, K. (1973) 'Uebersetzung und Sprachwissenschaft – eine Orientierung', *Babel* 19, 2, pp.75-86.

175. Feng, Z. W. (1982) 'Memoire pour une tentative de traduction multilingue du Chinois en Francais, Anglais, Japonais, Russe et Allemand', paper presented at the Ninth ICCL [COLING 82], Prague.

176. Francois, P. (1976-7) 'La banque de données terminologiques EURODICAUTOM de la Commission des Communautes Europeennes: Le resultat d'un travail pluridisciplinaire', *Comptes rendus du Colloque international de terminologie, Terminologies 76, Paris-La Defense* (1976), Association francaise de terminologie, Paris (1977).

177. Freigang, K-H. (1980) 'Zur Entwicklung der Homographenanalyse fuer das Englische', *LA Neue Folge,* Heft 3.2.

178. —— et al. (1979) 'Der Stand der Forschung auf dem Gebiet der maschinellen Uebersetzung', Saarland University, Saarbruecken.

179. Freigang, K-H. and Schmitz, K-D. (1982) 'Woerterbucherstellung, Woerterbuchsuche und Flexionsanalyse im Uebersetzungssystem Englisch-Deutsch', *Dokumentation* K2, SFB-100-K.

180. Fukushima, M. and Arita, H. (1983) 'A MAHA Machine Translation System from Japanese to English', technical note, Central Research Laboratory, Mitsubishi Electric Corp., Hyogo.

181. Ganeshsundaram, P. C. (1978) *Automatic Translation from English into Hindi using the PCG-Theory and a Scheme of Pre-Editing,* Series: AILA 1978 0124, Foreign Languages Section, Indian Institute of Science, Bangalore.

182. Geens, D. (1981) 'Machine Translation: Evaluation of the Eurotra Interface Mechanism', *Universite Libre de Bruxelle Rapport d'Activites de l'Institut de Phonetique* 15, pp.67-108.

183. Gerber, R. (1984) 'Etude des possibilités de coopération entre un système fondé sur des techniques de compréhension implicite (système logico-syntaxique) et un système fondé sur des techniques de compréhension explicite (système expert)', thèse de 3è cycle, Grenoble.

184. Gerhardt, T. C. (1980) Semantische Disambiguierung – technische Beschreibung', *LA Neue Folge,* Heft 3.2.
—— (1983) *SUSY – Handbuch fuer Semantische Disambiguierung,* Dokumentation A2/3.1, A2/3.2, SFB-100-A2..

185. Gerrard, M. (1976) 'Regular Use of Machine Translation of Russian at Oak Ridge National Laboratory', *AJCL* microfiche 46, pp.53-6.

186. Gervais, A. et DG de la Planification, de l'Evaluation et de la Verification (1980) *Rapport final d'evaluation du systeme pilote de traduction automatique TAUM-AVIATION,* Canada, Secretariat d'Etat.

187. Gibb, D. K. (1976) 'Interactive Analysis: A Synergistic Approach', *AJCL* 2, microfiche 48, p.45.

188. Glas, R. (1980) 'Disambiguierung deutscher Praepositionen', *LA Neue Folge,* Heft 3.2.

189. Gobeil, F. (1981) 'La traduction automatique au Canada', *L'Actualité terminologique* 14, 5.

190. Godden, K. S. (1982) 'Montague Grammar and Machine Translation between Thai and Engish', Ph D dissertation, University of Kansas.

191. Goetschalckx, J. (1974) 'Translation, Terminology, and Documentation in International Organizations', *Babel* 20, 4, pp.185-7.

192. —— (1979) 'Eurodicautom', in B. Snell (ed.), *Translating and the Computer,* pp.71-6, North-Holland, Amsterdam.

193. Goshawke, W. (1974) 'Spoken Languages Universal Numeric Translation (S L U N T)', mimeo.

194. Green, R. (1982) 'The M T Errors Which Cause Most Trouble to Posteditors', in V. Lawson (ed.), *Practical Experience of Machine Translation,* North-Holland, Amsterdam, pp.101-4.

195. Greenfield, C. C. and Serain, D. (1977) 'Machine-Aided Translation: From Terminology Banks to Interactive Translation Systems', Department of Computer Science, Carnegie-Mellon University, Pittsburgh, Pennsylvania, and Institut de Recherche en Informatique et Automatique, Paris, France, and E D R S.

196. Halliday, T. C. and Briss, E. A. (1977) 'The Evaluation and Systems Analysis of the S Y S T R A N Machine Translation System', final technical report R A D C-T R-76-399, November 1974-August 1976, Battelle Columbus Laboratories, Ohio, and N T I S report A D-A036 070/1G A.

197. Hardt, S. L. (1983) 'Automated Text Translation and the Organization of Conceptual and Lexical Information', *Proceedings of Trends and Applications 1983: Automating Intelligent Behavior – Applications and Frontiers,* I E E E, New York, pp.261-2.

198. Hauenschild, C., Huckert, E. and Maier, R. (1978) 'S A L A T: Entwurf eines automatischen Uebersetzungssystems', Arbeitspapier C E R 1, Sonderforschungsbereich 99, Teilprojekt A 2, Heidelberg University.

199. Hauenschild, C., Huckert, E. and Maier, R. (1979) 'S A L A T: Machine Translation via Semantic Representation', in R. Baeuerle, U. Egli and A. von Stechow (eds), *Semantics from Different Points of View,* Springer Verlag, Berlin, pp.324-52.

200. Hays, D. G. and Mathias, J., eds (1976) 'Summary Proceedings of the F B I S Seminar on Machine Translation, Rosslyn, Virginia', *AJCL* 2, microfiche 46, pp.1-96.

201. Henisz-Dostert, B. (1979) *A Review of the State of Machine Translation*, EDRS.

202. Henisz-Dostert, B., Macdonald, R. and Zarechnak, M. (1979) *Machine Translation*, Mouton, The Hague.

203. Heriard Dubreuil, S. (1979) 'Automatic Translation and Computer-Assisted Translation', *Inf. and Gestion* 107, pp.49-54.

204. Herkovits, A. (1973) *The Generation of French from a Semantic Representation*, Stanford University, California, Department of Computer Science.

205. Hitachi Ltd, Systems Development Laboratory (1981) *Research on Machine Translation in Hitachi Ltd*, Technical note, Tama-ku, Kawasaki.

206. Hoffman, T. (1976) 'Semantics in Aid of Automatic Translation', *Proceedings of the Sixth ICCL [COLING 76]*, Ottawa, and *AJCL* 2, microfiche 48, p.27.

207. —— (1978) 'Semantics in Aid of Automatic Translation', *Proceedings of the Seventh ICCL [COLING 78]*, Bergen.

208. Hubert, J. M. (1978) 'Constitution de lexiques multilingues pour traduction automatique. Problèmes posés dans le cas de TITUS', Communication au Congrès Européen de l'IAALD, Hambourg.

209. Hundt, M. G. (1982) 'Working with the Weidner Machine-Aided Translation System', in V. Lawson (ed.), *Practical Experience of Machine Translation*, North-Holland, Amsterdam, pp.45-51.

210. Hutchins, W. J. (1978) 'Progress in Documentation: Machine Translation and Machine-Aided Translation', *Journal of Documentation* 34, 2, pp.119-59.

211. —— (1979) 'Linguistic Models in Machine Translation', *University of East Anglia Papers in Linguistics* 9, pp.29-52.

212. —— (1982) 'The Evolution of Machine Translation Systems', in V. Lawson (ed.), *Practical Experience of Machine Translation*, North-Holland, Amsterdam, pp.21-37.

213. Hutton, F. C. (1976) 'Georgetown University MT System Usage', *AJCL* 2, microfiche 46, p.57.

214. Isabelle, P. et al. (1978) 'TAUM-AVIATION: description d'un système de traduction automatisée des manuels d'entretien en aéronautique', *Proceedings of the Seventh ICCL [COLING 78]*, Bergen.

215. Isabelle, P. (1981) *A Linguistic Description of the Taum-Aviation Computerized Translation System*, Groupe de recherche en traduction automatique, University of Montreal.

216. —— (1981) *Some Views on the Future of the TAUM Group and the TAUM-AVIATION System*, TAUM Group, University of Montreal, p.35.

217. Johnson, R. L. (1980) 'Contemporary Perspectives in Machine Translation', in S. Hanon and V. H. Pedersen (eds), *Human Translation Machine Translation* (Noter og Kommentarer 39), pp.133-47, Romansk Institut, Odense Universitut.

218. Jordan, S. R., Brown, A. F. R. and Hutton, F. C. (1976) 'Computerized Russian Translation at ORNL', in *Proceedings of the ASIS Annual Meeting, San Francisco,* p.163.

219. —— (1977) 'Computerized Russian Translation at ORNL', *ASIS Journal* 28, 1, pp.26-33.

220. —— (1977) *SLC Primer for Russian Translation Users,* Computer Sciences Division, Oak Ridge National Laboratory.

221. JPRS (1977) *Machine-Assisted Translation in West Germany* (translations of articles in German by various authors), NTIS document JPRS-68726.

222. Kay, M. (1973/5) 'Automatic Translation of Natural Languages', *Daedalus* 102, 3, Summer 1973, pp.217-30. Reprinted in E. Haugen and M. Bloomfield (eds), *Language as a Human Problem,* Lutterworth, Guildford, pp.219-32.

223. —— (1976) 'The Proper Place of Man and Machines in Translation', *AJCL,* microfiche 46.

224. —— (1982) 'Machine Translation', *AJCL* 8, 2, pp.74-8.

225. —— (1980) *The Proper Place of Men and Machines in Language and Translation,* technical report CSL-80-11, Xerox Palo Alto Research Center, California.

226. Kertesz, F. (1974) 'How to Cope with the Foreign-Language Problems: Experience Gained at a Multidisciplinary Laboratory', *ASIS Journal* 25, 2, pp.86-104.

227. King, M. (1981) 'Design Characteristic of a Machine Translation System', *Proceedings of the Seventh IJCAI, Vancouver,* vol.1, pp.43-6.

228. —— (1981) 'EUROTRA – A European System for Machine Translation', *Lebende Sprachen* 26, pp.12-14.

229. —— (1982) 'EUROTRA: An Attempt to Achieve Multilingual MT', in V. Lawson (ed.), *Practical Experience of Machine Translation,* North-Holland, Amsterdam, pp.139-47.

230. King, M. and Perschke, S. (1982) 'EUROTRA and its Objectives', *Multilingua* 1, 1, pp.27-32.

231. Kirschner, Z. (1982) 'On a Device in Dictionary Operations in Machine Translation', *Proceedings of the Ninth ICCL [COLING 82],* Prague, pp.157-60.

232. Kittredge, R. (1982) 'Variation and Homogeneity of Sublanguages', in R. Kittredge and J. Lehrberger (eds), *Sublanguage: Studies of Language in Restricted Semantic Domains,* De Gruyter, Berlin, pp.107-37.

233. —— (1983) *Sublanguage-Specific Computer Aids to Translation: A Survey of the Most Promising Application Areas,* Office of the State, Ottawa.

234. Kittredge, R., Ayotte, R., Stewart, G., Dansereau, J., Poulin, G., Ambrosi, A. and Bellert, I. (1973) *TAUM 73. Projet de Traduction Automatique de l'Universite de Montreal,* internal publication, pp.1-262.

235. Kittredge, R., Bourbeau, L. and Isabelle, P. (1976) 'Design and Implementation of an English-French Transfer Grammar', *Proceedings of the Sixth ICCL [COLING 76], Ottawa,* and *AJCL* 2, microfiche 48, p.69.

236. Knowles, F. E. (1979) 'Error Analysis of Systran Output: A Suggested Criterion for the "Internal" Evaluation of Translation Quality and a Possible Corrective for System Design', in B. M. Snell (ed.), *Translating and the Computer,* North-Holland, Amsterdam, pp.109-33.

237. —— (1979) 'Re-emergence of Interest in Machine Translation (MT)', paper presented at *BCS 79,* London.

238. —— (1982) 'The Pivotal Role of the Various Dictionaries in an MT System', in V. Lawson (ed.), *Practical Experience of Machine Translation,* North-Holland, Amsterdam, pp.149-62.

239. Korovina, T. I. (1978) 'Multilingual Word-for-Word Translation Using a Display', *Cybernetics* 14, 2, pp.282-3.

240. Kraass, K-H. (1974) 'Computer sind Dolmetscher der Zukunft', *Zeitschrift fuer Phonetik, Sprachwissenschaft und Kommunikationsforschung* 27, 4, pp.334-5.

241. Krollmann, F. (1974) 'Data Processing at the Translator's Service', *Babel* 20, 3, pp.121-9.

242. —— (1976) 'Translation Aids', *AJCL* 2, microfiche 48, p.58.

243. —— (1977) 'User Aspects of an Automatic Aid to Translation as Employed in a Large Translation Service', in Commission of the European Communities, *Overcoming the Language Barrier,* Verlag Dokumentation, Munich.

244. —— (1977) 'Benutzeraspekte beim Umgang mit einer maschinellen Uebersetzungshilfe aus der Sicht eines grossen Sprachendienstes', *Dritter Europaeischer Kongress ueber Dokumentationssysteme und -netze, Luxembourg,* vol.1, Munich, pp.201-14.

245. —— (1981) 'Computer Aids to Translation', *Meta* 26, 1, pp.85-94.

246. Kulagina, O. S. (1976) 'History and Present State of Machine Translation', *Cybernetics,* November-December, pp.937-44.

247. Labelle, F. (1981) *Tests et opérations de traduction,* TAUM Group, University of Montreal.

248. Landsbergen, J. (1982) 'Machine Translation Based on Logically Isomorphic Montague Grammars', *Proceedings of the Ninth ICCL [COLING 82]*, Prague, pp.175-81.

249. LATSEC (1975) 'Technical Proposal for Further Development of the SYSTRAN English-French Machine Translation System and Initial Development of the SYSTRAN French-English Translation System', Latsec Incorporated, La Jolla, California.

250. Lavorel, B. (1982) 'Experience in English-French Post-Editing', in V. Lawson (ed.), *Practical Experience of Machine Translation*, North-Holland, Amsterdam, pp.105-9.

251. Lawson, V. (1979) 'Tigers and Polar Bears, Or: Translating and the Computer', *The Incorporated Linguist* 18, pp.81-5.

252. —— (1980) 'Final Report on EEC Study Contract TH-21 (Feasibility Study on the Applicability of the Systran System of Computer-Aided Translation to Patent Texts)', Commission of the European Communities, Luxembourg, 1980.

253. —— (1981) 'Introduction: Help from the Computer', Machine Aids for Translators: *ASLIB Proceedings* 33, 7-8, pp.265-7.

254. —— (1982) 'Machine Translation and People', in V. Lawson (ed.), *Practical Experience of Machine Translation*, North-Holland, Amsterdam, pp.3-9.

255. —— ed. (1982) *Practical Experience of Machine Translation* (Proceedings of a conference held in London 1981), North-Holland, Amsterdam.

256. Lee, K-F. (1976) review of *Machine Translation,* by S-C. Loh et al., *ALLC Bulletin* 4, 3, pp.267-8.

257. Lehmann, W. P. (1978) 'Machine Translation', Encyclopedia of Computer Science and Technology, vol.10, pp.151-64, Dekker, New York.

258. Lehmann, W. P., Bennett, W. S. and Slocum, J. (1980) *Machine Translation: 1980,* unpublished paper, Linguistics Research Center, University of Texas at Austin.

259. Lehmann, W. P., Bennett, W. S., Slocum, J., Smith, H., Pfluger, S. M. V. and Eveland, S. A. (1981) 'The METAL System', final technical report RADC-TR-80-374, Linguistics Research Center, University of Texas at Austin, and NTIS report AO-97896.

260. Lehrberger, J. (1981) *The Linguistic Model: General Aspects,* TAUM Group, University of Montreal.

261. —— ed. (1981) *Possibilités d'extension du système TAUM-AVIA-TION,* TAUM Group, University of Montreal.

262. —— (1982) 'Automatic Translation and the Concept of Sub-language', in R. Kittredge and J. Lehrberger (eds), *Sublanguage: Studies of Language in Restricted Semantic Domains,* De Gruyter, Berlin, pp.81-106.

263. Lenders, W. (1975) 'Das Problem des Verstehens in der maschin-ellen Uebersetzung', *Abstracts of the Fourth International Congress on Applied Linguistics,* Stuttgart, p.22.

264. Lippmann, E. O. (1974) 'On Computational Linguistics and Computer-Aided Translation', *ATA Chronicle* 3, 5, pp.7-8.

265. —— (1975) 'On-Line Generation of Terminological Digests in Language Translation: An Aid in Terminology Processing', *IEEE Transactions on Professional Communications, PC*-18, no.4, pp.309-19.

266. —— (1976) 'Experimental On-Line Computer Aids for the Human Translator', *AJCL* 2, microfiche 46, pp.11-13.

267. —— (1977) 'Computer Aids for the Human Translator', report presented at the Eighth World Congress of FIT, Montreal.

268. —— (1980) 'Human Factors in the Design of Computer Aids for Translators', *ATA Chronicle* 9, 6, p.5.

269. Liu, Y-Q. (1980) 'Machine Translation in China – A Report', *ALLC Journal* 1, 2, p.60-6.

270. —— (1981) 'The System of Intermediate Constituents in Machine Translation from Foreign Languages into Chinese', paper presented at the Conference on Chinese Language Use, The Australian National University.

271. —— (1982) 'Aspects of Chinese Information Processing', paper presented at the Conference on International Cooperation in Chinese Bibliographical Automation, Canberra.

272. Ljudskanov, A. (1975) *Mensch und Maschine als Uebersetzer,* Niemeyer, Halle.

273. Locke, W. H. (1975) 'Machine Translation', Encyclopedia of Library and Information Science, vol.16, Marcel Dekker, New York, pp.414-44.

274. LOGOS (1973) 'Proposals to Systems Dimensions Limited for Operational English to French Machine Translation System', Logos Development Corporation, New Hampton, New York.

275. Loh, S-C. (1975) 'Final Report on Machine Translation Project', Department of Computer Science, Chinese University of Hong Kong.

276. —— (1975) 'Computer Translation of Chinese Journals', *AJCL* 3, microfiche 22, p.43.

277. —— (1975) 'Computer Translation of Chinese Scientific Journals', *ALLC Bulletin* 3, 3, p.258.

278. —— (1976) 'CULT: Chinese University Language Translator', in FBIS Seminar on Machine Translation, *AJCL,* microfiche 46, pp.46-50.

279. —— (1976) 'Machine Translation: Past, Present, and Future', *ALLC Bulletin* 4, 2, pp.105-14.

280. —— (1976) 'Translation of Three Chinese Scientific Texts into English by Computer', *ALLC Bulletin* 4, 2, pp.104-5.

281. Loh, S-C. and Kong, L. (1977) 'Automatische Uebersetzung chinesischer wissenschaftlicher Zeitschriften', *Dritter Europaeischer Kongress ueber Dokumentationssysteme und Dokumentationsnetze, Luxembourg,* vol.1, Munich, pp.563-78.

282. —— (1977) 'Computer Translation of Chinese Scientific Journals', in Commission of the European Communities, *Overcoming the Language Barrier,* Verlag Dokumentation, Munich, pp.631-46.

283. Loh, S-C., Kong, L. and Hung, H-S. (1978) 'Machine Translation of Chinese Mathematical Articles', *ALLC Bulletin* 6, 2, pp.111-20.

284. Loh, S-C. and Kong, L. (1979) 'An Interactive On-Line Machine Translation System (Chinese into English)', in B. M. Snell (ed.), *Translating and the Computer,* North-Holland, Amsterdam, pp.135-48.

285. Luckhardt, H-D. (1980) 'Weshalb SUSY mehrsprachig ist', *LA Neue Folge,* Heft 3.2, Saarbruecken.

286. —— (1980) 'Automatische Segmentierung von Saetzen', *LA Neue Folge,* Heft 3.2, Saarbruecken.

287. —— (1982) 'SATAN-Test. Beschreibung der Vorgehensweise und der Ergebnisse von Tests der deutschen Komponente des Saarbruecker automatischen Textanalysesystems SATAN', *LA Neue Folge,* Heft 6.

288. —— (1982) 'SUSY – Capabilities and Range of Application', *Multilingua* 1-4.

289. —— (1983) 'Probleme bei der Bewertung eines MU-Systems', *Sprache und Datenverarbeitung,* Heft 1981.

290. —— (1984) 'Erste Ueberlegungen zur Verwendung des Sub-language-Konzepts in SUSY', *Multilingua.*

291. —— (1984) 'Generation of Sentences from a Syntactic Deep Structure with a Semantic Component', in Bolc and McDonald (eds), *Natural Language Generation Systems.*

292. Luckhardt, H-D. and Maas, H. D. (1982) 'SUSY-Handbuch fuer Transfer und Synthese', *LA Neue Folge,* Heft 7, Saarbruecken.

293. Lytle, E. G. (1976) 'Automatic Language Processing Project', *AJCL* 2, microfiche 46, pp.14-23.

294. Lytle, E. G. and Packard, D. (1974) 'Junction Grammar as a Base for Automatic Language Processing', preprint of the Twelfth Annual Meeting of the Association for Computational Linguistics, Amherst, Massachusetts.

295. Lytle, E. G., Packard, D., Gibb, D., Melby, A. K. and Billings, F. H. (1975) 'Junction Grammar as a Base for Natural Language Processing', *AJCL* 3, microfiche 26, pp.1-77.

296. Maas, H. D. (1975) 'Automatische Textaufbereitung in der syntaktischen Analyse russischer und deutscher Saetze', *LA* 14, Saarbruecken.

297. —— (1976) 'Die Synthese deutscher Saetze im Rahmen der automatischen Uebersetzung', *Vorabdrucke des Internationalen Kolloquiums 'Automatische Lexikographie, Analyse und Uebersetzung'*, Saarbruecken, pp.30-5.

298. —— (1977) 'NOVERAD – Ein Programmsystem zur Analyse von Nominalphrasen', *LA*22, Saarbruecken.

299. —— (1977) 'Struktur der Nominalphrasen und ihre Darstellung im Computer', *LA* 22, Saarbruecken.

300. —— (1977) 'Struktur und Ergebnisse der Satzanalyse des Saarbruecker SATAN-Systems', *LA* 24, Saarbruecken.

301. —— (1977) 'Die transformationelle Synthese deutscher Saetze im Saarbruecker Verfahren der maschinellen Uebersetzung', *LA* 25, Saarbruecken.

302. —— (1977) 'Das Saarbruecker automatische Uebersetzungssystem SUSY', *Dritter Europaeischer Kongress ueber Dokumentationssysteme und -netze, Luxembourg*, vol.1, Munich, pp.527-35.

303. —— (1977) 'Maschinelle Uebersetzung mit dem Saarbruecker System SUSY', *Sprache und Datenverarbeitung* 1, 2.

304. —— (1977-8) 'The Saarbruecken Automatic Translation System (SUSY)', *Overcoming the Language Barrier*, pp.585-92, Verlag Dokumentation, Munich.

305. —— (1978) 'Zum Stand der automatischen Uebersetzung im Sonderforschungsbereich (SFB) 100. Das Uebersetzungssystem SUSY', in D. Krallmann (ed.), *Kolloquium zur Lage der linguistischen Datenverarbeitung*, Essen.

306. —— (1978) 'Das Saarbruecker Uebersetzungssystem SUSY', *Sprache und Datenverarbeitung*, pp.43-62.

307. —— (1980) 'Zur Entwicklung des Uebersetzungssystems SUSY und seiner einzelnen Komponenten', *LA Neue Folge*, Heft 3.1, Saarbruecken.

308. —— (1980) 'Zur Sprachunabhaengigkeit der Nominalanalyse', *LA Neue Folge*, Heft 3.2, Saarbruecken.

309. —— (1980) 'Zur Analyse deutscher Komposita und Derivationen', *LA Neue Folge*, Heft 3.2, Saarbruecken.

310. —— (1982) 'Repraesentation und Strategie bei der automatischen Analyse mit SUSY', *Sprache und Information*, Band 4.

311. —— (1983) 'SUSY I und SUSY II – verschiedene Analysestrategien in der maschinellen Uebersetzung', *Sprache und Datenverarbeitung*, Heft 1981.

312. Macklovitch, E. (1984) *Recent Canadian Experience in Machine Translation*, paper presented at the Cranfield Conference on MT.

313. Maegaard, B. (1976) 'The Recognition of Finite Verbs in French Texts', *ALLC Bulletin* 4, 1, pp.49-52.

314. Masterman, M. (1979) 'Essential Mechanism of Machine Translation', paper presented at BCS 79, London.

315. —— (1979) 'The Essential Skills to be Acquired for Machine Translation', in B. M. Snell (ed.), *Translating and the Computer*, North-Holland, Amsterdam, pp.159-80.

316. —— (1982) 'The Limits of Innovation in Machine Translation', in V. Lawson (ed.), *Practical Experience of Machine Translation*, North-Holland, Amsterdam, pp.163-86.

317. Mathias, J. (1973) 'The Chinese-English Translation Assistance Group and Its Computerized Glossary Project', *Federal Linguist* 5, 3-4, pp.7-13.

318. McNaught, J. (1981) 'Terminological Data Banks: A Model for a British Linguistic Data Bank (LDB)', Machine Aids for Translators: *ASLIB Proceedings* 33, 7-8.

319. Mel'cuk, I. A. (1974) 'Grammatical Meanings in Interlinguas for Automatic Translation and the Concept of Grammatical Meaning', in V. J. Rozencvejg (ed.), *Machine Translation and Applied Linguistics*, vol.I, Athenaion Verlag, Frankfurt am Main, pp.95-113.

320. Melby, A. K. (1973) 'Junction Grammar and Machine Assisted Translation', in A. Zampolli (ed.), *Computational and Mathematical Linguistics: Proceedings of the ICCL, Pisa*.

321. —— (1974) 'Forming and Testing Syntactic Transfers', MA thesis, Brigham Young University.

322. —— (1980) 'Design and Implementation of a Computer-Assisted Translation System', *British Computer Society Natural Language Translation Specialist Group Newsletter* 9, pp.7-19.

323. —— (1981) 'Linguistics and Machine Translation', in J. Copeland and P. Davis (eds), *The Seventh LACUS Forum 1980*, Hornbeam Press, Columbia, South Carolina.

324. —— (1981) 'Translators and Machines – Can They Co-operate?', *Meta* 26, 1, pp.23-34.

325. Melby, A. K., Smith, M. R. and Peterson, J. (1980) 'ITS: Interactive Translation System', *Proceedings of the Eighth ICCL [COLING 80]*, Tokyo, pp.424-9.

326. Meyer, E. (1976) 'The Storage and Retrieval of Alpha-Numeric Information', *Rechentechnische Datenverarbeitung* 13, 5, pp.43-7.

327. Morin, G. (1978) *SISIF: système d'identification, de substitution et d'insertion de formes*, TAUM Group, University of Montreal.

328. Morton, S. E. (1978) 'Designing a Multilingual Terminology Bank for United States Translators', *ASIS* 29, 6, pp.297-303.

329. —— (1978) 'Multilingual Terminology Banks Here and Abroad: Online Terminology Dissemination as an Aid to Translation', in E. H. Brenner (ed.), *The Information Age in Perspective: Proceedings of the ASIS Annual Meeting,* vol.15, pp.242-4.

330. Muraki, K. (1982) 'On a Semantic Model for Multi-lingual Paraphrasing', *Proceedings of the Ninth ICCL [COLING 82],* prague, pp.239-44.

331. Muraki, K. and Ichiyama, S. (1982) 'An Overview of Machine Translation Project at NEC Corporation', technical note, NEC Corporation, c and c Systems Research Laboratories.

332. Nagao, M. (1979) 'Machine Translation', *Information Processing Society of Japan* 20, 10, pp.896-902.

333. —— (1982) 'A Survey of Natural Language Processing and Machine Translation in Japan', in T. Kitagawa (ed.), *Computer Science and Technologies,* North-Holland, Amsterdam, pp.64-70.

334. —— (1983) 'Natural Language Processing and Machine Translation in Japan', *Digest of Papers: Spring COMPCON 83, Intellectual Leverage for the Information Society,* IEEE, New York, pp.306-7.

335. —— et al. (1981) 'On English Generation for a Japanese-English Translation System', Technical Report on Natural Language Processing 25, Information Processing of Japan.

336. Nagao, M., Tsujii, J., Mitamura, K., Hirakawa, H. and Kume, M. (1980) 'A Machine Translation System from Japanese into English: Another Perspective of MT Systems', *Proceedings of the Eighth ICCL [COLING 80],* Tokyo, pp.414-23.

337. Nagao, M., Tsujii, J., Yada, K. and Kakimoto, T. (1982) 'An English-Japanese Machine Translation System of the Titles of Scientific and Engineering Papers', *Proceedings of the Ninth ICCL [COLING 82],* Prague, pp.245-52.

338. Nancarrow, P. H. (1978) 'The Chinese University Language Translator (CULT) – A Report', *ALLC Bulletin* 6, 2, p.121.

339. Nedobejkine, N. (1976) 'Application du systeme ATEF à l'analyse morphologique de textes russes', in A. Zampolli (ed.), *Computational and Mathematical Linguistics: Proceedings of the ICCL, Pisa, 1973,* vol.2, Florence.

340. —— (1976) 'Niveaux d'interpretation dans une traduction multilingue: Application à l'analyse du russe', *Proceedings of the Sixth ICCL [COLING 76],* Ottawa, also in *AJCL* 2, microfiche 48, p.36.

341. Nikolova, B. and Nenova, I. (1982) 'Termservice – An Automated System for Terminology Services', *Proceedings of the Ninth ICCL [COLING 82],* Prague, pp.265-9.

342. Nishida, T. and Doshita, S. (1982) 'An English-Japanese Machine Translation System Based on Formal Semantics of Natural Language', *Proceedings of the Ninth ICCL [COLING 82]*, Prague, pp.277-82.

343. —— (1983) 'An Application of Montague Grammar to English-Japanese Machine Translation', *Proceedings of the ACL-NRL Conference on Applied Natural Language Processing*, Santa Monica, California, pp.156-65.

344. Nishida, F., Kiyono, M. and Doshita, S. (1981) 'An English-Japanese Machine Translation System Based on Formal Semantics of Natural Languages', technical report, Tokyo University.

345. Nishida, F. and Takamatsu, S. (1982) 'Japanese-English Translation through Internal Expressions', *Proceedings of the Ninth ICCL [COLING 82]*, Prague, pp.271-6.

346. Nishida, F., Takamatsu, S. and Kuroki, H. (1980) 'English-Japanese Translation through Case-Structure Conversion', *Proceedings of the 8th ICCL [COLING 80]*, Tokyo, pp.447-54.

347. Nitta, Y., Okajima, A., Yamano, F. and Ishihara, K. (1982) 'A Heuristic Approach to English-into-Japanese Machine Translation', *Proceedings of the Ninth ICCL,* Prague, pp.283-8.

348. Noel, J. (1975) 'Document Analysis Algorithms and Machine Translation Research', *Revue des Langues Vivantes* 41, 3, pp.237-60.

349. Nomura, H. and Shimazu, A. (1982) 'Machine Translation in Japan', technical note, Nippon Telegraph and Telephone Public Corporation, Musashino Electrical Communications Laboratory, Tokyo.

350. Nomura, H., Shimazu, A., Iida, H., Katagiri, Y., Saito, Y., Naito, S., Ogura, K., Yokoo, A. and Mikami, M. (1982) 'Introduction to LUTE (Language Understander, Translator & Editor)', technical note, Musashino Electrical Communication Laboratory, Research Division, Nippon Telegraph and Telephone Public Corporation, Tokyo.

351. Oh, Y-O. (1979) 'Maschinelle Uebersetzung der Sprachen Koreanisch und Deutsch: Syntaktische Basis fuer die kontrastive Linguistik', *Dhak Yonku (Language Research)* 15, pp.169-83.

352. Oubine, I. I. and Tikhomirov, B. D. (1982) 'Machine Translation Systems and Computer Dictionaries in the Information Service: Ways of Their Development and Operation', *Proceedings of the Ninth ICCL [COLING 82]*, Prague, pp.289-94.

353. Pankowicz, Z. L. (1978) 'Facts of Life in Assessment of Machine Translation', report on CETIL workshop on Evaluation Problems in Machine Translation, Luxembourg.

354. Pare, M. (1974) 'La banque de terminologie de l'Universite de Montreal', *Etudes francaises dans le monde* 2, 3, p.9.

355. —— (1975) 'Computerized Multilingual Word Banks can Provide Terminological Assistance to International Standards Organization', paper presented at the Symposium on International Cooperation in Terminology, Vienna.

356. —— (1976) 'Les banques automatisées de terminologies multilingues et les organismes de normalisation', *International Cooperation in Terminology,* First Infoterm Symposium, Vienna, 1975 (Infoterm Series, Vol.3), Munich, pp.224-34.

357. —— (1977) 'La banque de terminologie de l'Universite de Montreal', *Comptes rendus du Colloque international de terminologie, Terminologies 76, Paris-La Defense,* Association francaise de terminologie, Paris.

358. —— (1980) 'Y-a-t-il toujours une machine a traduire?', *Babel* 26, 2, pp.77-82.

359. Perschke, S. et al. (1974) Les travaux du CETIS dans le domaine de l'informatique documentaire', *Documentaliste* 1974 (numero special), pp.47-57.

360. Perschke, S., Fassone, G., Geoffrion, C., Kolar, W. and Fangmeyer, H. (1974) 'The SLC-II System Language Translator Package Concepts and Facilities', Commission of the European Communities, Luxembourg. Bericht EUR 5116e, Joint Nuclear Research Centre, Scientific Data Processing Centre, CETIS, Ispra, Italy.

361. Pertsov, N. V. (1973) 'Automatic Translation Research at Montreal University', *Nauchno-Tekh. Inf.* 2, 1, pp.36-44.

362. Perusse, D. (1983) 'Machine Translation', *ATA Chronicle* 12, 8, pp.6-8.

363. Petrick, S. R. (1976) 'Summary: FBIS Seminar on Machine Translation', *AJCL,* microfiche 46, pp.72-6.

364. Pigott, I. M. (1979) 'Theoretical Options and Practical Limitations of Using Semantics to Solve Problems of Natural Language Analysis and Machine Translation', in M. MacCafferty and K. Gray (eds), *The Analysis of Meaning; Informatics* 5, ASLIB, London, pp.239-68.

365. —— (1980) 'How Does SYSTRAN Translate?', paper presented to the BCS Natural Language Translation Group.

366. —— (1982) 'The Importance of Feedback from Translators in the Development of High-Quality Machine Translation', in V. Lawson (ed.), *Practical Experience of Machine Translation,* North-Holland, Amsterdam, pp.61-73.

367. Piotrowski, R. and Georgiev, H. (1974) 'La traduction automatique en USSR', *Revue Roumaine de Linguistique* 19, 1, pp.73-9.

368. Plum, T. (1974) 'The State of Machine Translation', *Translators and Translating: Selected Essays from the American Translators' Association,* State University of New York, Binghamton, pp.37-42.

369. Poulin, G. (1981) *SYDICAN: description du dictionnaire d'analyse*, TAUM Group, University of Montreal.

370. Quezel-Ambrunaz, M. (1978) 'ARIANE: Système interactif pour la traduction automatique multilingue, Version II, report G.3400.A, University of Grenoble.

371. —— (1979) 'Transfert en ARIANE-78: Le modele TRANSF'.

372. Ritzke, J. (1980) 'Zur Problematik der Disambiguierung franzoesischer Praepositionen', *LA Neue Folge*, Heft 3.2.

373. Roberts, A. H. and Zarechnak, M. (1974) 'Mechanical Translation', in T. A. Sebeok (ed.), *Current Trends in Linguistics*, vol.12: Linguistics and Adjacent Arts and Sciences, pt.4, Mouton, The Hague, pp.2825-68.

374. Rolling, L. N. (1978) 'The Facts about Automatic Translation', in *Proceedings of the Sixth Annual CAIS Conference on Information Science, Montreal*, Ottawa, pp.267-9.

375. —— (1979) 'The Second Birth of Machine Translation', paper given at Seventh Cranfield International Conference on Mechanised Information Storage and Retrieval Systems.

376. —— (1980) 'Automatic Translation Today and Tomorrow: The Computer is Cheaper than Human Translation', *Umsch. Wiss. und Tech*. 80, 11, pp.333-8.

377. Roos, M. (1981) 'A Report on Interaction of Human and Machine Translation: Harmony or Conflict?', University of Texas at Austin.

378. Rosenthal, J. (1979) 'Idiom Recognition for Machine Translation and Information Storage and Retrieval', PhD thesis, Georgetown University.

379. Rossi, F. (1982) 'The Impact of Posteditors' Feedback on the Quality of MT', in V. Lawson (ed.), *Practical Experience of Machine Translation*, North-Holland, Amsterdam, pp.113-17.

380. Rozencveig, V. J., ed. (1974) *Essays on Lexical Semantics*, Skriptor, Stockholm.

381. —— (1974) *Machine Translation and Applied Linguistics*, Athenaeum, Frankfurt.

382. Ruffino, J. R. (1982) 'Coping with Machine Translation', in V. Lawson (ed.), *Practical Experience of Machine Translation*, North-Holland, Amsterdam, pp.57-60.

384. Sager, J. C. (1979) 'Multilingual Communication: Chairman's Introductory Review of Translation and the Computer', in B. M. Snell (ed.), *Translating and the Computer*, North-Holland, Amsterdam, pp.1-25.

385. —— (1981) 'New Developments in Information Technology for Interlingual Communication', in Machine Aids for Translators: *ASLIB Proceedings* 33, 7-8.

386. —— (1982) 'Types of Translation and Text Forms in the Environment of Machine Translation', in V. Lawson (ed.), *Practical Experience of Machine Translation,* North-Holland, Amsterdam, pp.11-19.

387. Samuelsdorf, P. O. (1977) 'Relation of Machine Translation to Linguistics Exemplified by an English-German Translation Project', *Proceedings of the Third International Conference on Computers in the Humanities, Waterloo,* Ontario.

388. Sawai, S., Fukushima, H., Sugimoto, M. and Ukai, N. (1982) 'Knowledge Representation and Machine Translation', *Proceedings of the Ninth ICCL [COLING 82],* Prague, pp.351-6.

389. Sawai, S., Sugimoto, M. and Ukai, N. (1982) 'Knowledge Representation and Machine Translation', *Fujitsu Sci. Tech. Journal* 18, 1, pp.117-34.

390. Schmitz, K-D. (1980) 'Struktur und Updating des Analysewoerterbuchs', *LA Neue Folge,* Heft 3.2.

391. —— (1980) 'Die Woerterbuchsuche', *LA Neue Folge,* Heft 3.2.

392. —— (1982) 'PHRASEG – ein Verfahren zur Satzsegmentierung', *Dokumentation* κ5.1/κ5.2. sFB-100-κ.

393. Schulz, J. (1975) 'Der Computer hilft dem Uebersetzer: Ein Verfahren zur Abfrage eines auf Magnetband gespeicherten mehrsprachigen Fachwoerterbuchs', *Lebende Sprachen* 20, 4, pp.99-103.

394. —— (1976) 'Automatische Abfrage einer Terminologie-Datenbank', *Nachrichten fuer Dokumentation* 27, no.2, pp.62-6.

395. —— (1977) 'Eine Terminologiedatenbank fuer Uebersetzer: Abfragemoeglichkeiten im System TEAM', *Dritter Europaeischer Kongress ueber Dokumentationssysteme und -netze, Luxembourg,* vol.1, Munich, pp.97-140.

396. Schwab, W. (1981) 'Traduction et informatique: perspectives pour les années 80', *Meta* 26, 1, Montreal.

397. Schwanke, P. G. (1973) 'On Machine Translation: A Difference in Postulates', *Eco-logos* 19, 70, pp.3-7.

398. Seelbach, D. (1975) 'Maschinelle Uebersetzung von Key-Phrases', *Computerlinguistik und Dokumentation,* Munich, pp.132-46.

399. Seo, S. O. (1977) 'An Analysis of Korean by Junction Theory – with Reference to Machine Translation', *Dhak Yonku (Language Research)* 13, pp.139-63.

400. Serada, S. P. (1982) 'Practical Experience of Machine Translation', in V. Lawson (ed.), *Practical Experience of Machine Translation,* North-Holland, Amsterdam, pp.119-23.

401. Shaak, B. (1978) 'Machine Translation', *Comp. Bull.* 2, 16, pp.24-5.

402. Shalyapina, Z. M. (1980) 'Automatic Translation as a Model of the Human Translation Activity', *International Forum on Information and Documentation* 5, 2, pp.18-23.

403. —— (1980) 'Problems of Formal Representation of Text Structure from the Point of View of Automatic Translation', *Proceedings of the 8th ICCL [COLING 80]*, Tokyo, pp.174-82.

404. Shudo, K. (1973) 'On Machine Translation from Japanese into English for a Technical Field', *J. Inf. Process. Soc. Jap.* 14, 9, pp.661-8.

405. —— (1974) 'On Machine Translation from Japanese into English for a Technical Field', *Information Processing in Japan* 14, pp.44-50.

406. Sigurd, B. (1978) 'Machine Translation : State of the Art', in *Theory and Practice of Translation*, Peter Lang, Stockholm, pp.33-47.

407. Slocum, J. (1980) 'An Experiment in Machine Translation', *Proceedings of the Eighteenth Annual Meeting of the Association for Computational Linguistics, Philadelphia*, pp.163-7.

408. —— (1981) 'A Practical Comparison of Parsing Strategies for Machine Translation and Other Natural Language Processing Purposes', ph D thesis, University of Texas at Austin and University Microfilms International, Ann Arbor, Michigan.

409. —— (1981) 'The METAL Parsing System', working paper LRC-81-2, Linguistics Research Center, University of Texas at Austin.

410. —— (1981) 'A Practical Comparison of Parsing Strategies', *Proceedings of the Nineteenth Annual Meeting of the Association for Computational Linguistics, Stanford University*, pp.1-6.

411. —— (1981) 'The LRC Machine Translation System : An Application of Textual and Natural Language Processing Techniques to Translation Needs', report to the Siemens Corporation on current project, Linguistics Research Center, University of Texas at Austin.

412. —— (1982) 'The LRC Machine Translation System ; An Application of State-of-the-Art Text and Natural Language Processing Techniques', *Proceedings of the Ninth ICCL [COLING 82]*, Prague.

413. —— (1983) 'A Status Report on the LRC Machine Translation System', *Proceedings of the ACL-NRL Conference on Applied Natural Language Processing, Santa Monica, California*, pp.166-73.

414. —— (1983) 'MT: Its History, Current Status and Future Prospects', Siemens Communication Systems Inc. and the University of Texas.

415. —— (1984) 'How one might Automatically Identify and Adapt to a Sublanguage', paper presented at the Workshop on Sublanguages, New York University.

416. —— (1984) 'Machine Translation: Its History, Current Status, and Future Prospects', *Proceedings of the Tenth International Conference on Computational Linguistics (COLING)*, Stanford University.

417. Slocum, J. and Bennett, W. S. (1982) 'The LRC Machine Translation System: An Application of State-of-the-Art Text and Natural Language Processing Techniques to the Translation of Technical Manuals', working paper LRC-82-1, Linguistics Research Center, University of Texas at Austin.

418. Snell, B. M., ed. (1979) *Translating and the Computer* [Proceedings of a seminar held in London, 1978], North-Holland, Amsterdam.

419. —— (1980) 'Electronic Translation?', *ASLIB Proceedings* 32, 4, pp.179-86.

420. Somers, H. L. (1981) 'Bede – The CCL/UMIST Machine Translation System: Rule-Writing Formalism (3rd revision)', report no.81-5, Centre for Computational Linguistics, University of Manchester Institute of Science and Technology.

421. —— (1983) 'Investigating the Possibility of a Microprocessor-based Machine Translation System', *Proceedings of the Conference on Applied Natural Language Processing,* Santa Monica, California, pp.149-55.

422. Somers, H. L. and McNaught, J. (1980) 'The Translator as a Computer User', *The Incorporated Linguist* 19, pp.49-53.

423. Stachowitz, R. A. (1973) *Voraussetzungen fuer maschinelle Uebersetzung: Probleme, Loesungen, Aussichten,* Athenaeum Skripten, Frankfurt am Main.

424. —— (1976) 'Beyond the Feasibility Study: Lexicographic Progress', in A. Zampolli (ed.), *Computational and Mathematical Linguistics: Proceedings of the ICCL, Pisa,* vol.1, Florence.

425. —— (1976) 'Beyond the Feasibility Study: Syntax and Semantics', in A. Zampolli (ed.), *Computational and Mathematical Linguistics: Proceedings of the ICCL, Pisa, 1973,* vol.2, Florence.

426. Stahl, G. (1980) 'Idées directrices dans l'analyse syntaxique automatique de l'allemand', internal report DRET.

427. Staples, C. O. (1983) 'The Logos Intelligent Translation System', paper distributed in connection with the Logos exhibition at IJCAI, Karlsruhe.

428. Stegentritt, E. (1980) 'Die Komplementanalyse im Saarbruecker Uebersetzungssystem, dargestellt am Franzoesischen', *LA Neue Folge,* Heft. 3.2.

429. Steiff, S. (1979) 'The Development of the "Titus" Four-Language Automatic Translation Method', *Inf. and Doc.* 4, pp.20-6.

430. Stessel, J. P. (1973) 'Une methode practique de decodage des messages meteorologiques', Institut Royal Meteorologique de Belgique, Brussels, and NTIS.

431. Stiegler, A. D. (1981) 'Machine Aids for Translators: *ASLIB Proceedings* 33, 7-8.

432. Straub, J. R. and Rogers, C. A. (1979) 'Computer Analysis of Basic English as a First Step in Machine Translation', in R. Trappl, F. Hanika and F. R. Pichler (eds), *Progress in Cybernetics and Systems Research*, vol.5, pp.491-4.

433. Stutzman, W. J. (1976) 'Organizing Knowledge for English-Chinese Translation', *Proceedings of the Sixth ICCL [COLING 76]*, Ottawa.

434. —— (1976) 'Organizing Knowledge for English-Chinese Translation', *AJCL* 2, microfiche 48, p.28.

435. Sugita, S. (1973) 'Mechanical Translation Between English and Japanese', in S. Gould (ed.), *Proceedings of the First International Symposium on Computers and the Chinese Input/Output Systems, Taipei, Taiwan*, Academia Sinica, vol.13, pp.555-72.

436. Tanke, E. (1975) 'Electronic Data Processing in the Service of Translators, Terminologists, and Lexicographers', *Philips Terminology Bulletin* 4, 2/3, pp.3-19.

437. —— (1975) 'Future Developments – Three Lectures given at the Seminar on Computer Aids to Human Translation, Eindhoven', *Philips Terminology Bulletin* 4, 2/3, pp.38-48.

438. —— (1979) 'Implementing Machine Aids to Translation', in B. M. Snell (ed.), *Translating and the Computer*, North-Holland, Amsterdam, pp.45-70.

439. TAUM (1973) 'Le système de traduction automatique de l'Universite de Montreal (TAUM)', *Meta* 18, pp.227-89.

440. —— (1977) *Projet Aviation*, rapport d'étape, University of Montreal (mimeo).

441. —— (1979) 'Presentation de la chaîne de traduction informatisée TAUM/AVIATION', University of Montreal.

442. Thiel, M. (1980) 'Mehrsprachige Woerterbuchsuche: Wortanalyse', *LA Neue Folge*, Helft 3.2, Saarbruecken.

443. —— (1980) 'Verfahren der Flexionsanalyse', *LA Neue Folge*, Heft 3.2, Saarbruecken.

444. —— (1980) 'Erkennung fester Syntagmen', *LA Neue Folge*, Heft 3.2.

445. —— (1980) 'Verbalgruppen- und Komplementanalyse', *LA Neue Folge*, Heft 3.2.

446. —— (1983) 'Systemarchitektur von SUSY unter benutzerspecifischem Aspekt', *Sprache und Datenverarbeitung*, Heft 1981.

447. —— (1983) 'LICA – Sprach- und texttypspezifische Parametersaetze', *Dokumentation* A2/6, SFB-100-A2.

448. Thome, G. (1980) 'Probleme der Teilsatzerkennung im Englischen', *LA Neue Folge*, Heft 3.2.

449. Thouin, B. (1976) 'Système informatique pour la generation morphologique de langues naturelles en états finis', *AJCL* 1976, no.2, microfiche 48, p.56, also in *Proceedings of the Sixth ICCL [COLING 76]*, Ottawa.

450. —— (1979) 'The Future of Computer Translation', *CIPS Review* 3, 6, pp.18-19.

451. —— (1982) 'The Meteo System', in V. Lawson (ed.), *Practical Experience of Machine Translation*, North-Holland, Amsterdam, pp.39-44.

452. Toma, P. (1974) 'Computer Translation: In Its Own Right', *Kommunikationsforschung und Phonetik: IKP-Forschungsberichte*, vol.50, Buske, Hamburg, pp.155-64.

453. —— (1976) 'An Operational Machine Translation System', in R. W. Brislin (ed.), *Translation: Applications and Research*, Gardner Press, New York, pp.247-59.

454. —— (1976) 'SYSTRAN', in FBIS Seminar on Machine Translation, *AJCL*, microfiche 46, pp.40-5.

455. —— (1977) 'SYSTRAN: A Third-Generation Machine Translation System', *Sprache und Datenverarbeitung* 1, 1, pp.38-46.

456. —— (1977) 'SYSTRAN as a Multi-lingual Machine Translation System', in Commission of the European Communities, *Overcoming the Language Barrier*, vol.1, Verlag Dokumentation, Munich, pp.569-81.

457. Toma, P., Carlson, J. A., Stoughton, D. R. and Ryan, J. P. (1973) 'Machine-Aided Editing', final report RADC-TR-73-368, LATSEC Inc., La Jolla, California.

458. Toma, P., Garrett, P., Kozlik, L., Perwin, D. and Starr, C. (1974) 'Some Semantic Considerations in Russian-English Machine Translation', final report RADC-TR-74-189, LATSEC Inc., La Jolla, California, and NTIS Report AD-787-671.

459. Toma, P. and Kozlik, L. A. (1973) 'Rand Corporation Data in SYSTRAN', final report RADC-TR-73-262, 2 vols, LATSEC Inc., La Jolla, California.

460. Toma, P., Kozlik, L. A. and Perwin, D. G. (1974) 'Optimization of SYSTRAN System', final technical report RADC-TR-73-155-rev, LATSEC Inc., La Jolla, California, and NTIS report AD-777-850.

461. Troike, R. C. (1976) 'The future of MT', FBIS seminar, *AJCL* 3, microfiche 51, pp.47-9.

462. —— (1976) 'The View from the Center: The Future of MT', *The Linguistic Reporter* 18, 9, p.2.

463. Tsujii, J. (1982) 'The Transfer Phase in an English-Japanese Translation System', *Proceedings of the Ninth ICCL [COLING 82]*, Prague, pp.383-90.

464. Tucker, A. B., Jr. (1984) 'A Perspective on Machine Translation: Theory and Practice', *Communications of the ACM* 27, 4, pp.322-9.

465. Tucker, A. and Nirenburg, S. (1984) 'Machine Translation: A Contemporary View', *Annual Review of Info, Science and Terminology,* vol.19.

466. Tucker, A. B., Vasconcellos, M. and Leon, M. (1980) 'PAHO Machine Translation System: Introduction and Users' Manual', Pan American Health Organization, Washington, DC.

467. Uchida, H. and Sugiyama, K. (1979) 'Automated Translation of Japanese Kana Input into Mixed Kana-Kanji Output', *Fujitsu Science and Technology Journal* 15, 2, pp.21-43.

468. —— (1980) 'A Machine Translation System from Japanese into English Based on Conceptual Structure', *Proceedings of the 8th ICCL [COLING 80],* Tokyo, pp.455-62.

469. Usui, T. (1982) 'An Experimental Grammar for Translating English to Japanese', technical report, Department of Computer Sciences, University of Texas at Austin.

470. Van Slype, G. (1978) 'Second Evaluation of the SYSTRAN Automatic Translation System', final report, Commission of the European Communities, Luxembourg.

471. —— (1982) 'Economic Aspects of Machine Translation', in V. Lawson (ed.), *Practical Experience of Machine Translation,* North-Holland, Amsterdam, pp.79-93.

472. —— (1979) 'Evaluation by the EEC Commission of the SYSTRAN Automatic Translation System, 1978 Version', *Inf. and Doc.* 4, pp.27-35.

473. —— (1979) 'Evaluation du système de traduction automatique SYSTRAN anglais-francais, version 1978, de la Commission des communautés Europeennes', *Babel* 25, 3, pp.157-62.

474. —— (1979) 'Critical Study of Methods for Evaluating the Quality of Machine Translation', final report, Commission of the European Communities, Brussels/Luxembourg.

475. Van Slype, G. and Pigott, I. (1979) 'Description du système de traduction automatique SYSTRAN de la Commission des Communautés Europeennes', *Documentaliste* 16, 4, pp.150-9.

476. Vasconcellos, M. (1983) Management of the Machine Translation Environment: Interaction of Functions at the Pan American Health Organization', unpublished manuscript, Pan American Health Organization, Washington, DC.

477. Vauquois, B. (1975) 'La traduction automatique à Grenoble', *Documents de Linguistique Quantitative* 24, Dunod, Paris.

478. —— (1976) 'Automatic Translation – A Survey of Different Approaches', *Statistical Methods in Linguistics* 1, pp.127-35.

479. —— (1976) 'Les procédés formels de représentation des structures profondes', in *Preprints of the International Colloquium 'Automatische Lexikographie, Analyse und Uebersetzung', Saarbruecken,* pp.67-9.

480. —— (1978) 'L'évolution des logiciels et des modèles linguistiques pour la traduction automatisee', *TA Informations* 19.

481. —— (1979) 'Aspects of Mechanical Translation in 1979', Conference for Japan IBM Scientific Program; Groupe d'Etudes pour la Traduction Automatique; Institut de Recherches et Mathematiques Avancées, University of Grenoble.

482. —— (1981) 'L'informatique au service de la traduction', *Meta* 26, 1, pp.8-17.

483. —— (1983) 'Automatic Translation', *Proceedings of the Summer School 'The Computer and the Arabic Language',* Rabat, Ch.9.

484. Vernimb, C. O. (1977) 'The European Network for Scientific, Technical, Economic, and Social Information', *Nachrichten fuer Dokumentation* 28, 1, pp.11-18.

485. Wang, W. S. Y. (1976) 'Chinese-English Machine Translation', *AJCL* 2, 1976, microfiche 46, p.24.

486. —— et al. (1975) 'Chinese-English Machine Translation System', final technical report RADC-TR-75-109, Department of Linguistics, University of California at Berkeley, and NTIS Report AD-A011-715.

487. Wang, W. S. Y. and Chan, S. W. (1974) 'Development of Chinese-English Machine Translation System', final technical report RADC-TR-74-22, University of California at Berkeley.

488. Wang, W. S. Y., Chan, S. W. and Robyn, P. (1976) 'Chinese-English Machine Translation System', final technical report RADC-TR-76-21, Department of Linguistics, University of California at Berkeley.

489. Wang, W. S. Y., Chan, S. W. and T'sou, B. K. (1973) 'Chinese Linguistics and the Computer', *Linguistics* 118, pp.89-117.

490. Weber, H-J. (1976) 'Semantische Merkmale zur Identifikation von Satz- und Textstrukturen', *Vorabdrucke des Internationalen Kolloquiums 'Automatische Lexikographie, Analyse und Uebersetzung', Saarbruecken,* pp.70-9.

491. Weickenmeier, E. (1980) 'Das Saarbruecker Verfahren zur Homographenanalyse', *LA Neue Folge,* Heft 3.2.

492. —— (1980) 'Strategie der Nominalanalyse', *LA Neue Folge,* Heft 3.2.

493. Weidner. 'The Weidner System Version II', Weidner Communications Corporation, undated technical report.

494. Weissenborn, J. (1976) 'Zur Strategie der automatischen Analyse des Franzoesischen im Hinblick auf die maschinelle Uebersetzung', *Vorabdrucke des Internationalen Kolloquiums 'Automatische Lexikographie, Analyse und Uebersetzung', Saarbruecken*, pp.80-8.

495. Wheeler, P. (1983) 'The Errant Avocado (Approaches to Ambiguity in SYSTRAN Translation)', *Newsletter* 13, Natural Language Translations Specialist Group, BCS.

496. Whitelock, P. J. and Kilby, K. J. (1983) 'An In-depth Study of MT Techniques', final report, Science and Engineering Research Council, Social Studies Research Council, Joint Commission, contract no.GR/C 01276.

497. Widmann, R. L. (1975) 'Trends in Computer Applications to Literature', *Computers and the Humanities* 9, 5, pp.231-5.

498. Wiechowski, M. J. (1973) 'Translation al machina (Interlingua) (Machine Translation)', *Eco-logos* 19, 68, pp.3, 14, 19-20.

499. Wilks, Y. (1973) 'The Stanford Machine Translation Project', in R. Rustin (ed.), *Natural Language Processing: Courant Computer Science Symposium, no.8, 1971*, Algorithmics Press, New York, pp.243-90.

500. —— (1977) 'Four Generations of Machine Translation Research and Prospects for the Future', paper presented at the NATO Symposium, Venice.

501. Wilks, Y. and LATSEC Inc. 'Comparative Translation Quality Analysis', final report on contract F33657-77-C-0695.

502. Wilss, W. (1977) *Uebersetzungswissenschaft: Probleme und Methoden*, Kleet, Stuttgart.

503. Witkam, A. P. M. (1983) 'DLT-distributed Language Translation – A Multilingual Facility for Videotex Information Networks', BSO (Buro voor Systemontwikkeling), Utrecht, Netherlands.

504. Xyzyx Information Corporation (1974) 'Computer Aided Language Translation System, English to French', Canoga Park, California.

505. Yang, C. J. (1981) 'High Level Memory Structures and Text Coherence in Translation', *Proceedings of the Seventh IJCAI, Vancouver*, vol.1, pp.47-9.

506. Young, M. E. (1978) 'Machine Translation (A Bibliography with Abstracts)', NTIS report PS-78/0448/7GA.

507. —— (1981) 'Machine Translation 1964-May 1981 (Citations from the NTIS Data Base)', NTIS report PB81-806507.

508. Zachary, W. W. (1978) 'Machine Aids to Translation: A Concise State of the Art Bibliography', *AJCL* 3, microfiche 77, pp.34-40.

509. —— (1979) 'A Survey of Approaches and Issues in Machine-Aided Translation Systems', in *Computers and the Humanities* 13, 1, pp.13-28.

510. —— (1980) 'Translation Databases: Their Content, Use, and Structure', in J. Raben and G. Marks (eds), *Data Bases in the Humanities and Social Sciences,* North-Holland, Amsterdam, pp.217-21.

511. Zarechnak, M. (1976) 'Russian-English System, Georgetown University', in FBIS Seminar on Machine Translation, *AJCL,* microfiche 46, pp.51-2.

512. —— (1979) 'The History of Machine Translation', in B. Henisz-Dostert, R. R. Macdonald and M. Zarechnak, *Machine Translation,* Mouton, The Hague, pp.1-87.

513. —— (1980) 'Machine Translation: Past, Present, and Future', *ATA Chronicle* 6, 7, pp.10-12.

514. Zingel, H-J. (1974) 'Computer spricht vier Sprachen: Der Textildokumentation gelingt erstmals automatische Uebersetzung wissenschaftlicher Texte', *VDI-Nachrichten* 16, pp.2-6.

515. —— (1975) 'Das TITUS-System – die Fachdatenbank Textil – entstanden durch internationale Kooperation, internationales Analysesystem und automatische Uebersetzung', *Bericht von der Tagung der Deutschen Gesellschaft fuer Dokumentation,* Muenchen.

516. —— (1975) 'TITUS, Eine viersprachige Datenbank des VDI ueber Dokumentationen der Chemiefaser, Textil und Bekleidungs-industrie', paper presented at the IBM-Seminar Dialogverarbeitung mit Bildschirmen in der Chemiefaser-, Textil- und Bekleidungs-industrie, Sindelfingen.

517. —— (1975) 'TITUS, Internationale Dokumentation der Textil-industrie und Bekleidungsindustrie', *Nachrichten duer Dokumentation* 26, 2, pp.77-8.

518. —— (1978) 'Experiences with Titus II', *Int. Classif.* 5, 1, pp.33-7.

Software
519. Aho, A. V. and Ullman, J. (1972) *The Theory of Parsing, Translation and Compiling,* Vol.1, Prentice-Hall, New Jersey.

520. Ashman, B. D. (1982) 'Production Systems and Their Application to Machine Translation: Transfer Report', report no.82-9, Centre for Computational Linguistics, University of Manchester Institute of Science and Technology.

521. Bachut, D. (1983) 'ATLAS manuel d'utilisation', Contract ADI Nu83/175.
Bachut, D. and Verastégui, N. (1983) 'VISULEX manuel d'exploitation sous CMS', Document ADI.

522. —— (1984) 'Software Tools for the Environment of a Computer-Aided Translation System', *Proceedings of COLING-84,* Stanford.

523. Backus, J. (1978) 'Can Programming be Liberated from the von Neumann Style?', *CACM,* August.

524. Boitet, C. (1979) 'Automatic Production of CF and CS Analyzers using a General Tree-Transducer', paper presented at the Second International Colloquium on Machine Translation, Lexicography, and Analysis, Saarbruecken.

525. —— (1982) *Le point sur ARIANE-78 début 1982*, vol.I, part I: *Le logiciel*, GETA-Champollion CAP SOGETI.

526. Boitet, C. and Nedobejkine, N. (1982) 'Russian-French Machine Translation at Grenoble: A General Software used for Implementing a Particular Linguistic Strategy', *Linguistics*.

527. Case, R. and Berglund, L. (1976) 'FTD Edit System', final technical report, January 1974-June 1976, Operating Systems Inc., Woodland, California.

528. Chappuy (1983) 'Formalisation de la description des niveaux d'interprétation des langues naturelles', thèse de 3è cycle, Grenoble.

529. Chauché, J. (1974) 'Transducteurs et arborescences: Etude et realisation de systemes appliqués aux grammaires transformationnelles', Thèse Docteur Es-Sciences Mathematiques, University of Grenoble, France.

530. —— (1976) 'Vers un modèle algorithmique pour le traîtement automatique des langues naturelles', *AJCL* 2, microfiche 48, p.25.

531. Clemente-Salazar (1982) 'Etudes et algorithmes liés à une nouvelle structure de données en TA: les E-graphes', Thèse de Docteur-ingénieur, USMG & INPG, Grenoble, and *Proceedings of COLING-82*, Prague.

532. Colmerauer, A. (1971) 'Les systemes-Q: un formalisme pour analyser et synthetiser des phrases sur ordinateur', TAUM Group, University of Montreal.

533. Davis, R., Buchanan, B. G. and Shortliffe, E. H. (1977) 'Production Rules as a Representation for a Knowledge-based Consultation System', in *Artificial Intelligence* 8, North-Holland, pp.15-45.

534. Davis, R. and King, J. (1975) *An Overview of Production Systems*, Stanford Artificial Intelligence Laboratory, Memo AIM-271.

535. Gérin-Lajoie, R. (1981) *Vérification des manipulations d'arbres en LEXTRA*, MA thesis, University of Montreal.

536. Hoare, C. and Wirth, N. (1973) 'An Axiomatic Definition of the Programming Language Pascal', *Acta Infomatica* vol.2, no.4, pp.335-5.

537. Johnson, S. C. (1975) 'YACC – Yet Another Compiler Compiler', *CSTR* 432, Bell Labs, Murray Hill, NJ.

538, Keil, G. C. (1979) 'Machine Translation Software', Paper Summaries, British Computer Society 79. Paper presented at BCS 79, London.

539. Langley P. and Nicholas, D. (1983) 'A Framework for Exploring Production System Architectures', draft paper, Carnegie Mellon University.

540. Liskov, B. H. and Berzins (1979) 'An Appraisal of Program Specifications', in P. Wegner (ed.), *Research Directions in Software Technology*, MIT, London.

541. Luther, D. A., Montgomery, C. and Case, R. M. (1977) 'An Interactive Text-Editing System in Support of Russian Translation by Machine', *IFIPS National Computer Conference Proceedings*, vol.46, AFIPS Press, Montvale, New Jersey, pp.789-90.

542. Magnenat-Thalmann, N. (1982) 'Choosing an Implementation Language for Automatic Translation', *Comput. Lang.* 7, 3-4, pp.161-70.

543. Marcus, R. S. and Reintjes, J. F. (1981) 'A Translating Computer Interface for End-User Operation of Heterogeneous Retrieval Systems, I: Design', *ASIS* 32, 4, pp.287-303.

544. —— (1981) 'A Translating Computer Interface for End-User Operation of Heterogeneous Retrieval Systems, II: Evaluations', *ASIS* 32, 4, pp.304-17.

545. Martin, W. A., Church, K. W. and Patil, R. S. (1981) 'Preliminary Analysis of a Breadth-First Parsing Algorithm: Theoretical and Experimental Results', paper presented at the University of Texas Symposium on Modelling Human Parsing Strategies.

546. Mathias, J. (1973) 'Some Computer Functions for Machine-Aided Translation', in S. Gould (ed.), *Proceedings of the First International Symposium on Computers and the Chinese Input/Output Systems, Taipei, Taiwan*, Academia Sinica, vol.13, pp.589-92.

547. McCarthy, J. (1960) 'Recursive Functions of Symbolic Expressions and their Computation by Machine', *CACM* 3, 4.

548. Melby, A. K. (1981) 'A Suggestion-Box Translator Aid', *Proceedings of the DLLS Symposium*, Brigham Young University, Dept. of Linguistics.

549. —— (1982) 'Multi-level Translation Aids in a Distributed System', *Ninth ICCL [COLING 82]*, Prague, pp.215-20.

550. —— (1983) 'Computer-Assisted Translation Systems: The Standard Design and a Multi-Level Design', *Proceedings of the ACL-NRL Conference on Applied Natural Language Processing, Santa Monica, California*, pp.174-7.

551. —— (1984) 'Recipe for a Translator Work Station', *Multilingua* vol.3-4, Mouton Publishers (based on the current paper with permission).

552. Mozota (1984) 'Un formalisme d'expressions pur la spécification du contrôle dans les systèmes de production', Thèse de 3è cycle, Grenoble.

553. Pratt, V. R. (1973) 'A Linguistics Oriented Programming Language', *Proceedings of the third IJCAI, Stanford University*, pp.372-81.

554. —— (1975) 'LINGOL – A Progress Report', *Advance Papers of the Fourth International Joint Conference on Artificial Intelligence, Tbilisi, Georgia, USSR,* pp.422-8.

555. Quezel-Ambrunaz, A. and Guillaume, P. (1976) 'Analyse automatique de textes par un système d'états finis', in A. Zampolli (ed.), *Computational and Mathematical Linguistics: Proceedings of the ICCL, Pisa, 1973,* vol.2, Florence.

556. Richards, M. and Whitby-Strevens, C. (1980) *BCPL – The Language and Its Compiler,* Cambridge University Press.

557. Rosner, M. (1983) 'Production Systems', in M. King (ed.), *Parsing Natural Language,* Academic Press, London, pp.35-58.

558. Scott, D. (1973) 'Models for Various Type-Free in P. Suppes et al. (eds), *Logic, Methodology, and Philosophy of Science IV,* North-Holland, pp.157-87.

559. Stewart, G. (1978) 'Spécialisation et compilation des ATN: REZO', paper presented at COLING-78, Bergen.

560. Ter-Mikaelyan, T. M. and Urutyan, R. L. (1974) 'General Description of the 'Garni' Computer and of the Process of Realization of a Translation Algorithm. Edited machine translation of Vychislitelnyi Tsentr, Erevan. Trudy (USSR) n7 p9-23 1972, by C. T. Ostertag', Foreign Technology Division, Wright-Patterson Air Force Base, Ohio.

561. Varile, G. B. (1983) 'Charts: A Data Structure for Parsing', in M. King (ed.), *Parsing Natural Language,* Academic Press, London, pp.73-88.

562. Verastegui, N. (1982) 'Etude du parallélisme appliqué à la traduction automatisée par ordinateur. STAR-PALE: un système parallèle', Thèse de Docteur Ingénieur, USMG & INPG, Grenoble, and *Proceedings of COLING-82,* Prague.

563. Wychoff, S. K. (1979) 'Computer-Assisted Translation: Mainframe to Microprocessor', *COMPCON 79 Proceedings: Using Microprocessors – Extending Our Reach,* Washington, DC, pp.472-6.

Linguistics and Computational Linguistics

564. Association for Computational Linguistics (1983) *Proceedings of the First Conference of the European Chapter of the Association for Computational Linguistics,* Pisa.

565. Borer, H. (1984) *Parametric Syntax: Case Studies in Semitic and Romance Languages,* Foris Publications.

566. Bresnan, J. W. (1977) 'A Realistic Transformational Grammar', in Halle, Bresnan and Miller (eds), *Linguistic Theory and Psychological Reality,* MIT Press.

567. —— (1982) *The Mental Representation of Grammatical Relations,* MIT Press.

568. —— (1982) 'Polyadicity', in J. Bresnan (ed.), *The Mental Representation of Grammatical Relations*, MIT Press.

569. Burton, R. (1976) *Semantic Grammar*, BBN Report 3453.

570. Chomsky, N. (1957) *Syntactic Structures*, Mouton, The Hague.

571. —— (1965) *Aspects of the Theory of Syntax*, MIT Press.

572. —— (1972) *Studies in Semantics in Generative Grammar*, Mouton, The Hague.

573. —— (1977) *Essays on Form and Interpretation*, North-Holland.

574. —— (1980) 'On Binding', *Linguistic Inquiry* 11.

575. —— (1981) *Lectures on Government and Binding*, Foris Publications.

576. —— (1982) *Some Concepts and Consequences of the Theory of Government and Binding*, MIT Press.

577. Chomsky, N. and Lasnik, H. (1977) 'Filters and Control', *Linguistic Inquiry* 8, pp.425-504.

578. Curry, H. B. (1961) 'Some Logical Aspects of Grammatical Structure', in *Structure of Language and Its Mathematical Aspects* (American Mathematical Society, Rhode Island), pp.56-68.

579. Damerau, F. J. (1981) 'Operating Statistics for the Transformational Question Answering System', *AJCL* 7 (1), pp.30-42.

580. Dowty, D. R. (1982) 'Grammatical Relations and Montague Grammar', in P. Jacobson and G. K. Pullum (eds), *The Nature of Syntactic Representation*, Synthese Language Library, Reidel, Dordrecht, pp.79-130.

581. Dowty, D. R., Wall, R. E. and Peters, S. (1981) *Introduction to Montague Semantics*, Reidel, Dordrecht.

582. Fillmore, C. (1968) 'The Case for Case', in E. Bach and R. Harms (eds), *Universals in Linguistic Theory*, New York, 1-90.

583. Gawron, J. M., King, J., Lamping, J., Loebner, E., Paulson, A., Pullum, G., Sag, I. and Wasow, T. (1982) 'The GPSG Linguistics System', in *Proceedings of the 20th Annual Meeting of the Association for Computational Linguistics*, pp.74-81.

584. Gazdar, G. (1981) 'Unbounded Dependencies and Coordinate Structure', in *Linguistic Inquiry*, 12 (2), pp.155-84.

585. —— (1982) 'Phrase Structure Grammar', in P. Jacobson and G. Pullum (eds), *The Nature of Syntactic Representation*, Reidel, Dordrecht.

586. —— (1983) 'Phrase Structure Grammars and Natural Languages', *Proceedings of the Eighth International Joint Conference on Artificial Intelligence, Karlsruhe*, vol.1, pp.556-65.

587. Grice, H. (1975) 'Logic and Conversation', in P. Cole and J. Morgan (eds), *Syntax and Semantics*, vol.3: *Speech Acts*, Academic Press, pp.41-58.

588. Grishman R. (1984) 'An Introduction to Computational Linguistics', draft, New York University.

589. Harman, G. H. (1963) 'Generative Grammars without Transformation Rules: a Defense of Phrase Structure', *Language* 39.

590. Hays, D. G. (1964) 'Dependency Grammar, a Formalism and some Observations', *Language* 40, pp.511-25.

591. Hayes, P. (1976) 'Semantic Markers and Selectional Restrictions, in Charniak and Wilks (eds), *Computational Semantics,* Amsterdam, pp.41-54.

592. Hayes, P. J. and Mouradian, G. V. (1981) 'Flexible Parsing', *American Journal of Computational Linguistics* 7, 4, pp.232-42.

593. Heidorn, G. E., Jensen, K., Miller, L. A., Byrd, R. J. and Chodorow, M. S. (1982) 'The EPISTLE Text-critiquing System', *IBM Systems Journal* 21 (3).

594. Hendrix, G. G., Sacerdoti, E. D., Sagalowicz, D. and Slocum, J. (1978) 'Developing a Natural Language Interface to Complex Data', *ACM Transactions on Database Systems,* 3 (2), pp.105-47.

595. Jackendoff, R. (1972) *Semantic Interpretation in Generative Grammar,* MIT Press.

596. —— (1977) *X-bar Syntax: A Study of Phrase Structure,* MIT Press.

597. Janssen, T. M. V. (1980) 'On Problems Concerning the Quantification Rules in Montague Grammar', in C. Rohrer (ed.), *Time, Tense and Quantifiers,* Max Niemeyer Verlag, Tuebingen, pp.113-34.

598. Jensen, K. and Heidorn, G. E. (1982) 'The Fitted Parse: 100 per cent Parsing Capability in a Syntactic Grammar of English', research report RC-9729 (no.42958), Computer Sciences Department, IBM Thomas J. Watson Research Center, Yorktown Heights, New York.

599. Kaplan, R. M. and Bresnan, J. (1982) 'Lexical Functional Grammar: A Formal System for Grammatical Representation', in J. Bresnan (ed.), *The Mental Representation of Grammatical Relations,* MIT Press.

600. Katz, J. and Fodor, J. A. (1963) 'The Structure of a Semantic Theory', *Language* 39, pp.170-210.

601. Kay, M. (1973) 'The MIND System', in R. Rustin (ed.), *Natural Language Processing,* Algorithmics Press, New York, pp.155-88.

602. King, M., ed. (1983) *Parsing Natural Language,* Academic Press, London.

603. —— (1983) 'Transformational Parsing', in M. King (ed.), *Parsing Natural Language,* Academic Press, London.

604. Kittredge, R. and Lehrberger, J., eds (1982) *Sublanguage. Studies of Language in Restricted Semantic Domains,* Berlin and New York.

605. Landsbergen, J. (1981) 'Adaptation of Montague Grammar to the Requirements of Parsing', in J. A. G. Groenendijk, T. M. V. Janssen and M. B. J. Stokhof (eds), *Formal Methods in the Study of Language,* Part 2, MC Tract 136, Mathematical Centre, Amsterdam, pp.399-420.

606. Lyons, J. (1968) *Introduction to Theoretical Linguistics,* Cambridge University Press.

607. Marcus, M. (1980) *A Theory of Syntactic Recognition for Natural Language,* MIT Press.

608. Miller, L. A., Heidorn, G. E. and Jensen, K. (1980) 'Text-Critiquing with the EPISTLE System: An Author's Aid to Better Syntax', research report RC 8601 (no.37554), Behavioral Sciences and Linguistics Group, Computer Sciences Dept., IBM Thomas J. Watson Research Center, Yorktown Heights, New York.

609. Newmeyer, F. J. (1980) *Linguistic Theory in America,* Academic Press, New York.

610. Partee, B. H. (1973/6) 'Some Transformational Extensions of Montague Grammar', *Journal of Philosophical Logic* 2, 1973, pp.509-34. Reprinted in B. H. Partree (ed.), *Montague Grammar,* Academic Press, New York, pp.51-76.

611. Paxton, W. H. (1977) 'A Framework for Speech Understanding', technical note 142, AI Center, SRI International, Menlo Park, California.

612. Petrick, S. R. (1973) 'Transformational Analysis', in R. Rustin (ed.), *Natural Language Processing,* Algorithmics Press, New York, pp.27-41.

613. Radford, A. (1981) 'A Transformational Syntax', Cambridge University Press.

614. Ritchie, G. (1983) 'Semantics in Parsing', in M. King (ed.), *Parsing Natural Language,* pp.199-217.

615. Robinson, J. J. (1982) 'DIAGRAM: A Grammar for Dialogues', *CACM* 25 (1).

616. Ross, H. (1967) 'Constraints on Variables in Syntax', MIT dissertation, distributed by Indiana University Linguistics Club.

617. Sager, N. (1981) *Natural Language Information Processing,* Addison-Wesley, Reading, Massachusetts.

618. Stowell, T. (1981) 'Origin of Phrase Structure', MIT dissertation, to be published by MIT Press.

619. Thomason, R. D., ed. (1974) *Formal Philosophy, Selected Papers of Richard Montague,* Yale University Press, New Haven.

620. Wehrli, E. (1983) 'A Modular Parser for French', in *Proceedings of the Eighth International Joint Conference on Artificial Intelligence,* William Kaufmann Inc., pp.686-9.

621. —— (1984) 'A Government and Binding Parser for French', Working Paper no.48, ISSCO, University of Geneva.

622. Weischedel, R. M. and Black, J. E. (1980) 'If the Parser Fails', *Proceedings of the 18th Annual Meeting of the ACL, University of Pennsylvania.*

623. Wierzbicka, A. (1972) *Semantic Primitives,* Athenaum.

624. —— (1980) *Lingua Mentalis: The Semantics of Natural Language,* Academic Press, Sydney.

625. Winograd, T. (1983) *Language as a Cognitive Process,* Addison-Wesley.

626. Woods, W. A. (1970) 'Transition Network Grammars for Natural Language Analysis', *CACM* 13 (10), pp.591-606.

627. —— (1975) 'Syntax, Semantics, and Speech', BBN Report 3067, Bolt, Beranek, and Newman Inc., Cambridge, Massachusetts.

628. —— (1980) 'Cascaded ATN Grammars', *AJCL* 6 (1), pp.1-12.

629. Zwicky, A. and Sadock, J. (1975) 'Ambiguity Tests and How to Fail Them', in J. Kimball (ed.), *Syntax and Semantics,* Academic Press, New York, pp.1-36.

Artificial Intelligence
630. Barr, A. and Feigenbaum, E. A. (1981) *The Handbook of Artificial Intelligence, vol.*I, William Kaufmann.

631. Berwick, R. (1983) 'Transformational Grammar and Artificial Intelligence: A Contemporary View', *Cognition and Brain Theory* 6, 4.

632. Carbonell, J. G., Cullingford, R. E. and Gershman, A. V. (1978) 'Towards Knowledge-based Machine Translation', *Abstracts, Seventh ICCL [COLING 78], Bergen,* and NTIS Report RR-146.

633. —— (1981) 'Steps Toward Knowledge-based Machine Translation', *IEEE Transactions on Pattern Analysis and Machine Intelligence (PAMI)* 3, 4, pp.376-92.

634. Charniak, E. (1973) 'Jack and Jane in Search of a Theory of Knowledge', *IJCAI,* pp.115-24.

635. Charniak, E. and Wilks, Y., eds (1976) *Computational Semantics,* Amsterdam.

636. Cullingford, R. E. (1978) 'Script Application: Computer Understanding of Newspaper Stories', Research Report No.116, CS Dept., Yale University.

637. Dreyfus, H. (1972) *What Computers Can't Do.* Harper & Row, New York.

638. Dyer, M. G. (1983) *In-depth Understanding: A Computer Model of Integrated Processing for Narrative Comprehension,* Cambridge, Massachusetts.

639. Erman, L., Hayes-Roth, F., Lesser, V. R. and Reddy, D. R. (1980) 'The Hearsay-II Speech-Understanding System: Integrating Knowledge to Resolve Uncertainty', *Computing Surveys* 12(2), pp.213-53.

640. Hauenschild, C. and Pause, P. E. (1983) 'Faktoren-Analyse zur Modellierung des Textverstehens', *Linguistische Berichte* 88, pp.101-20.

641. Hayes-Roth, F. et al. (1983) *Building Expert Systems*, London.

642. Hobbs, J. R. (1976) 'Pronoun Resolution', Research Report 76-1, Department of Computer Sciences, City College, City University of New York.

643. King, M. (1981) 'Semantics and Artificial Intelligence in Machine Translation', *Sprache und Datenverarbeitung*, pp.5-8.

644. Lytinen, S. L. and Schank, R. (1982) *Representation and Translation*, Yale Research Report 234.

645. Metzing, D., ed. (1980) *Frame Conceptions and Text Understanding*, Berlin.

646. Minsky, M. (1975) 'A Framework for Representing Knowledge', in Winston (ed.), *The Psychology of Computer Vision*, New York, pp.211-77.

647. Schank, R., ed. (1975) *Conceptual Information Processing*, Amsterdam.

648. —— (1980) Language and Memory, in *Cognitive Science* 4, pp.243-84.

649. —— (1982) *Dynamic Memory: A Theory of Reminding and Learning in Computers and People*, Cambridge.

650. Schank, R. and Abelson, R. (1975) 'Scripts, Plans and Knowledge', *IJCAI* 4, pp.151-7.

651. —— (1977) *Scripts, Plans, Goals and Understanding*, Hillsdale.

652. Small, S. (1980) 'Word Expert Parsing: A Theory of Distributed Word-Based Natural Language Understanding', Technical Report 954, cs Dept., University of Maryland.

653. Wilks, Y. (1973) *Natural Language Inference*, Stanford University, California, Department of Computer Science.

654. —— (1973) 'An Artificial Intelligence Approach to Machine Translation', in R. C. Schank and K. M. Colby (eds), *Computer Models of Thought and Language*, Freeman, San Francisco, pp.114-51.

655. —— (1975) 'A Preferential, Pattern-seeking Semantics for Natural Language Inference', *Artificial Intelligence* 6, pp.55-76.

656. —— (1976) 'Parsing English II', in Charniak and Wilks, *Computational Semantics*, pp.155-84.

657. —— (1976) 'Semantics and World Knowledge in MT', *AJCL* 2, microfiche 48, pp.67-9.

658. —— (1977) 'Frames for Machine Translation', *New Scientist* 76, pp.802-3.

659. —— (1978) *Making Preference more Active,* DAI Research Report 32, University of Essex.

660. —— (1979) 'Machine Translation and Artificial Intelligence', in B. M. Snell (ed.), *Translating and the Computer,* North-Holland, Amsterdam, pp.27-43.

661. —— (1979) 'Machine Translation and Artificial Intelligence', paper presented at BCS 79, London.

662. —— (1980) 'Frames, Semantics and Novelty', in Metzing (ed.), *Frame Conceptions and Text Understanding,* pp.134-63.

663. —— (1983) 'Deep and Superficial Parsing', in King (ed.), *Parsing Natural Language,* pp.219-46.

664. Winograd, T. (1972) *Understanding Natural Language,* Edinburgh University Press and Academic Press, New York.

665. Winston, P. H. (1984) *Artificial Intelligence,* Addison-Wesley.

INDEX

abstracts, 186
accuracy, 92, 107-8, 135, 149, 347-9
  evaluation of, 92, 99-100, 270-1;
    difficulty of, 92, 95, 98, 149
  expectations, 92, 320, 347
  in SUSY, 220
  in SYSTRAN, 192
  in TAUM-AVIATION, 270-1
  in TAUM-METEO, 265
  in WEIDNER, 179
  see also errors
adjunction grammars, 138
ADL *see under* ALPS
agreement, 178, 179
AI *see* Artificial Intelligence
ALEX *see under* LOGOS
ALPAC (Automatic Language
  Processing Advisory Committee) report, 18, 19-20, 22-4,
  26, 29
  effect, 20, 22, 91, 274, 319
  reply to, 23-4
ALPS, 26, 30, 134, 172-5, 177
  ADL (Automatic Dictionary
    Look-up), 174-5
  CTS (Computer Translation
    System), 174-5
  SDL (Selective Dictionary Look-
    up), 174
  see also Brigham Young Uni-
    versity
ambiguity, 34, 35-6, 71, 96-8, 271-3
  and anaphora, 74-5, 79-80
  and translation, 33-4, 96
  and world knowledge, 71, 74,
    75-6, 79-80, 85
  handling in WEIDNER, 177-8,
    180
  in English, 107

resolution *see* disambiguation
tests for, 262
types, 74-5
see also homographs
anaphora, 34, 74-5, 78, 79-80
  resolution, 86-8, 95-8, 99-103
antecedent, 207, 208
arcs (in charts), 119
ARIANE-78, 135, 139, 140-5, 146,
  151, 152, 278, 318
  portable version, 144-5
Arthern, P. J., 92
article, 149
Artificial Intelligence (AI), 28,
  71-2, 106, 151, 274
  and MT, 35-6, 74, 75-90, 146, 153
Artsruni, George, 3
ASM360, 139
assessment *see* evaluation
ATALA, 150
ATEF *see under* GETA: German-
  to-French system
ATLAS, 142, 151
ATNs (Augmented Transition
  Networks), 138, 268, 327, 328,
  333
Automatic Language Processing
  Advisory Committee *see*
  ALPAC

backtracking, 114
Bacon, Francis, 93-4
Bar-Hillel, Y., 7, 9, 17-18, 45, 46,
  47, 51, 145-6
batch systems, 116
BBN, 138
binding theory, 65
Birkbeck College, 5, 14
Bloomfield, 43, 44-5